Lecture Notes
in Computational Science
and Engineering

85

Editors:

Timothy J. Barth
Michael Griebel
David E. Keyes
Risto M. Nieminen
Dirk Roose
Tamar Schlick

For further volumes:
http://www.springer.com/series/3527

James Blowey • Max Jensen

Editors

Frontiers in Numerical Analysis - Durham 2010

 Springer

Editors
James Blowey
Max Jensen
Durham University
Department of Mathematical Sciences
South Road
DH1 3LE Durham
United Kingdom
j.f.blowey@durham.ac.uk
m.p.j.jensen@durham.ac.uk

ISSN 1439-7358
ISBN 978-3-642-23913-7   e-ISBN 978-3-642-23914-4
DOI 10.1007/978-3-642-23914-4
Springer Heidelberg Dordrecht London New York

Library of Congress Control Number: 2011944196

Mathematics Subject Classification (2010): 65N30, 65N15, 65N25, 65N55, 78M10, 92C35

Cover image: By courtesy of Carlo D' Angelo

Printed on acid-free paper

Springer is part of Springer Science+Business Media (www.springer.com)

Preface

The Twelvefth LMS-EPSRC Summer School in Computational Mathematics and Scientific Computation was held at the University of Durham, UK, from the 25th to the 31st of July 2010. This was the fourth of these schools to be held in Durham, having previously been hosted by the University of Lancaster and the University of Leicester. The purpose of the summer school was to present high quality instructional courses on topics at the forefront of computational mathematics and scientific computing research to postgraduate students.

This volume presents written contributions from each of the speakers. In all cases, these contributions are more comprehensive versions of the lecture notes which were distributed to participants during the meeting.

At the time of writing it is now more than two years since we first contacted the guest speakers and during that period they have given significant portions of their time to making the summer school, and this volume, a success. We would like to thank all four of them for the care which they took in the preparation and delivery of their lectures. The main speakers were Daniele Boffi, Susanne Brenner, Professor Peter Monk and Paolo Zunino.

Instrumental to the school were the two tutors who ran a very successful tutorial programme (Richard Norton & Angela Mihai). There was also a successful programme of contributed talks from six students in the afternoon. The UKIE section of SIAM contributed prizes for the best talks given by graduate students. The invited speakers took on the bulk of the task of judging these talks. After careful and difficult consideration the prizes were awarded to Edward Tucker (Imperial College) and Alexander Raisch (Bonn University.) The general quality of the student presentations was impressively high promising a vibrant future for the subject.

The audience consisted of forty-three research students from within the UK and Europe. As always, one of the most important aspects of the summer school was proving a forum for UK numerical analysts, both young and old, to meet for an extended period and exchange ideas.

Durham
June 2011

James F. Blowey
Max P.J. Jensen

Acknowledgements

We thank the LMS and the Engineering and Physical Sciences Research Council for their financial support which covered all the costs of the main speakers, tutors, plus the accommodation costs of the participants.

We would also like to thank the Grey College for their hospitality in hosting the participants, the Maths Secretaries for their support and our families for supporting our efforts.

Contents

Contributors

Daniele Boffi Dipartimento di Matematica, Università degli Studi di Pavia, via Ferrata 1, I-27100 Pavia, Italy, daniele.boffi@unipv.it

Susanne C. Brenner Department of Mathematics and Center for Computation & Technology, Louisiana State University, Baton Rouge, LA 70803, USA, brenner@math.lsu.edu

Erik Burman Department of Mathematics, University of Sussex, Falmer, Brighton, BN1 9RF United Kingdom, E.N.Burman@sussex.ac.uk

Qiang Chen Department of Mathematical Sciences, University of Delaware, Newark, DE 19716, USA, qchen@math.udel.edu

Francesca Gardini Dipartimento di Matematica, Università degli Studi di Pavia, via Ferrata 1, I-27100 Pavia, Italy, francesca.gardini@unipv.it

Lucia Gastaldi Dipartimento di Matematica, Università di Brescia, via Valotti 9, I-25133 Brescia, Italy, lucia.gastaldi@ing.unibs.it

Peter Monk Department of Mathematical Sciences, University of Delaware, Newark, DE 19716, USA, monk@math.udel.edu

Paolo Zunino MOX - Department of Mathematics, Politecnico di Milano, P.zza Leonardo da Vinci 32, 20133 Milano, Italy, paolo.zunino@polimi.it

Some Remarks on Eigenvalue Approximation by Finite Elements

Daniele Boffi, Francesca Gardini, and Lucia Gastaldi

Abstract The aim of this paper is to supplement the results of Boffi (Acta Numer. 19:1–120, 2010) with some additional remarks. In particular we deal with three distinct topics: we review some tutorial examples in one dimension and provide numerical codes for them; we analyze the case of multiple eigenvalues and show some numerical; we review a posteriori error analysis for eigenvalue problems.

1 Introduction

A recent survey [21] reports on the state of the art of the approximation of symmetric and compact eigenvalue problem by the finite element method. The aim of this paper is to supplement it with some additional theoretical results, application examples, and numerical codes.

This paper deals with three different topics. The first topic is considered in Sects. 2, 3, and 4 where some preliminary examples reported in [21, Part 1] are revisited in more detail. With a didactic purpose, particular emphasis is put on basic one dimensional examples in the standard and in the mixed Galerkin setting. Each example is completed by appropriate Matlab codes.

The second topic, discussed in Sect. 5, deals with the approximation of multiple eigenvalues. It is a common practice to restrict the analysis of eigenvalue/eigenfunction convergence to the case of simple modes and to only state the results in the case

D. Boffi (✉) · F. Gardini
Dipartimento di Matematica, Università degli Studi di Pavia, via Ferrata 1, I-27100 Pavia, Italy
e-mail: daniele.boffi@unipv.it; francesca.gardini@unipv.it; http://www-dimat.unipv.it/boffi; http://www-dimat.unipv.it/gardini

L. Gastaldi
Dipartimento di Matematica, Università di Brescia, via Valotti 9, I-25133 Brescia, Italy
e-mail: lucia.gastaldi@ing.unibs.it; http://www.ing.unibs.it/~gastaldi

J. Blowey and M. Jensen (eds.), *Frontiers in Numerical Analysis – Durham 2010*, Lecture Notes in Computational Science and Engineering 85, DOI 10.1007/978-3-642-23914-4_1, © Springer-Verlag Berlin Heidelberg 2012

of multiplicities higher than one. Here we want to make it precise how the analysis goes in the case of multiple eigensolutions and, in particular, to focus on the case when an eigenspace can contain eigenfunctions of variable smoothness. The typical example of this situation is given by a double eigenvalue with one singular and one smooth eigenfunction. We will recall the theoretical estimates for this particular situation and we will confirm them with appropriate numerical experiments.

Finally, Sect. 6 is devoted to a fundamental topic which could not be covered in [21]: a posteriori error control for eigenvalue approximations. We will review the main ideas behind adaptive mesh refinement for the approximation of eigenvalue problems in the case of standard elliptic and mixed formulations.

2 Variationally Posed Eigenvalue Problems

In this paper we deal with the finite element approximation of *symmetric* and *compact* eigenvalue problem arising from partial differential equations. We introduce in this section our setting and recall some basic results.

The convergence analysis for eigenvalue problems usually consists of two parts: in the first step one shows that all continuous eigensolutions are approximated by the correct number of discrete eigenmodes (counted according to their multiplicities) and that no spurious eigenvalue is present; in the second step error estimates are looked for, which provide the order of convergence for eigenvalues and eigenfunctions. In this section we focus on the first step. The question of the rate of convergence will be detailed in Sect. 5

Let V and H be real Hilbert spaces. We suppose $V \subset H$ with dense and continuous embedding. Let $a : V \times V \to \mathbb{R}$ and $b : H \times H \to \mathbb{R}$ be symmetric and continuous bilinear forms, and consider the problem: find $\lambda \in \mathbb{R}$ and $u \in V$, with $u \neq 0$, such that

$$a(u, v) = \lambda b(u, v) \quad \forall v \in V. \tag{1}$$

The Galerkin discretization of problem (1) is based on a finite dimensional space $V_h \subset V$ and reads: find $\lambda_h \in \mathbb{R}$ and $u_h \in V_h$, with $u_h \neq 0$, such that

$$a(u_h, v) = \lambda_h b(u_h, v) \quad \forall v \in V_h. \tag{2}$$

The convergence analysis of the eigensolutions of (2) to those of (1) is usually performed with the introduction of suitable solution operators. We assume that for any $f \in H$ there exists a unique $Tf \in V$ and a unique $T_h f \in V_h$ such that

$$a(Tf, v) = b(f, v) \quad \forall v \in V \tag{3}$$

and

$$a(T_h f, v) = b(f, v) \quad \forall v \in V_h. \tag{4}$$

This is the case, for instance, when a is V-elliptic and b is equivalent to a scalar product in H. Unless otherwise expressly written, we will assume that we are in this setting. We are then given two self-adjoint operators from H into itself and we assume that

$$T : H \to H \text{ is compact.} \tag{5}$$

It is clear that, being a finite rank operator, T_h is compact as well.

Our main question about the convergence of the eigenvalue and the absence of spurious modes is indeed equivalent to the convergence in norm of T_h to T (see [21] for more details and for a formal definition of convergence): we will then discuss sufficient and necessary conditions for obtaining

$$\|T - T_h\|_{\mathscr{L}(H)} \to 0 \quad \text{when } h \to 0. \tag{6}$$

It can be seen that the eigenvalue convergence is also ensured by a convergence in the norm of V

$$\|T - T_h\|_{\mathscr{L}(V)} \to 0 \quad \text{when } h \to 0, \tag{7}$$

which can of course hold true only under the additional hypothesis that T is compact in $\mathscr{L}(V)$.

The most elegant theorem that proves (6) has been stated in this framework by Kolata in [58], although results in this direction were known and used before by many authors (see [8], for instance). The starting point is the standard Galerkin orthogonality which reads

$$T_h = P_h T, \tag{8}$$

where $P_h : V \to V_h$ is the elliptic projection associated to the bilinear form a. The next theorem (see [58] and [21, Theorem 7.6]) is often referred to by saying that compactness turns *pointwise* into *uniform* convergence.

Theorem 1. *If T is compact from H to V and P_h converges strongly (i.e., pointwise) to the identity operator from V to H, then T_h converges to T (uniformly) in the norm of $\mathscr{L}(H)$ (see (6)).*

Remark 1. It should be noted that the compactness hypothesis of Theorem 1 is stronger than (5). This is however needed when using the representation $T_h = P_h T$, since P_h is naturally defined in V and not in H. There is another option which consists of assuming T to be compact in $\mathscr{L}(V)$ and P_h converging strongly to I in $\mathscr{L}(V)$: this implies the convergence in norm (7).

Remark 2. The convergence in norm (6) can often be obtained by examining directly the error estimates linking (4) and (3). In many applications it is possible to get estimates of the form

$$\|Tf - T_h f\|_V \le C h^k \|f\|_H.$$

3 One Dimensional Examples

In this section we consider some one dimensional examples. We start with the standard Laplace eigenvalue problem which fits pretty well the theory presented in Sect. 2. Then we consider eigenvalue problems in mixed form and show how the theory of Sect. 2 should be changed in order to deal with this setting. In particular, an analogue of Theorem 1 cannot be proved in the case of the Laplace eigenvalue problem in mixed form due to a lack of compactness.

The examples are discussed in detail and particular emphasis is given to the numerical results (including the source code for Matlab computations).

3.1 Standard Laplace Eigenvalue Problem

Given the interval $\Omega =]0, \pi[$ we look for eigenvalues λ and eigenfunctions u with $u \neq 0$ such that

$$
\begin{cases}
-u''(x) = \lambda u(x) & \text{in } \Omega \\
u(0) = u(\pi) = 0.
\end{cases}
\tag{9}
$$

The exact eigenvalues are given by $\lambda = 1, 4, 9, 16, \ldots$ and the eigenspaces are generated by $\sin(kx)$ for $k = 1, 2, 3, 4, \ldots$.

This problem fits the setting of Sect. 2 with the following choices:

$$
V = H_0^1(\Omega)
$$

$$
H = L^2(\Omega)
$$

$$
a(u, v) = \int_0^\pi u'(x)v'(x)\mathrm{d}x
$$

$$
b(u, v) = \int_0^\pi u(x)v(x)\mathrm{d}x.
$$

Let us consider the conforming approximation of (9) by *continuous piecewise linear* finite elements. It is well-known that the matrix form of the discrete problem is given by

$$
A\mathbf{x} = \lambda M \mathbf{x}
$$

where the stiffness matrix A is

$$
a_{ij} = \frac{1}{h} \cdot
\begin{cases}
2 & \text{for } i = j \\
-1 & \text{for } |i - j| = 1 \\
0 & \text{otherwise,}
\end{cases}
$$

Table 1 Eigenvalues computed using the code in Listing 1.1 for different values of n

Exact	Computed (rate)				
	$n = 8$	$n = 16$	$n = 32$	$n = 64$	$n = 128$
1	1.0129	1.0032 (2.0)	1.0008 (2.0)	1.0002 (2.0)	1.0001 (2.0)
4	4.2095	4.0517 (2.0)	4.0129 (2.0)	4.0032 (2.0)	4.0008 (2.0)
9	10.0803	9.2631 (2.0)	9.0652 (2.0)	9.0163 (2.0)	9.0041 (2.0)
16	19.4537	16.8382 (2.0)	16.2067 (2.0)	16.0515 (2.0)	16.0129 (2.0)
25	33.2628	27.0649 (2.0)	25.5059 (2.0)	25.1257 (2.0)	25.0314 (2.0)
36	51.3724	40.3212 (1.8)	37.0525 (2.0)	36.2610 (2.0)	36.0651 (2.0)
49	69.5582	57.0672 (1.3)	50.9572 (2.0)	49.4840 (2.0)	49.1206 (2.0)
64		77.8147	67.3528 (2.0)	64.8266 (2.0)	64.2059 (2.0)
81		103.0473	86.3943 (2.0)	82.3258 (2.0)	81.3299 (2.0)
100		133.0513	108.2597 (2.0)	102.0237 (2.0)	100.5030 (2.0)
DOF	7	15	31	63	127

Fig. 1 The second eigenfunction computed and plotted using the code in Listing 1.1 ($n = 8$ and $k = 2$)

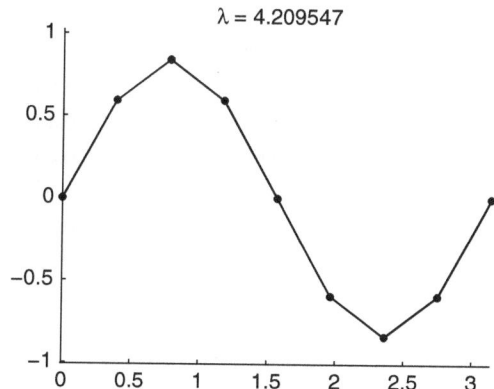

$\lambda = 4.209547$

and the mass matrix M is

$$m_{ij} = h \cdot \begin{cases} 2/3 & \text{for } i = j \\ 1/6 & \text{for } |i - j| = 1 \\ 0 & \text{otherwise}, \end{cases}$$

with $i, j = 1, \ldots, n-1$, where n is the number of subdivisions of the interval $[0, \pi]$ (the dimensions of A and M equal the number of internal nodes).

In Listing 1.1 we report a simple Matlab code that solves our problem, which displays the first ten eigenvalues, and plots a specific eigenfunction. The result of the computations for successively refined meshes is included in Table 1. Second order convergence can be clearly appreciated (this is compatible with the error estimates which will be made precise in Sect. 5.1). Figure 1 shows the second eigenfunction for $n = 16$ (this is the plot which is generated with the parameters reported in Listing 1.1).

Table 2 Eigenvalues computed using the code in Listing 1.2 for different values of n

Exact	Computed (rate)				
	$n = 8$	$n = 16$	$n = 32$	$n = 64$	$n = 128$
1	1.0000	1.0000 (4.0)	1.0000 (4.0)	1.0000 (4.0)	1.0000 (4.0)
4	4.0020	4.0001 (4.0)	4.0000 (4.0)	4.0000 (4.0)	4.0000 (4.0)
9	9.0225	9.0015 (3.9)	9.0001 (4.0)	9.0000 (4.0)	9.0000 (4.0)
16	16.1204	16.0082 (3.9)	16.0005 (4.0)	16.0000 (4.0)	16.0000 (4.0)
25	25.4327	25.0307 (3.8)	25.0020 (3.9)	25.0001 (4.0)	25.0000 (4.0)
36	37.1989	36.0899 (3.7)	36.0059 (3.9)	36.0004 (4.0)	36.0000 (4.0)
49	51.6607	49.2217 (3.6)	49.0148 (3.9)	49.0009 (4.0)	49.0001 (4.0)
64	64.8456	64.4814 (0.8)	64.0328 (3.9)	64.0021 (4.0)	64.0001 (4.0)
81	95.7798	81.9488 (4.0)	81.0659 (3.8)	81.0042 (4.0)	81.0003 (4.0)
100	124.9301	101.7308 (3.8)	100.1229 (3.8)	100.0080 (3.9)	100.0005 (4.0)
DOF	15	31	63	127	255

Fig. 2 The second eigenfunction computed and plotted using the code in Listing 1.2 ($n = 4$ and $k = 2$). The dashed line represents the linear part of the solution

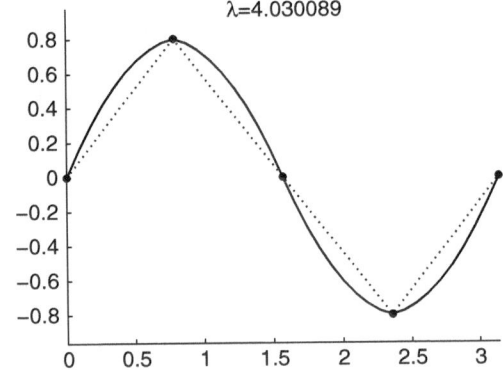

As an additional example, in Listing 1.2 we report a code that solves the same problem using continuous piecewise quadratic finite elements. The matrix construction has been made using a hierarchical approach: in the interval $[x_i, x_{i+1}]$ a quadratic function is seen as the sum of its affine part with matching values at the endpoints and a quadratic bubble. It turns out that the stiffness matrix A is a 2×2 block matrix of size $2n - 1$ (n being the number of subintervals) in which only the two blocks on the main diagonal are different from zero: the block A_{11} of size $n-1$ is equal to the stiffness matrix of the previous case and the block A_{22} of size n is given by the contribution of the bubbles. The mass matrix M has a block structure too, but now the off-diagonal terms are non zero since there are contributions coming from the interaction between affine functions and bubbles.

The results in Table 2 show that the eigenvalues are approximated with fourth order accuracy. A direct comparison with Table 1 confirms that quadratic elements provide much more accurate results even with less degrees of freedom. Figure 2 shows the second eigenfunctions computed with the code reported in Listing 1.2: the linear part is drawn with a dashed line and the sum of the linear part and the

Fig. 3 Exponential
convergence of eigenvalues
computed with the code
reported in Listing 1.3

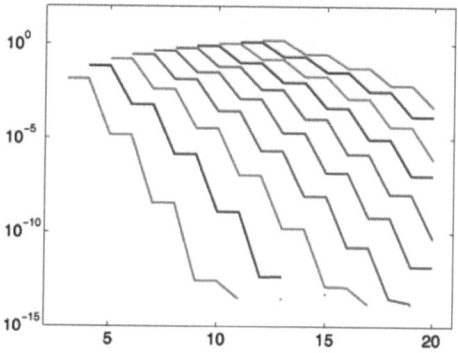

bubble (i.e., the full solution) is represented by the solid line. This plot compares to Fig. 1, where the same eigenfunctions was computed with piecewise linear elements. It is apparent that (even on such coarse meshes) quadratic elements provide a more accurate result using the same number of degrees of freedom (7).

The results shown so far concern the approximation of problem (9) using the h version of the finite element method. This means that a better approximation is obtained by successive refinements of the mesh and error estimates are given in terms of the meshsize h, which is supposed to tend to zero. We conclude the discussion of this example with a remark on the p version of the finite element methods, that is the mesh is kept fixed and a better approximation is obtained by raising the order of the polynomials. The code reported in Listing 1.3 computes the eigenvalues of problem (9) using p-th order polynomials on a mesh which is composed of a single element (i.e., using a plain spectral method).

The stiffness and mass matrix are evaluated using the arguments presented in [30, Sect. 3.8]. It is interesting to analyze the results of some computations. Since we are expecting exponential convergence, we plot the errors using a semilog scale in Fig. 3. The x axis represents the polynomial order (ranging from 3 to 20) and the y axis reports the logarithm of the error. The errors in the approximation of the eigenvalues are reported from the left to the right (this means, in particular, that the most left line corresponds to the error in the approximation of the first eigenvalue $\lambda = 1$). It can be observed a typical behavior of spectral approximations: the convergence is exponential and is dependent on the parity of the polynomial order. Table 3, for instance, shows the computed values of the fifth discrete eigenvalue approximating $\lambda = 25$. The polynomial order varies from 7 to 20 (the fifth eigenvalue shows up only when the order is at least 7 and the dimension of the system is 5). It is apparent that the approximation improves only every other step (when the order is raised from even to odd in this case, but the situation is the opposite if an even eigenvalue is considered, see Fig. 3). It is also clear that the accuracy of the discrete values is not very good when the order is 7 or 8 but, after the convergence has started, it is very fast (more than 10 digits with 17 degrees of freedom, while we needed over 100 degrees of freedom to get only 3 digits with quadratic elements, see Table 2).

Table 3 Fifth eigenvalue
computed with the code
in Listing 1.3 for different
values of p

p	DOF	Computed
7	5	35.5593555378041
8	6	35.5593555378041
9	7	25.7779168651921
10	8	25.7779168651921
11	9	25.0306605127133
12	10	25.0306605127132
13	11	25.0004945052929
14	12	25.0004945052929
15	13	25.0000037734250
16	14	25.0000037734250
17	15	25.0000000156754
18	16	25.0000000156756
19	17	25.0000000000389
20	18	25.0000000000389

Listing 1.1 Matlab code for the 1D Laplace eigenvalue problem: piecewise linear elements

```
clear all
close all
k=2; % eigenfunction to be plotted
n=8; % number of subdivisions
kev=4; % number of eigenvalues to compute
a=0; b=pi; % interval endpoints
h=(b-a)/n; % mesh size
x=linspace(a,b,n+1); % mesh nodes
%
% stiffness matrix A
%
e=ones(n-1,1);
A=spdiags([-e 2*e -e]/h,-1:1,n-1,n-1);
%
% mass matrix M
%
M=spdiags([1/6*e 2/3*e 1/6*e]*h,-1:1,n-1,n-1);
%
% compute solution and sort eigenmodes
%
[v,d]=eigs(A,M,min(kev,length(M)),'SM');
[ev,I]=sort(diag(d));
ef=v(:,I);
%
% display first 10 eigenvalues
%
ev(1:min(length(ev),10))
```

```
%
% plot selected eigenfunction
%
vector =[0; ef (: ,k );0];
c=min( vector ); d=max( vector );
c=c-(d-c )*.1; d=d+(d-c )*.1;
set (0 ,' defaultaxesfontsize ' ,15)
set (0 ,' defaulttextfontsize ' ,15)
figure
hold on
plot (x, vector ,'-k')
plot (x, vector ,'.k')
axis ([a b c d])
set ( findobj ('type ','line '),'linewidth ' ,2 ,...
    'markersize ' ,20)
lambda=sprintf ('%0.7g',ev(k ));
title ([ '\lambda=' lambda ])
```

Listing 1.2 Matlab code for the 1D Laplace eigenvalue problem: piecewise quadratic elements

```
clear all
close all
k=2; % eigenfunction to be plotted
n=4; % number of subdivisions
kev=4; % number of eigenvalues to compute
a=0; b=pi; % interval endpoints
h=(b-a)/n; % mesh size
x=linspace (a,b,n+1); % mesh nodes
%
% stiffness matrix A (contributions coming from P1)
%
e=ones (n ,1);
A11=spdiags ([-e 2*e -e ]/h, -1:1 ,n-1,n-1);
%
% stiff. matrix A (contributions coming from bubbles)
%
A22=spdiags (16/3*e/h ,0 ,n,n );
%
% stiff. matrix A: assembly
%
A=[A11, zeros (n-1,n); zeros (n,n-1),A22 ];
%
% mass matrix M (contributions coming from P1)
%
```

```
M11=spdiags([1/6*e 2/3*e 1/6*e]*h,-1:1,n-1,n-1);
%
% mass matrix M (contributions coming from bubbles)
%
M22=spdiags(8/15*e*h,0,n,n);
%
% mass matrix M (interactions P1/bubbles)
%
M12=spdiags([e e]/3*h,[0 1],n-1,n);
%
% mass matrix M: assembly
%
M=[M11,M12;M12',M22];
%
% compute solution and sort eigenmodes
%
[v,d]=eigs(A,M,min(kev,length(M)),'SM');
[ev,I]=sort(diag(d));
ef=v(:,I);
%
% display first 10 eigenvalues
%
ev(1:min(length(ev),10))
%
% plot selected eigenfunction
%
% P1 component
%
vector=[0;ef(1:n-1,k);0];
%
% bubble component
%
bubble=ef(n:end,k);
%
c=min(vector); d=max(vector);
c=c-(d-c)*.1; d=d+(d-c)*.1;
figure
hold on
set(0,'defaultaxesfontsize',15)
set(0,'defaulttextfontsize',15)
%
kk=10; % number of points for bubble recostructions
for i=1:n
  for j=1:kk
```

```
        xx=linspace(0,h,kk);
        plot([x(i)+xx],[bubble(i)*4/h^2*xx.*(h-xx)+...
            (vector(i)+(vector(i+1)-vector(i))/h*xx)],...
            '-k')
    end
end
%
plot(x,vector,':k')
plot(x,vector,'.k')
axis([a b c d])
set(findobj('type','line'),'linewidth',2,...
    'markersize',20)
lambda=sprintf('%0.7g',ev(k));
title(['\lambda=' lambda])
```

Listing 1.3 Matlab code for the 1D Laplace eigenvalue problem: spectral method

```
clear all;
p=10; % order of the polynomial
a=0; b=pi; % interval endpoints
%
% stiffness matrix A (modal basis in [-1 1])
%
A=eye(p+1);
A(1:2,1:2)=[1/2 -1/2;-1/2 1/2];
%
% mass matrix M (modal basis in [-1 1])
%
diagonal=zeros(p,1);
for k=2:p
    diagonal(k+1)=2/(2*k-3)/(2*k+1);
end
M=diag(diagonal);
%
diagonal2=zeros(p,1);
for k=2:p-2
    diagonal2(k+2)=-1/(2*k+1)/sqrt((2*k-1)*(2*k+3));
end
M=M+diag(diagonal2(2:p),-2)+diag(diagonal2(2:p),2);
%
M(1:2,1:4)=[2/3 1/3 1/sqrt(6) -1/3/sqrt(10);...
            1/3 2/3 1/sqrt(6) 1/3/sqrt(10)];
M(3:4,1:2)=[1/sqrt(6) 1/sqrt(6);...
            -1/3/sqrt(10) 1/3/sqrt(10)];
```

```
%
% solve for Dirichlet boundary conditions
% and rescale interval
%
ev=sort(eig(A(3:p,3:p),M(3:p,3:p)))*4/(b-a)/(b-a);
ev(1:min(10,p-2))
```

3.2 Laplace Eigenvalue Problem in Mixed Form

The standard mixed formulation of problem (9) is: given $\Sigma = H^1(\Omega)$ and $U = L^2(\Omega)$, find $\lambda \in \mathbb{R}$ and $u \in U$, with $u \neq 0$, such that for some $s \in \Sigma$

$$
\begin{cases}
\displaystyle\int_0^\pi s(x)t(x)\mathrm{d}x + \int_0^\pi u(x)t'(x)\mathrm{d}x = 0 & \forall t \in \Sigma \\
\displaystyle\int_0^\pi s'(x)v(x)\mathrm{d}x = -\lambda \int_0^\pi u(x)v(x)\mathrm{d}x & \forall v \in U.
\end{cases}
\tag{10}
$$

Its Galerkin discretization is based on discrete subspaces $\Sigma_h \subset \Sigma$ and $U_h \subset U$ and reads: find $\lambda_h \in \mathbb{R}$ and $u_h \in U_h$ with $u_h \neq 0$, such that for some $s_h \in \Sigma_h$ it holds

$$
\begin{cases}
\displaystyle\int_0^\pi s_h(x)t(x)\mathrm{d}x + \int_0^\pi u_h(x)t'(x)\mathrm{d}x = 0 & \forall t \in \Sigma_h \\
\displaystyle\int_0^\pi s_h'(x)v(x)\mathrm{d}x = -\lambda_h \int_0^\pi u_h(x)v(x)\mathrm{d}x & \forall v \in U_h.
\end{cases}
$$

The matrix form of the problem is

$$
\begin{pmatrix} A & B^T \\ B & 0 \end{pmatrix} \begin{pmatrix} \mathsf{x} \\ \mathsf{y} \end{pmatrix} = -\lambda \begin{pmatrix} 0 & 0 \\ 0 & M \end{pmatrix} \begin{pmatrix} \mathsf{x} \\ \mathsf{y} \end{pmatrix},
$$

where A is the mass matrix in Σ_h, M is the mass matrix in U_h, and B is the matrix defined as follows:

$$
b_{jk} = \int_0^\pi \varphi_k'(x)\psi_j(x)\mathrm{d}x,
$$

where $\{\varphi_k\}$ and $\{\psi_j\}$ are bases in Σ_h and U_h, respectively.

Example 1 (P1-P1 element). We start with the simplest choice of finite element spaces: continuous piecewise linears for both Σ_h and U_h. The corresponding Matlab code is reported in Listing 1.4. It should be noted that the homogeneous Dirichlet boundary conditions are enforced in a natural way through the formulation (10), so that the matrices A, B, and M do not include boundary conditions.

Table 4 Eigenvalues computed using the code in Listing 1.4 for different values of n

Exact	Computed (rate)				
	$n = 8$	$n = 16$	$n = 32$	$n = 64$	$n = 128$
	0.0000	−0.0000	0.0000	0.0000	−0.0000
1	1.0001	1.0000 (4.1)	1.0000 (4.0)	1.0000 (4.0)	1.0000 (4.0)
4	3.9660	3.9981 (4.2)	3.9999 (4.0)	4.0000 (4.0)	4.0000 (4.0)
	7.4257	8.5541	8.8854	8.9711	8.9928
9	8.7603	8.9873 (4.2)	8.9992 (4.1)	9.0000 (4.0)	9.0000 (4.0)
16	14.8408	15.9501 (4.5)	15.9971 (4.1)	15.9998 (4.0)	16.0000 (4.0)
25	16.7900	24.5524 (4.2)	24.9780 (4.3)	24.9987 (4.1)	24.9999 (4.0)
	38.7154	29.7390	34.2165	35.5415	35.8846
36	39.0906	35.0393 (1.7)	35.9492 (4.2)	35.9970 (4.1)	35.9998 (4.0)
49		46.7793	48.8925 (4.4)	48.9937 (4.1)	48.9996 (4.0)

In [21, Table 4.1] it has already been observed that this method does not provide reliable results. Table 4 shows that the correct eigenvalues are approximated with fourth order accuracy, while several other *spurious* modes are present. Figure 4 shows the first two spurious modes, corresponding to the value $\lambda = 0$ and to a discrete value which seems to converge to $\lambda = 9$ (i.e., this mode is spurious in the sense of a wrong multiplicity). The eigenfunction u is plotted in the left part of the figure, while the corresponding component s is on the right. An example of correct eigenfunction is shown in Fig. 5: this is the eigenfunction obtained exactly with the parameters of Listing 1.4 and should be compared with Figs. 1 and 2. For the sake of completeness, we report in Fig. 6 the eigenfunction corresponding to the discrete eigenvalue approximating the *correct* continuous eigenvalue $\lambda = 9$ (i.e., the fifth discrete mode).

Example 2 (P1-P0 element). We now describe a *convergent* mixed scheme for which no spurious mode is present. Since the space U is $L^2(\Omega)$, there is no need to consider a finite element approximation U_h made of continuous functions. For reasons which are clear from the abstract theory, it is natural to consider the following choice: continuous piecewise linear elements for the approximation of Σ and discontinuous piecewise constants for the approximation of U. The corresponding Matlab code is reported in Listing 1.5.

As already observed in [21, Sect. 4.2], the numerical results are pretty much related to the ones of the standard Galerkin approximation of the Laplace eigenvalue problem: the eigenvalues are reported in Table 5 where second order of convergence is clearly detected. As far as the number of degrees of freedom is concerned, Table 5 shows the dimension of the space U_h, since in the solution procedure the variable s can be eliminated (see Listing 1.5).

Figure 7 shows the second eigenfunction computed using the code in Listing 1.5 and should be compared with Fig. 1.

Example 3 (P2-P0 element). We conclude this section about one-dimensional examples with the discussion of the P2-P0 scheme for the approximation of the

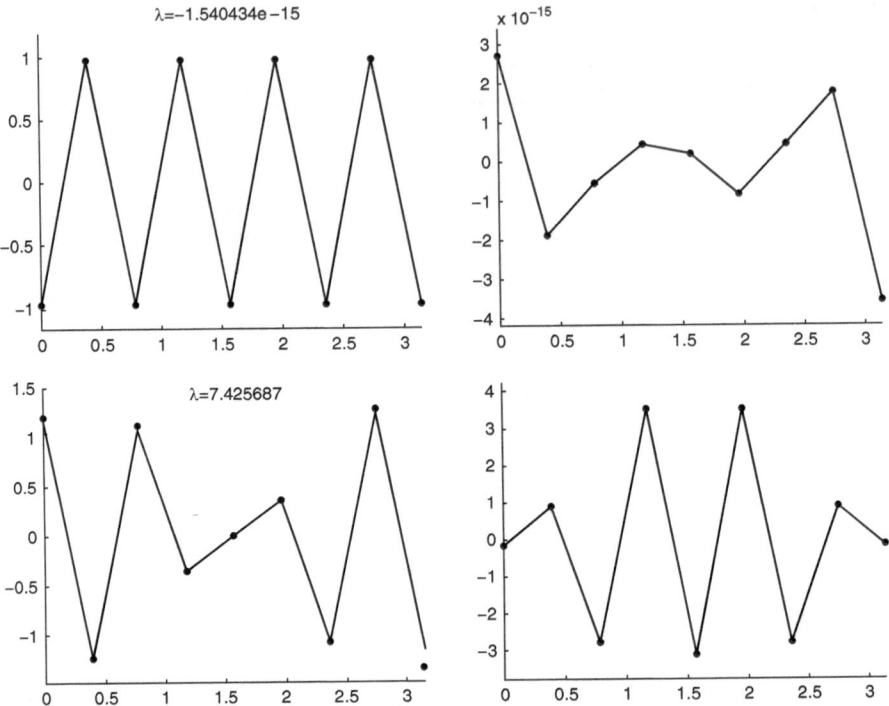

Fig. 4 Some spurious modes computed with the code in Listing 1.4: the first (top) and fourth (bottom) eigenfunction. The left subplots correspond to the component u and the right ones to s. Notice that the function in the top right subplot is zero up to machine precision

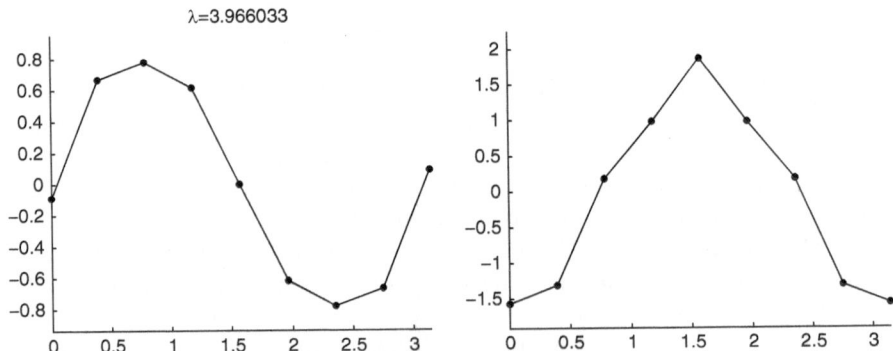

Fig. 5 An approximation of the second eigenfunction of problem (10) (the component u on the left side and the component s on the right) computed with the code in Listing 1.4 for $n = 8$ and $k = 3$

mixed problem (10). We started from the P1-P1 element (see Example 1) which is affected by spurious modes and moved to the P1-P0 element (see Example 2) which is nicely convergent. In this last example, we shall demonstrate the bad behavior of the P2-P0 element where we use continuous piecewise quadratic elements for

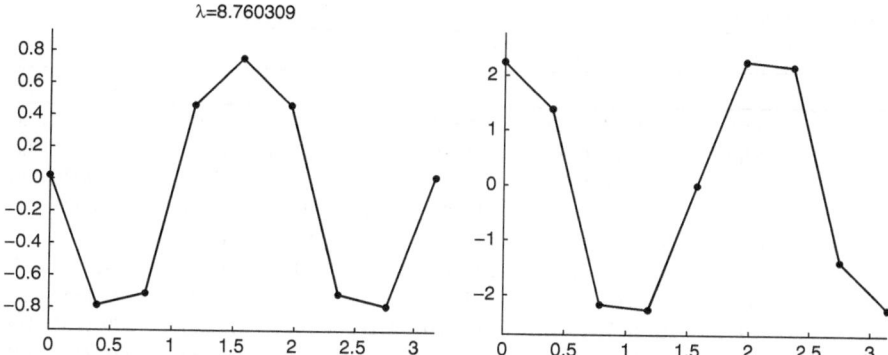

Fig. 6 An approximation of the third eigenfunction of problem (10) (the component u on the left side and the component s on the right) computed with the code in Listing 1.4 for $n = 8$ and $k = 5$

Table 5 Eigenvalues computed using the code in Listing 1.5 for different values of n

Exact	Computed (rate)				
	$n = 8$	$n = 16$	$n = 32$	$n = 64$	$n = 128$
1	1.0129	1.0032 (2.0)	1.0008 (2.0)	1.0002 (2.0)	1.0001 (2.0)
4	4.2095	4.0517 (2.0)	4.0129 (2.0)	4.0032 (2.0)	4.0008 (2.0)
9	10.0803	9.2631 (2.0)	9.0652 (2.0)	9.0163 (2.0)	9.0041 (2.0)
16	19.4537	16.8382 (2.0)	16.2067 (2.0)	16.0515 (2.0)	16.0129 (2.0)
25	33.2628	27.0649 (2.0)	25.5059 (2.0)	25.1257 (2.0)	25.0314 (2.0)
36	51.3724	40.3212 (1.8)	37.0525 (2.0)	36.2610 (2.0)	36.0651 (2.0)
49	69.5582	57.0672 (1.3)	50.9572 (2.0)	49.4840 (2.0)	49.1206 (2.0)
64	77.8147	77.8147 (0.0)	67.3528 (2.0)	64.8266 (2.0)	64.2059 (2.0)
81		103.0473	86.3943 (2.0)	82.3258 (2.0)	81.3299 (2.0)
100		133.0513	108.2597 (2.0)	102.0237 (2.0)	100.5030 (2.0)
DOF	8	16	32	64	128

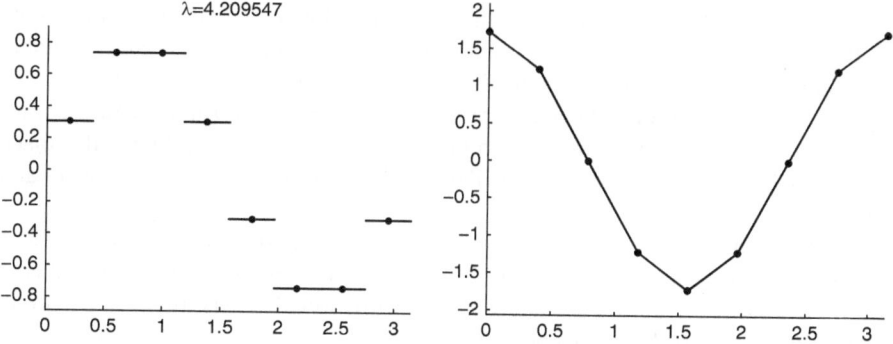

Fig. 7 The second eigenfunction computed and plotted using the code in Listing 1.5 ($n = 8$ and $k = 2$). The component u is on the left side and the component s on the right

Table 6 Eigenvalues computed using the code in Listing 1.6 for different values of n

Exact	Computed (rate with respect to 6λ)				
	$n = 8$	$n = 16$	$n = 32$	$n = 64$	$n = 128$
1	5.7061	5.9238 (1.9)	5.9808 (2.0)	5.9952 (2.0)	5.9988 (2.0)
4	19.8800	22.8245 (1.8)	23.6953 (1.9)	23.9231 (2.0)	23.9807 (2.0)
9	36.7065	48.3798 (1.6)	52.4809 (1.9)	53.6123 (2.0)	53.9026 (2.0)
16	51.8764	79.5201 (1.4)	91.2978 (1.8)	94.7814 (1.9)	95.6925 (2.0)
25	63.6140	113.1819 (1.2)	138.8165 (1.7)	147.0451 (1.9)	149.2506 (2.0)
36	71.6666	146.8261 (1.1)	193.5192 (1.6)	209.9235 (1.9)	214.4494 (2.0)
49	76.3051	178.6404 (0.9)	253.8044 (1.5)	282.8515 (1.9)	291.1344 (2.0)
64	77.8147	207.5058 (0.8)	318.0804 (1.4)	365.1912 (1.8)	379.1255 (1.9)
81		232.8461	384.8425 (1.3)	456.2445 (1.8)	478.2172 (1.9)
100		254.4561	452.7277 (1.2)	555.2659 (1.7)	588.1806 (1.9)
DOF	8	16	32	64	128

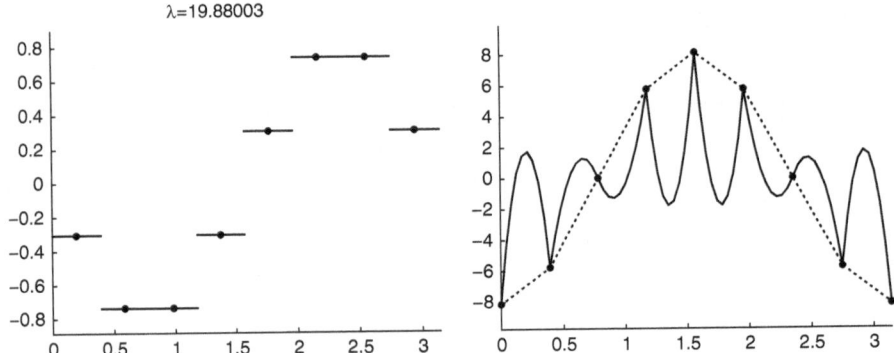

Fig. 8 The second eigenfunction computed and plotted using the code in Listing 1.6 ($n = 8$ and $k = 2$). The component u is on the left side and the component s on the right

the approximation of Σ and discontinuous piecewise constants for U. The final message will be that, as expected, when dealing with mixed schemes the choice of the discrete spaces has to be made very carefully in order to meet suitable assumptions. For a discussion on this issues the reader is referred to Sect. 4.

A Matlab code for this element is presented in Listing 1.6. As in Listing 1.2 we use a hierarchical approach in order to implement the P2 element, that is the space of quadratic elements is presented as the sum of continuous piecewise linears and local bubbles.

In [21, Sect. 4.4] it has been shown that this scheme does not converge. More precisely, the discrete eigenvalues computed with the code in Listing 1.6 converge up to second order to wrong values which correspond to six times the correct eigenvalues. Table 6 shows this behavior.

Moreover, it can be shown that the eigenspaces are good approximations of the correct ones as far as the component u is concerned, while the component s is badly approximated. Figure 8, for instance, shows the eigenfunctions corresponding to the second eigenvalue (which converges to six times $\lambda = 4$). It can be observed that u_h is a good approximation of the space generated by $\sin(2x)$; on the other hand, the *linear* part of s_h is a good approximation of the space generated by $2\cos(2x)$, while the *bubble* part of s provides a *spurious* component of the solution. This fact is a consequence of the lack in the ellipticity in the kernel property (see Sect. 4).

Listing 1.4 Matlab code for the 1D Laplace eigenvalue problem in mixed form: P1-P1 element

```
clear all
close all
k=3; % eigenfunction to be plotted
n=8; % number of subdivisions
kev=5; % number of eigenvalues to compute
a=0; b=pi; % interval endpoints
h=(b-a)/n; % mesh size
x=linspace(a,b,n+1); % mesh nodes
%
% mass matrix A (no boundary conditions)
%
e=ones(n+1,1);
A=spdiags([1/6*e 2/3*e 1/6*e]*h,-1:1,n+1,n+1);
A(1,1)=1/3*h; A(n+1,n+1)=1/3*h;
%
% the matrix B (no boundary conditions)
%
B=spdiags([e/2 -e/2],[-1 1],n+1,n+1);
B(1,1)=1/2; B(n+1,n+1)=-1/2;
%
% mass matrix M is equal to A
%
%
% compute solution and sort eigenmodes
%
Schur=A\B';
[v,d]=eigs(B*Schur,A,min(kev,length(A)),.1);
[ev,I]=sort(diag(d));
ef=v(:,I);
%
% display first 10 eigenvalues
%
ev(1:min(10,length(ev)))
%
% plot selected eigenfunction
%
```

```
set(0,'defaultaxesfontsize',15)
set(0,'defaulttextfontsize',15)
%
% the component "u"
%
figure(1)
clf
hold on
vector=ef(:,k);
c=min(vector); d=max(vector);
c=c-(d-c)*.1; d=d+(d-c)*.1;
plot(x,vector,'-k')
plot(x,vector,'.k')
set(findobj('type','line'),'linewidth',2,...
    'markersize',20)
axis([a b c d])
lambda=sprintf('%0.7g',ev(k));
title(['\lambda=' lambda])
%
% the component "s"
%
figure(2)
clf
hold on
vector=-Schur*ef(:,k);
c=min(vector); d=max(vector);
c=c-(d-c)*.1; d=d+(d-c)*.1;
plot(x,vector,'-k')
plot(x,vector,'.k')
set(findobj('type','line'),'linewidth',2,...
    'markersize',20)
axis([a b c d])
```

Listing 1.5 Matlab code for the 1D Laplace eigenvalue problem in mixed form: P1-P0 element

```
clear all
close all
k=2; % eigenfunction to be plotted
n=8; % number of subdivisions
kev=4; % number of eigenvalues to compute
a=0; b=pi; % interval endpoints
h=(b-a)/n; % mesh size
x=linspace(a,b,n+1); % mesh nodes
%
```

```
% mass  matrix  A
%
e=ones(n+1,1);
A=spdiags([1/6*e  2/3*e  1/6*e]*h,-1:1,n+1,n+1);
A(1,1)=1/3*h;  A(n+1,n+1)=1/3*h;
%
% the  matrix  B
%
B=spdiags([-e  e],[0  1],n,n+1);
%
% mass  matrix  M
%
M=spdiags(e*h,0,n,n);
%
% compute  solution  and  sort  eigenmodes
%
Schur=A\B';
[v,d]=eigs(B*Schur,M,min(kev,length(M)),'SM');
[ev,I]=sort(diag(d));
ef=v(:,I);
%
% display  first  10  eigenvalues
%
ev(1:min(10,length(ev)))
%
% plot  selected  eigenfunction
%
set(0,'defaultaxesfontsize',15)
set(0,'defaulttextfontsize',15)
%
% the  component  "u"
%
figure(1)
clf
hold on
c=min(ef(:,k));  d=max(ef(:,k));
c=c-(d-c)*.1;  d=d+(d-c)*.1;
for  j=1:n
   value=ef(j,k);
   plot([x(j)  x(j+1)],[value,value],'-k')
   plot((x(j)+x(j+1))/2,value,'.k')
end
set(findobj('type','line'),'linewidth',2,...
    'markersize',20)
```

```
axis([a b c d])
lambda=sprintf('%0.7g',ev(k));
title(['\lambda=' lambda])
%
% the component "s"
%
figure(2)
clf
hold on
vector=-Schur*ef(:,k);
c=min(vector); d=max(vector);
c=c-(d-c)*.1; d=d+(d-c)*.1;
plot(x,vector,'-k')
plot(x,vector,'.k')
set(findobj('type','line'),'linewidth',2,...
    'markersize',20)
axis([a b c d])
```

Listing 1.6 Matlab code for the 1D Laplace eigenvalue problem in mixed form: P2-P0 element

```
clear all
close all
k=2; % eigenfunction to be plotted
n=8; % number of subdivisions
kev=4; % number of eigenvalues to compute
a=0; b=pi; % interval endpoints
h=(b-a)/n; % mesh size
x=linspace(a,b,n+1); % mesh nodes
%
% mass matrix A (contributions coming from P1)
%
e=ones(n+1,1);
A11=spdiags([1/6*e 2/3*e 1/6*e]*h,-1:1,n+1,n+1);
A11(1,1)=1/3*h; A11(n+1,n+1)=1/3*h;
%
% mass matrix A (contributions coming from bubbles)
%
A22=spdiags(8/15*e*h,0,n,n);
%
% mass matrix A (interactions P1/bubbles)
%
A12=spdiags([e e]/3*h,[-1 0],n+1,n);
A=[A11,A12;A12',A22];
%
```

```
% the matrix B
%
B=spdiags([-e e],[0 1],n,n+n+1);
%
% mass matrix M
%
M=spdiags(e*h,0,n,n);
%
% compute solution and sort eigenmodes
%
Schur=A\B';
[v,d]=eigs(B*Schur,M,min(kev,length(M)),'SM');
[ev,I]=sort(diag(d));
ef=v(:,I);
%
% display first 10 eigenvalues
%
ev(1:min(length(ev),10))
%
% plot selected eigenfunction
%
set(0,'defaultaxesfontsize',15)
set(0,'defaulttextfontsize',15)
%
% the component "u"
%
figure(1)
clf
hold on
for j=1:n
  value=ef(j,k);
  plot([x(j) x(j+1)],[value,value],'-k')
  plot((x(j)+x(j+1))/2,value,'.k')
end
set(findobj('type','line'),'linewidth',2,...
    'markersize',20)
c=min(ef(:,k)); d=max(ef(:,k));
c=c-(d-c)*.1; d=d+(d-c)*.1;
axis([a b c d])
lambda=sprintf('%0.7g',ev(k));
title(['\lambda=' lambda])
%
% the component "s"
%
```

```
figure(2)
clf
hold on
vector=-Schur*ef(:,k);
%
% P1 component
%
linear=vector(1:n+1);
c=min(linear);  d=max(linear);
c=c-(d-c)*.1;  d=d+(d-c)*.1;
%
% bubble component
%
bubble=vector(n+2:end);
%
kk=10; % number of points for bubble recostructions
%
for  i=1:n
   for  j=1:kk
      xx=linspace(0,h,kk);
      plot([x(i)+xx],[bubble(i)*4/h^2*xx.*(h-xx)+...
            (linear(i)+(linear(i+1)-linear(i))/h*xx)],...
            '-k')
   end
end
%
plot(x,linear,':k')
plot(x,linear,'.k')
axis([a b c d])
set(findobj('type','line'),'linewidth',2,...
      'markersize',20)
```

4   Eigenvalue Problems in Mixed Form

The examples presented in Sect. 3 confirm that Galerkin discretizations of eigen-value problems in mixed form present a different behavior from standard Galerkin approximations of variationally posed eigenvalue problems. In particular, Sect. 3.1 shows that standard Galerkin approximations of the Laplace eigenvalue problem are optimally convergent as soon as we choose a reasonable approximating space. This is the important consequence of Theorem 1. Let us define $T : L^2 \to L^2$ using the source problem associated to (9): given $f \in L^2$ let $Tf \in H^1$ be the solution of the problem

$$\int_0^\pi (Tf)'(x)v'(x)\mathrm{d}x = \int_0^\pi f(x)v(x)\mathrm{d}x.$$

The discrete operator $T_h : L^2 \to L^2$ can be defined analogously using the discrete source problem. It turns out that the operator T is compact from L^2 into H^1 and that the elliptic projection onto standard finite element spaces converges pointwise to the identity operator from H^1 to L^2. Hence, we can apply Theorem 1 with the choice $H = L^2$ and $V = H^1$ in order to conclude that T_h converges to T in the space $\mathscr{L}(L^2)$.

On the other hand, Examples 1, 2 and 3 show that in the case of mixed approximation we need to impose suitable compatibility assumptions between the two approximating finite element spaces. This fact is not surprising since we are used to the classical inf-sup conditions [29] for mixed finite elements, but we shall see that for eigenvalue problems the situation is different from that of the source problem.

4.1   Inf-Sup Conditions

We now discuss Examples 1, 2, and 3 in respect to the classical inf-sup conditions.

It is well-known that for the well-posedness of the *source* problem associated with (10) the following two conditions are sufficient and, in a suitable sense, necessary: the ellipticity in the discrete kernel

$$\|t_h\|_{L^2}^2 \geq \alpha \|t_h\|_{H^1}^2 \quad \forall t_h \in \mathbb{K}_h, \tag{11}$$

where $\mathbb{K}_h = \{t_h \in \Sigma_h : \int_0^\pi v(x)t_h'(x)\mathrm{d}x = 0 \ \forall v \in U_h\}$, and the inf-sup condition

$$\inf_{v_h \in U_h} \sup_{t_h \in \Sigma_h} \frac{\int_0^\pi v_h(x)t_h'(x)\mathrm{d}x}{\|v_h\|_{L^2}\|t_h\|_{H^1}} \geq \beta. \tag{12}$$

Remark 3. Conditions (11) and (12) should be changed to use the divergence operator instead of the first derivative and $H(\mathrm{div})$ instead of H^1 in the multidimensional case.

We start by showing that the P1-P0 element discussed in Example 2 satisfies both conditions (11) and (12).

The ellipticity in the discrete kernel is a trivial consequence of the fact that the derivative of a function in P1 is an object of P0, hence functions in $\mathbb{K}_h$ have vanishing derivative (take $v = t'$ in the definition of $\mathbb{K}_h$).

The inf-sup condition can be proved by constructing a Fortin operator (see [29, Prop. II.2.8]). We need to find $\Pi_h : H^1 \to \Sigma_h$ such that

$$\int_0^\pi (s'(x) - (\Pi_h s)'(x))v(x)\mathrm{d}x = 0 \quad \forall v \in U_h$$

and $\|\Pi_h s\|_{H^1} \leq C\|s\|_{H^1}$. It can be easily observed that the standard *nodal* interpolation operator $s \mapsto s_I$ satisfies the required properties; indeed, for each element $[x_i, x_{i+1}]$ we have

$$\int_{x_i}^{x_{i+1}} s_I'(x)\mathrm{d}x = s_I(x_{i+1}) - s_I(x_i) = s(x_{i+1}) - s(x_i) = \int_{x_i}^{x_{i+1}} s'(x)\mathrm{d}x.$$

Remark 4. The P1-P0 element is the one dimensional counterpart of the well-known Raviart–Thomas element

Moving to the P1-P1 element discussed in Example 1, it has been observed several times in the literature that it does not satisfies the inf-sup condition (see, for instance, [11]). What is surprising about the bad behavior reported in Table 4 is that the P1-P1 element is convergent for the corresponding *source* problem when the solution is smooth and the eigenfunctions of the Laplace problem are analytic. It turns out that the approximation of eigenvalue problems does not follow the same lines as the approximation of the corresponding source problem: there are several spurious modes even if the regularity of the eigenfunctions is not an issue. To be more precise, the zero frequency reported in Table 4 was expected and is a consequence of the lack of the inf-sup condition: there is a function $u_h \in U_h$ with $u_h \neq 0$ such that $\int_0^\pi u_h(x)t'(x)\mathrm{d}x = 0$ for all $t \in \Sigma_h$; such a function is the eigenfunction shown in Fig. 4 (top left); on the other hand, the other spurious modes cannot be predicted from the standard theory of mixed finite elements.

Let us conclude this section with the analysis of the P2-P0 element presented in Example 3 which can be seen as a modification of the P1-P0 element with an enrichment of the space Σ_h. The hierarchical construction of the matrices in Listing 1.6 shows explicitly the nature of the enrichment which consists of a single quadratic bubble in each element. Since the P1-P0 element satisfies the inf-sup condition (12), the P2-P0 satisfies the inf-sup condition as well (the supremum is taken over a larger space). On the other hand the enrichment of the space Σ_h implies a modification of the discrete kernel $\mathbb{K}_h$ for the definition of the ellipticity condition (11). In particular, all the element bubbles are elements of $\mathbb{K}_h$ (they vanish at the endpoints of the interval, hence their first derivatives have zero mean value) and, moreover, are functions for which the uniform ellipticity (11) does not hold as it can be easily observed by a standard scaling argument or by explicit computation. Figure 8 confirms that the bad behavior of the method is due to the presence of the bubbles which pollute the discrete solution. For similar consideration related to the corresponding source problem the reader is referred to [26].

4.2 Convergence of Eigenvalue Problems in Mixed Forms

From the discussion of Sect. 4.1 it might seem that the two main conditions for the stability of the mixed source problem, namely the ellipticity in the discrete kernel (11) and the inf-sup condition (12), are sufficient for the convergence of the mixed eigenvalue problem as well. On the other hand, the analysis of the P1-P1 element should warn the reader: in that case the approximation of the source problem is convergent when the solution is smooth enough, while the approximation of the eigenvalue problem presents spurious modes (even though all exact eigenfunctions are smooth).

The fundamental results contained in [22, 23] state that the natural conditions for the good approximation and the absence of spurious modes in the approximation of eigenvalue problems in mixed form *are not* the classical inf-sup conditions. We recall in this setting the conditions introduced in [22] (see also [21, Part 3]).

In order to convince the reader that the eigenvalue problem has a substantially different nature from the source problem, we try to repeat the argument of Theorem 1 in the framework of mixed approximations: we will show that the situation is now more complicated. Roughly speaking, Theorem 1 says that a suitable compactness assumption turns pointwise convergence into uniform convergence. In order to use a similar argument in this framework, we need to introduce a suitable solution operator T and to show a suitable compactness property. Since the solution of the source problem corresponding to (10) has two components s and u, we firstly have to choose how to define the solution operator. A first (and, as we shall see, wrong) possibility is to define $T_{\Sigma U} : L^2 \times L^2 \to L^2 \times L^2$ as follows:

$$(f, g) \overset{\text{cutoff}}{\longmapsto} (0, g) \overset{T_2}{\longmapsto} (s, u),$$

where the operator $T_2 : L^2 \times L^2 \to L^2 \times L^2$ corresponds to $T_2(f, g) = (s, t) \in H^1 \times L^2$ solution of the following source problem

$$
\begin{cases}
\displaystyle\int_0^\pi s(x)t(x)\mathrm{d}x + \int_0^\pi u(x)t'(x)\mathrm{d}x = \int_0^\pi f(x)t(x)\mathrm{d}x & \forall t \in H^1 \\
\displaystyle\int_0^\pi s'(x)v(x)\mathrm{d}x = -\int_0^\pi g(x)v(x)\mathrm{d}x & \forall v \in L^2.
\end{cases}
\tag{13}
$$

The discrete operator $T_{\Sigma U,h}$ can be defined analogously using the same cutoff function and the discrete source problem. In order to fit the framework of Theorem 1 one needs to introduce a suitable projection operator. This can be done by using the source mixed problem: let $Q_h : H^1 \times L^2 \to \Sigma_h$ and $R_h : H^1 \times L^2 \to U_h$ be defined starting from $(s, u) \in H^1 \times L^2$ in order to satisfy the following equations

$$\int_0^\pi Q_h(s,u)(x)t(x)\mathrm{d}x + \int_0^\pi R_h(s,u)(x)t'(x)\mathrm{d}x$$

$$= \int_0^\pi s(x)t(x)\mathrm{d}x + \int_0^\pi u(x)t'(x)\mathrm{d}x \quad \forall t \in H^1$$

$$\int_0^\pi (Q_h(s,u))'(x)v(x)\mathrm{d}x = \int_0^\pi s'(x)v(x)\mathrm{d}x \quad \forall v \in L^2.$$

Taking $P_h = (Q_h, R_h)$, it is clear that we have $T_{\Sigma U,h} = P_h T_{\Sigma U}$, so that we might think of adapting Theorem 1 to this situation.

Unfortunately, the compactness assumption on $T_{\Sigma U}$ *does not* hold. Indeed, taking $H = L^2 \times L^2$ and $V = H^1 \times L^2$, we would need $T_{\Sigma U}$ compact from $L^2 \times L^2$ in $H^1 \times L^2$ which is in conflict with the fact that in (13) the derivative of s is equal to $-g$. As a possible workaround, we can try to use the second comment contained in Remark 1. Unfortunately, for the same reason as before $T_{\Sigma U}$ is not compact from $H^1 \times L^2$ in $H^1 \times L^2$ either.

The correct approach for the definition of the solution operator has been introduced in [22]. Being interested in an eigenvalue problem which involves the eigenfunction u, the natural choice is to define $T : L^2 \to L^2$ as follows: given $g \in L^2$ find $s \in H^1$ and $Tg \in U_h$ such that

$$\begin{cases} \displaystyle\int_0^\pi s(x)t(x)\mathrm{d}x + \int_0^\pi Tg(x)t'(x)\mathrm{d}x = 0 \quad \forall t \in H^1 \\ \displaystyle\int_0^\pi s'(x)v(x)\mathrm{d}x = -\int_0^\pi g(x)v(x)\mathrm{d}x \quad \forall v \in L^2. \end{cases} \tag{14}$$

The discrete operator T_h can be defined analogously using the discrete source problem. In [22] sufficient (and, in a suitable sense, necessary) conditions for the convergence of T_h to T have been introduced. These conditions, which in particular imply the good approximation of Problem (10), are:

the *weak approximability* of H^2, that is

$$\int_0^\pi v(x)t_h'(x)\mathrm{d}x \le \rho(h)\|t_h\|_{L^2}\|v\|_{H^2} \quad \forall v \in H^2 \ \forall t_h \in \mathbb{K}_h, \tag{15}$$

where here and in the next two properties $\rho(h)$ denotes a quantity that tends to zero as h goes to zero;

the *strong approximability* of H^2, that is for all $v \in H^2$ there exists $v_I \in \mathbb{K}_h$ such that

$$\|v - v_I\|_{H^1} \le \rho(h)\|v\|_{H^2}; \tag{16}$$

the *Fortid* condition, that is there exists a bounded Fortin operator $\Pi_h : H^1 \to \Sigma_h$ converging in norm to the identity

$$\|t - \Pi_h t\|_{L^2} \le \rho(h)\|t\|_{H^1} \quad \forall t \in H^1. \tag{17}$$

For the sake of completeness, we recall that a bounded Fortin operator satisfies

$$\int_0^\pi (s'(x) - (\Pi_h s)'(x)) v(x) dx = 0 \quad \forall s \in H^1 \ \forall v \in U_h$$

and $\|\Pi_h s\|_{H^1} \leq C \|s\|_{H^1}$.

Coming back to the three mixed methods discussed in Examples 1, 2, and 3, it can be seen that all of them satisfy the strong approximability property, while only the P1-P0 element satisfies the remaining two properties. Indeed, we have already shown in Sect. 4.1 that it satisfies the ellipticity in the kernel property (which implies the weak approximability property) and we proved that the interpolation operator is a Fortin operator (which easily implies the Fortid property). On the other hand, the P1-P1 element does not satisfy the Fortid property (it is known not to meet the inf-sup condition) and the P2-P0 element does not satisfy the weak approximability condition.

5 Error Estimates for Multiple Eigenvalues

This part of the paper is devoted to a priori error estimates for the eigenvalue problem in variational formulation. After a brief review of the fundamental results on spectral approximation mainly based on the theory of Babuška and Osborn [12], in the second section, we focus on the problem of error estimates for multiple eigenvalues. The main result of this section, due to Knyazev and Osborn [57], shows that the eigenvalue errors depend mainly on the approximability of the corresponding eigenspace. We end this section with some numerical results.

5.1 Fundamental Results on Spectral Approximation

In this section, we recall the error estimates, collected in [21], which show how the eigenvalues and eigenfunctions of T are approximated by those of T_h and then how they apply to the case of variationally posed eigenproblems. Throughout this section we assume that X is a Hilbert space with inner product (u, v) and that $T : X \to X$ is a compact self-adjoint positive linear operator. Let $T_h : X \to X$ be a family of compact self-adjoint positive linear operators of finite rank. We assume that T_h converges uniformly to T, that is

$$\|T - T_h\|_{\mathscr{L}(X)} \to 0 \quad \text{as } h \to 0. \tag{18}$$

Let μ be an eigenvalue of T of algebraic multiplicity m. Since T is self-adjoint and X is a Hilbert space, then the ascent of μ is 1. Let $\mu_{i,h}$ for $i = 1, \ldots, m$ be the eigenvalues, repeated according to their multiplicity, of T_h converging to μ. We denote by $E \subseteq X$ the eigenspace associated to μ and by E_h the direct sum of all the eigenspaces associated to the eigenvalues $\mu_{i,h}$. Then it holds

$$\hat{\delta}(E, E_h) \leq C \|(T - T_h)_{|E}\|_{\mathscr{L}(X)}, \tag{19}$$

where $\hat{\delta}(E, F)$ represents the gap between Hilbert subspaces and is defined by

$$\delta(E, F) = \sup_{\substack{u \in E \\ \|u\|_X = 1}} \inf_{v \in F} \|u - v\|_X, \quad \hat{\delta}(E, F) = \max(\delta(E, F), \delta(F, E)). \tag{20}$$

Let us introduce some notation and observations which will be useful later on. For nonzero functions u and v, if $E = \text{span}\{u\}$, we write $\delta(u, F)$ instead of $\delta(E, F)$ and if $E = \text{span}\{u\}$ and $F = \text{span}\{v\}$, we write $\delta(u, v)$ for $\delta(E, F)$. We have $0 \leq \delta(E, F) \leq 1$ and $\delta(E, F) = 0$ if and only if $E \subseteq F$. If $\dim E = \dim F < \infty$ then $\delta(E, F) = \delta(F, E)$.

In the remainder of the section we shall have $\dim E \leq \dim F$, then the following Lemma holds true, see [28, Lemma 3.4].

Lemma 1. *Let $\{\phi_i, \ i = 1, \ldots, \dim E\}$ form an orthogonal basis for the subspace E. Then*

$$\delta^2(E, F) \leq \sum_i \delta^2(\phi_i, F).$$

If P and Q are the orthogonal projections onto E and F, respectively, then $\delta(E, F)$ equals the largest singular value of the operator $(I - Q)P$ and

$$\delta(E, F) = \|(I - Q)P\|_X. \tag{21}$$

Let us now go back to the approximation of eigenvalues and eigenfunctions. We have the following error estimate for the eigenvalues.

Theorem 2. *Let $\{\phi_1, \ldots, \phi_m\}$ be a basis of the eigenspace E associated to the eigenvalue μ. Then, for $i = 1, \ldots, m$*

$$|\mu - \mu_{i,h}| \leq C \left(\sum_{j,k=1}^{m} |((T - T_h)\phi_j, \phi_k)| + \|(T - T_h)_{|E}\|_{\mathscr{L}(X)}^2 \right). \tag{22}$$

Moreover, we have the following estimate for the eigenfunctions.

Theorem 3. *Let $\{\mu_h\}$ be a sequence of discrete eigenvalues of T_h converging to a non-zero eigenvalue μ of T. Consider a sequence $\{u_h\}$ of unit vectors in the eigenspace E_h associated to μ_h. Then there exists an eigenfunction $u(h)$ associated to the eigenvalue μ of T such that*

$$\|u(h) - u_h\|_X \leq C \|(T - T_h)_{|E}\|_{\mathscr{L}(X)}.$$

We first observe that Theorems 2 and 3 also hold in the case of operators which are not self-adjoint. In such a case, one has to take into account the concept of ascent multiplicity of $\mu - T$.

As a corollary of the above theorems we obtain the error estimates for eigenvalues and eigenfunctions of problem (1). In order to embed the operator T defined in Sect. 2 (see (3)) into the abstract setting of this section, we assume that $T : V \to V$ is compact, hence we can apply the results of Theorems 2 and 3 with $X = V$ and $\lambda = 1/\mu$. This is the case, for instance, if $V \subset H$ is compact; the convergence (18) is a consequence of Theorem 1 (see Remark 1).

Corollary 1. *Let λ be an eigenvalue of problem* (1) *and let $\lambda_{i,h}$, for $i = 1, \ldots, m$ be the eigenvalues of problem* (2) *converging to λ. Then we have for $i = 1, \ldots, m$:*

$$|\lambda - \lambda_{i,h}| \le C \sup_{\substack{u \in E \\ \|u\|=1}} \inf_{v \in V_h} \|u - v\|_V^2. \tag{23}$$

Corollary 2. *Let $\{\lambda_h\}$ be a sequence of discrete eigenvalues of* (2) *converging to an eigenvalue λ of* (1). *Consider a sequence $\{u_h\}$ of unit vectors in the eigenspace E_h associated to λ_h. Then there exists an eigenfunction $u(h)$ associated to the eigenvalue λ of* (1) *such that*

$$\|u(h) - u_h\|_V \le C \sup_{\substack{u \in E \\ \|u\|=1}} \inf_{v \in V_h} \|u - v\|_V.$$

We notice here that one can deduce analogous results using the less strong assumption (6). We refer to [21, Sect. 10], for instance, for a discussion on the error estimates for the Laplace eigenproblem.

From the last theorems, we infer that the rate of convergence of multiple eigenvalues depends on the rate of approximability of the corresponding eigenspace, hence on the approximation rate of the least regular eigenfunction. On the other hand, the numerical experiments reported in [12, Sect. 10] show different rates of convergence for the approximate eigenvalues in the presence of eigenfunctions having different approximabilities. In the next section this point will be addressed following the ideas of Knyazev and Osborn [57] in the case of variationally posed eigenproblems.

5.2 Error Estimates for Ritz-Galerkin Approximation of Multiple Eigenvalues

This section is devoted to sharp error estimates in the case of multiple eigenvalues which take into account the possibility that a multiple eigenvalue might be associated to eigenfunctions with different regularities. In particular, we shall see that in this case it is possible to identify discrete eigenvalues converging to the multiple eigenvalue with different rates of convergence which take into account the regularities of the corresponding eigenfunctions. More precisely, the rate of convergence towards an eigenvalue λ of multiplicity m associated to eigenfunctions with different regularities depends on the regularity of the eigenspace which is approximated by the sequence of the discrete eigenvectors associated to the m eigenvalues $\lambda_{i,h}$ for $1 \leq i \leq m$ which converge to λ.

The results we are going to report here are based on the theory developed in [57].

Let us go back to the setting of Sect. 2. Let V_h a finite dimensional subspace of V with $\dim V_h = N$. Let $T : V \to V$ and $T_h : V \to V$ be defined, respectively, in (3) and in (4). Then we have that T_h is the Ritz approximation of $T : V \to V$ since

$$T_h = P_h T \tag{24}$$

where $P_h : V \to V_h$ is the elliptic projection onto V_h defined as follows: for all $u \in V$, $P_h u \in V_h$ is such that

$$a(P_h u - u, v) = 0 \quad \forall v \in V_h. \tag{25}$$

We assume that the bilinear form $a : V \times V \to \mathbb{R}$ is coercive and continuous in V, that is

$$\alpha \|u\|_V^2 \leq a(u, u) \quad \forall u \in V, \quad \text{with } \alpha > 0,$$
$$|a(u, v)| \leq C \|u\|_V \|v\|_v \quad \forall u, v \in V, \quad \text{with } C > 0 \tag{26}$$

Since $\sqrt{a(u, u)}$ is an equivalent norm in V, in this section we will use the following norm in V

$$\|u\|_V = \sqrt{a(u, u)}. \tag{27}$$

Notice that the elliptic projection P_h results in an orthogonal projection with respect to this norm.

Let us denote by $0 < \lambda_1 \leq \lambda_2 \leq \ldots$ the eigenvalues of (1) and by $0 < \lambda_{1,h} \leq \lambda_{2,h} \leq \cdots \leq \lambda_{N,h}$ those of (2), both repeated according to their algebraic multiplicity. Moreover, we denote by u_i the eigenfunction associated to the eigenvalue λ_i and by $u_{i,h}$ the discrete eigenfunction associated to $\lambda_{i,h}$, that is

$$a(u_i, v) = \lambda_i b(u_i, v) \quad \forall v \in V$$
$$a(u_{i,h}, v) = \lambda_{i,h} b(u_{i,h}, v) \quad \forall v \in V_h.$$

From now on we assume that $b(u_i, u_j) = \delta_{ij}$ and $b(u_{i,h}, u_{j,h}) = \delta_{ij}$ for $i, j = 1, \ldots, N$. Notice that this implies

$$a(u_i, u_i) = \lambda_i$$
$$a(u_i, u_j) = 0 \quad i \neq j$$

and

$$a(u_{i,h}, u_{i,h}) = \lambda_{i,h}$$

$$a(u_{i,h}, u_{j,h}) = 0 \quad i \neq j.$$

Let $E_{1,\ldots,i} \subset V$ (resp. $E_{1,\ldots,i,h} \subset V_h$) denote the span of the first i eigenvectors $u_1, \ldots, u_i$ (resp. $u_{1,h}, \ldots, u_{i,h}$) and let $P_{1,\ldots,i}$ (resp. $P_{1,\ldots,i,h}$) be the elliptic projection onto $E_{1,\ldots,i}$ (resp. $E_{1,\ldots,i,h}$), that is

$$a(u - P_{1,\ldots,i}u, v) = 0 \quad \forall v \in E_{1,\ldots,i}$$

$$a(u - P_{1,\ldots,i,h}u, v) = 0 \quad \forall v \in E_{1,\ldots,i,h}.$$

We recall here the following characterization of the eigenvalues by means of the Rayleigh quotient:

$$\lambda_1 = \min_{\substack{v \in V \\ v \neq 0}} \frac{a(v,v)}{b(v,v)}, \lambda_i = \min_{\substack{v \in \left(\oplus_{j=1}^{i-1} E_j\right)^{\perp} \\ v \neq 0}} \frac{a(v,v)}{b(v,v)},$$

$$\lambda_{1,h} = \min_{\substack{v \in V_h \\ v \neq 0}} \frac{a(v,v)}{b(v,v)}, \lambda_{i,h} = \min_{\substack{v \in \left(\oplus_{j=1}^{i-1} E_{j,h}\right)^{\perp} \\ v \neq 0}} \frac{a(v,v)}{b(v,v)}, \tag{28}$$

Moreover, the i-th eigenvalue λ_i of (1) and the i-th discrete eigenvalue $\lambda_{i,h}$ of (2) satisfy:

$$\lambda_i = \min_{E \in V^{(i)}} \max_{v \in E} \frac{a(v,v)}{b(v,v)}, \quad \lambda_{i,h} = \min_{E \in V_h^{(i)}} \max_{v \in E} \frac{a(v,v)}{b(v,v)} \tag{29}$$

where $V^{(i)}$ and $V_h^{(i)}$ denote the set of all subspaces of V, respectively V_h with dimension equal to i (see, for instance, [21, Prop. 7.2]).

Notice that as a consequence of (29) we have that

$$\lambda_i \leq \lambda_{i,h} \quad \text{for } i = 1, \ldots, N. \tag{30}$$

In the case of self-adjoint operators and of their Ritz approximation, one can make more precise the statement of Theorem 2 with the following result, proved in [55], which shows that the error for an eigenvalue depend on the approximability of all previous eigenvectors.

Theorem 4. *For $i = 1, \ldots, N$ we have*

$$0 \leq \frac{\lambda_{i,h} - \lambda_i}{\lambda_{i,h}} \leq \delta^2(E_{1,\ldots,i}, V_h) = \|(I - P_h)P_{1,\ldots,i}\|_{\mathscr{L}(V)}^2. \tag{31}$$

Proof. We have $\delta(E_{1,\ldots,i}, V_h) \leq 1$. If $\delta(E_{1,\ldots,i}, V_h) = 1$ then (31) is obviously true. Hence, let us suppose that

$$\delta(E_{1,\ldots,i}, V_h) < 1.$$

Since dim $E_{1,...,i} = i \le \dim V_h = N < +\infty$, we can apply (21) and obtain

$$\delta(E_{1,...,i}, V_h) = \|(I - P_h)P_{1,...,i}\|_{\mathscr{L}(V)},$$

from which, thanks to [53, Theorem 6.34, Cap. I], we deduce that P_h provides a one-to-one map from $E_{1,...,i}$ to $P_h E_{1,...,i}$. Therefore dim $P_h E_{1,...,i} = \dim E_{1,...,i} = i$.

Let us take $\bar{u} \in P_h E_{1,...,i}$ such that $\|\bar{u}\|_V = 1$ and

$$\lambda(\bar{u}) = \max_{w \in P_h E_{1,...,i}, w \neq 0} \lambda(w)$$

where $\lambda(w)$ is the Rayleigh quotient defined by

$$\lambda(w) = \frac{a(w, w)}{b(w, w)}.$$

We recall that thanks to the coercivity assumption (26) and to the norm definition (27) we also have $a(\bar{u}, \bar{u}) = \|\bar{u}\|_V^2 = 1$.

We consider the following orthogonal decomposition of $\bar{u}$ in V

$$\bar{u} = u + v \quad \text{for } u \in E_{1,...,i}, \; v \in E_{1,...,i}^\perp. \tag{32}$$

Hence, we have $a(v, w) = 0$ for all $w \in E_{1,...,i}$ and consequently $a(v, Tw) = 0$, since $E_{1,...,i}$ is an invariant subspace of T. By the definition of T we also get that $b(v, w) = a(v, Tw) = 0$ for all $w \in E_{1,...,i}$.

The definition (32) of v yields

$$\|v\|_V = \delta(\bar{u}, E_{1,...,i}) \le \delta(P_h E_{1,...,i}, E_{1,...,i})$$
$$= \delta(E_{1,...,i}, P_h E_{1,...,i}) = \delta(E_{1,...,i}, V_h) < 1, \tag{33}$$

since $P_h E_{1,...,i}$ and $E_{1,...,i}$ have the same dimension.

From (33) we obtain that $u \neq 0$ so that $\lambda(u)$ is well defined. We now show that

$$\lambda(u) \le \lambda_i \le \lambda_{i,h} \le \lambda(\bar{u}). \tag{34}$$

We already know that $\lambda_i \le \lambda_{i,h}$ (see (30)). By the definition of $\bar{u}$ and the min-max characterization of the eigenvalues (28) the last inequality holds true. It remains to prove the first one. Since $u \in E_{1,...,i}$, we have $u = \sum_{j=1}^{i} \alpha_j u_j$ and

$$\lambda(u) = \frac{a(u, u)}{b(u, u)} = \frac{\sum_{j=1}^{i} \alpha_j^2 a(u_j, u_j)}{\sum_{j=1}^{i} \alpha_j^2 b(u_j, u_j)} = \frac{\sum_{j=1}^{i} \alpha_j^2 \lambda_j}{\sum_{j=1}^{i} \alpha_j^2} \le \lambda_i.$$

due to the orthogonalities of the eigenfunctions.

We observe that

$$\lambda(\bar{u}) = \frac{a(u, u) + a(v, v)}{b(u, u) + b(v, v)}$$

If $v = 0$, then $\lambda(\bar{u}) = \lambda(u)$. If $v \neq 0$, a direct calculation gives

$$\frac{1}{\lambda(u)} - \frac{1}{\lambda(\bar{u})} = \left(\frac{1}{\lambda(\bar{u})} - \frac{1}{\lambda(v)}\right) \frac{a(v, v)}{a(u, u)} \leq \frac{1}{\lambda(\bar{u})} \frac{a(v, v)}{a(u, u)} \leq \frac{1}{\lambda_{i,h}} \frac{a(v, v)}{a(u, u)}.$$

It is now easy to see that

$$0 \leq \frac{1}{\lambda_i} - \frac{1}{\lambda_{i,h}} \leq \frac{1}{\lambda(u)} - \frac{1}{\lambda(\bar{u})} \leq \frac{1}{\lambda_{i,h}} \frac{a(v, v)}{a(u, u)},$$

which implies

$$\frac{\lambda_{i,h}}{\lambda_i} \leq \frac{a(v, v)}{a(u, u)} + 1 = \frac{1}{a(u, u)}$$

and

$$\frac{\lambda_{i,h} - \lambda_i}{\lambda_{i,h}} = 1 - \frac{\lambda_i}{\lambda_{i,h}} \leq 1 - a(u, u) = a(v, v) = \|v\|_V.$$

This inequality together with (33) concludes the proof of the theorem. ☐

We see that the error estimate for the eigenvalue in Theorem 4 depends on the approximability properties of all previous eigenvectors, while in Corollary 1 it depends on the approximability properties of eigenspace associated to the eigenvalue of interest. On the other hand we see that the estimate (31) does not depend on any undetermined constant. Let us consider an eigenvalue λ_p with multiplicity $m > 1$, then from the above theorem it is easy to derive the following result.

Corollary 3. *Assume that*

$$\lambda_{p-1} < \lambda_p = \dots \lambda_{p+m-1} < \lambda_{p+m}, \tag{35}$$

with $p + m - 1 \leq N$. Then for any index $i = p, \dots, p + m - 1$ we have

$$0 \leq \frac{\lambda_{i,h} - \lambda_p}{\lambda_{i,h}} \leq \inf_{\substack{E_{1,\dots,p-1} \subset E_{1,\dots,i} \subseteq E_{1,\dots,p+m-1} \\ \dim E_{1,\dots,i} = i}} \delta^2(E_{1,\dots,i}, V_h)$$

$$= \delta^2(E_{1,\dots,p+m-1}, V_h). \tag{36}$$

Corollary 3 provides different estimates for every eigenvalue, but it requires approximability of all previous eigenvectors.

The following lemma suggests that the rate of convergence of the error for the multiple eigenvalue does not necessarily depend on the approximability of all the associated eigenfunctions.

Lemma 2. *For $i = 1, \ldots, N$, the following relation holds true*

$$0 \leq \frac{\lambda_{i,h} - \lambda_i}{\lambda_{i,h}} = \frac{1}{\lambda_i} \|(I - P_{i,h})u_i\|_V^2 - \frac{\lambda_i}{\lambda_{i,h}} a((I - P_i)u_{i,h}, T(I - P_i)u_{i,h})$$

where P_i and $P_{i,h}$ are the elliptic projection onto $\mathrm{span}\{u_i\}$ and $\mathrm{span}\{u_{i,h}\}$, respectively.

Proof. The proof is quite simple. Using the definition of the Rayleigh quotient and the equality $b(u_i, u_i) = b(u_{i,h}, u_{i,h}) = 1$, we have $\lambda_{i,h} = a(u_{i,h}, u_{i,h})$ and $1 = b(u_{i,h}, u_{i,h}) = a(T_h u_{i,h}, u_{i,h})$, hence

$$0 \leq \frac{\lambda_{i,h} - \lambda_i}{\lambda_i} = \frac{\lambda_{i,h}}{\lambda_i} - 1 = \frac{1}{\lambda_i} a(u_{i,h}, u_{i,h}) - a(T_h u_{i,h}, u_{i,h})$$

$$= \frac{1}{\lambda_i} a((I - P_i)u_{i,h}, (I - P_i)u_{i,h}) + \frac{1}{\lambda_i} a(u_{i,h}, P_i u_{i,h}) - a(T u_{i,h}, u_{i,h})$$

$$= \frac{1}{\lambda_i} \|(I - P_i)u_{i,h}\|_V^2 + \frac{1}{\lambda_i} a(u_{i,h}, P_i u_{i,h})$$
$$\quad - a(T(I - P_i)u_{i,h}, u_{i,h}) - a(T P_i u_{i,h}, u_{i,h})$$

$$= \frac{1}{\lambda_i} \|(I - P_i)u_{i,h}\|_V^2 + a\left(u_{i,h}, \frac{1}{\lambda_i} P_i u_{i,h} - T P_i u_{i,h}\right)$$

$$= \frac{1}{\lambda_i} \|(I - P_i)u_{i,h}\|_V^2 - a((I - P_i)u_{i,h}, T(I - P_i)u_{i,h})$$

thanks to $\left(\frac{1}{\lambda_i} I - T\right) P_i u_{i,h} = 0$ and $a(T(I - P_i)u_{i,h}, P_i u_{i,h}) = a((I - P_i)u_{i,h}, T P_i u_{i,h}) = 0$. We conclude the proof by observing that

$$\frac{1}{\lambda_{i,h}} \|(I - P_i)u_{i,h}\|_V^2 = \frac{1}{\lambda_i} \|(I - P_{i,h})u_i\|_V^2,$$

so that

$$0 \leq \frac{\lambda_{i,h} - \lambda_i}{\lambda_{i,h}} = \frac{\lambda_i}{\lambda_{i,h}} \frac{\lambda_{i,h} - \lambda_i}{\lambda_i}$$

$$= \frac{1}{\lambda_{i,h}} \|(I - P_i)u_{i,h}\|_V^2 - \frac{\lambda_i}{\lambda_{i,h}} a((I - P_i)u_{i,h}, T(I - P_i)u_{i,h}).$$

$\square$

Since T is positive, we immediately obtain using (21)

$$0 \leq \frac{\lambda_{i,h} - \lambda_i}{\lambda_i} \leq \frac{1}{\lambda_i} \|(I - P_{i,h})u_i\|^2 \leq \frac{1}{\lambda_i} \delta^2(u_i, u_{i,h}).$$

Lemma 2 shows that the estimate on the i-th eigenvalue depends explicitly on both the continuous and the discrete associated eigenfunctions u_i and $u_{i,h}$.

The next theorem, proved in [57], provides an estimate which does not depend explicitly on the approximate eigenfunction $u_{i,h}$, but only on the approximability properties of u_i in the discrete space V_h.

Theorem 5. *Let us fix an index i with $1 \le i \le N$ such that*

$$\min_{j=1,\dots,i-1} |\lambda_{j,h} - \lambda_i| \neq 0, \tag{37}$$

then

$$0 \le \frac{\lambda_{i,h} - \lambda_i}{\lambda_{i,h}} \le \frac{\|(I - P_h + P_{1,\dots,i-1,h})u_i\|_V^2}{\|u_i\|_V^2}$$

$$\le \left(1 + \max_{j=1,\dots,i-1} \frac{\lambda_{j,h}^2 \lambda_j^2}{|\lambda_{j,h} - \lambda_i|^2} \|(I - P_h)TP_{1,\dots,i-1,h}\|_{\mathscr{L}(V)}^2 \right) \delta^2(u_i, V_h), \tag{38}$$

where $P_{1,\dots,i-1,h}$ is the elliptic projection onto $E_{1,\dots,i-1,h} = \mathrm{span}\{u_{1,h}, \dots, u_{i-1,h}\}$ defined as follows: for $u \in V$, $P_{1,\dots,i-1,h}u \in E_{1,\dots,i-1,h}$ such that

$$a(u - P_{1,\dots,i-1,h}u, v) = 0 \qquad \forall v \in E_{1,\dots,i-1,h},$$

Proof. The proof of the first inequality in (5) follows the same lines as that of Theorem 31. Let $i > 1$, since the case $i = 1$ is already covered by Theorem 4.

Let $\tilde{u}_i = u_i/\|u_i\|_V$. Since the operator $I - P_h + P_{1,\dots,i-1,h}$ is an orthogonal projection in V with respect to the norm defined in (27) and $\tilde{u}_i$ is normalized, we have $\|(I - P_h + P_{1,\dots,i-1,h})\tilde{u}_i\|_V \le 1$. We assume that $\|(I - P_h + P_{1,\dots,i-1,h})\tilde{u}_i\|_V < 1$, since otherwise the inequality (38) is obviously true.

Let $E_i = \mathrm{span}\{u_i\}$, then $\dim E_i = 1$ and also $\dim(P_h - P_{1,\dots,i-1,h})E_i = 1$. Then by [53, Theorem 6.34, Cap. I] we obtain

$$\delta(E_i, (P_h - P_{1,\dots,i-1,h})E_i) = \delta(u_i, (P_h - P_{1,\dots,i-1,h})u_i) = \|(I - P_h + P_{1,\dots,i-1,h})\tilde{u}_i\|_V < 1.$$

Let us take $\bar{u} \in (P_h - P_{1,\dots,i-1,h})E_i$ such that $\|\bar{u}\|_V = 1$ and consider its orthogonal decomposition with respect to the norm of V as follows:

$$\bar{u} = u + v \quad \text{with } u \in E_{1,\dots,i}, \ v \in (E_{1,\dots,i})^\perp,$$

so that

$$a(v, w) = a(\bar{u} - u, w) = 0 \quad \forall w \in E_{1,\dots,i}.$$

Hence

$$\|v\|_V = \|\bar{u} - u\|_V = \inf_{w \in E_{1,\dots,i}} \|\bar{u} - w\|_V$$

$$= \delta(\bar{u}, E_{1,\dots,i}) \le \delta(\bar{u}, u_i) = \delta((P_h - P_{1,\dots,i-1,h}u_i, u_i) < 1. \tag{39}$$

This implies that $u \neq 0$ and that $\lambda(u)$ is defined. As in Theorem 4, if we show that

$$\lambda(u) \leq \lambda_i \leq \lambda_{i,h} \leq \lambda(\bar{u}), \tag{40}$$

then we get

$$0 \leq \frac{\lambda_{i,h} - \lambda_i}{\lambda_{i,h}} \leq \|v\|_V,$$

and this would conclude the proof of the theorem.

We prove only the last inequality in (40), since the others are quite standard.

We observe that the operator $P_h - P_{1,\ldots,i-1,h}$ is the projection onto the subspace of V_h spanned by the eigenfunctions $u_{j,h}$ with $j = i, \ldots, N$. Hence, by definition, we have

$$\bar{u} \in \left(\bigoplus_{j=1}^{i-1} E_{j,h} \right)^{\perp},$$

then (28) yields

$$\lambda(\bar{u}) = \frac{a(\bar{u}, \bar{u})}{b(\bar{u}, \bar{u})} \geq \min_{\substack{w \in \left(\oplus_{j=1}^{i-1} E_{j,h} \right)^{\perp} \\ w \neq 0}} \frac{a(w, w,)}{b(w, w)} = \lambda_{i,h}.$$

It remains to obtain the second line of (38). We have

$$\|(I - P_h + P_{1,\ldots,i-1,h})\tilde{u}_i\|_V^2 = \|(I - P_h)\tilde{u}_i\|_V^2 + \|P_{1,\ldots,i-1,h}\tilde{u}_i\|_V^2.$$

In order to use [56, Theorem 3.2], we observe that $P_{1,\ldots,i-1,h}TP_{1,\ldots,i-1,h|E_{1,\ldots,i-1,h}} = T_{h|E_{1,\ldots,i-1,h}}$, hence the spectrum of $P_{1,\ldots,i-1,h}TP_{1,\ldots,i-1,h|E_{1,\ldots,i-1,h}}$ is the set of the eigenvalues $1/\lambda_{j,h}$ for $j = 1, \ldots, i-1$ of T_h. Then we have

$$\|P_{1,\ldots,i-1,h}\tilde{u}_i\|_V \leq \frac{\|(I - P_h)TP_{1,\ldots,i-1,h}\|_{\mathscr{L}(V)}}{d_h} \|(I - P_h)\tilde{u}_i\|_V$$

where

$$d_h = \min_{j=1,\ldots,i-1} \left| \frac{1}{\lambda_{j,h}} - \frac{1}{\lambda_i} \right| = \min_{j=1,\ldots,i-1} \frac{|\lambda_{j,h} - \lambda_i|}{|\lambda_{j,h}\lambda_i|}.$$

$\square$

Remark 5. Let us assume that each $\lambda_{i,h}$ converges to λ_i, then the assumption (37) reads:

$$\min_{j=1,\ldots,i-1} |\lambda_{j,h} - \lambda_i| \approx \lambda_i - \lambda_{i-1}.$$

Notice that this quantity enters in the denominator of (38), hence the constant in the second line of (38) increases as $\lambda_i - \lambda_{i-1}$ becomes smaller.

In order to better explain the result presented in Theorem 5, let us consider the case of an eigenvalue λ_p with multiplicity $m = 2$ so that

$$\lambda_{p-1} < \lambda_p = \lambda_{p+1} < \lambda_{p+2}.$$

We can choose $i = p$ or $i = p + 1$. If $i = p$ then the denominator in (38) approximates $\lambda_p - \lambda_{p-1}$ hence it is strictly positive. For $i = p + 1$, instead, we have

$$\min_{j=1,\ldots,p} |\lambda_{j,h} - \lambda_{p+1}| \approx |\lambda_{p,h} - \lambda_{p+1}|$$

and it tends to 0 as $h \to 0$.

We can make the result of Theorem 5 more precise in the following corollary.

Corollary 4. *Assume that the eigenvalue λ_p with $p > 1$ has multiplicity $m > 1$ so that (35) holds with $p + m - 1 \leq N$ and that (37) holds true for $i = p$. Let $E_{p,\ldots,p+m-1}$ be the corresponding eigenspace, then*

$$0 \leq \frac{\lambda_{p,h} - \lambda_p}{\lambda_{p,h}} \leq \min_{u \in E_{p,\ldots,p+m-1}, \, \|u\|_V = 1} \|(I - P_h + P_{1,\ldots,p-1,h})u\|_V^2$$

$$\leq \left(1 + \max_{j=1,\ldots,p-1} \frac{\lambda_{j,h}^2 \lambda_p^2}{|\lambda_{j,h} - \lambda_p|^2} \|(I - P_h)T P_{1,\ldots,p-1,h}\|_{\mathscr{L}(V)}^2 \right) \min_{\substack{u \in E_{p,\ldots,p+m-1} \\ \|u\|_V = 1}} \delta^2(u, V_h)$$

$$= \left(1 + \max_{j=1,\ldots,p-1} \frac{\lambda_{j,h}^2 \lambda_p^2}{|\lambda_{j,h} - \lambda_p|^2} \|(I - P_h)T P_{1,\ldots,p-1,h}\|_{\mathscr{L}(V)}^2 \right) \delta^2(E_{p,\ldots,p+m-1}, V_h).$$

Notice that Corollary 4 gives an estimate containing only the gap between the eigenspace spanned by the m eigenfunctions associated to λ_p and the discrete space V_h, even if the approximability of the previous eigenfunctions still appears in the constant.

The final result of this section provides an error estimate for λ_p which can take into account the case of eigenfunctions associated to a multiple eigenvalue with different approximability properties. In addition, this estimate also covers the case of clustered eigenvalues.

Theorem 6. *Let i and q be fixed with $1 \leq i \leq N$ and $1 \leq q \leq i$. Let us denote by $E_{i-q+1,\ldots,i}$ the q-dimensional invariant subspace corresponding to eigenvalues $\lambda_{i-q+1} \leq \cdots \leq \lambda_i$ and by $P_{i-q+1,\ldots,i}$ the elliptic projection onto $E_{i-q+1,\ldots,i}$. If*

$$\min_{j=1,\ldots,i-q} |\lambda_{j,h} - \lambda_i| \neq 0 \tag{41}$$

then the following error estimate holds true

$$0 \leq \frac{\lambda_{i,h} - \lambda_i}{\lambda_{i,h}} \leq \|(I - P_h + P_{1,\ldots,i-q,h})P_{i-q+1,\ldots,i}\|^2_{\mathscr{L}(V)}$$

$$\leq \left(1 + \max_{j=1,\ldots,i-q} \frac{\lambda_{j,h}^2 \lambda_i^2}{|\lambda_{j,h} - \lambda_i|^2} \|(I - P_h)TP_{1,\ldots,i-q,h}\|^2_{\mathscr{L}(V)}\right)\|(I - P_h)P_{i-q+1,\ldots,i}\|^2_{\mathscr{L}(V)}$$

$$(42)$$

where $P_{1,\ldots,i-q,h}$ is the elliptic projection onto $E_{1,\ldots,i-q,h} = \mathrm{span}\{u_{1,h},\ldots,u_{i-q,h}\}$.

Proof. The proof is similar to that of Theorem 5. The operators $I - P_h + P_{1,\ldots,i-q,h}$ and $P_{i-q+1,\ldots,i}$ are orthogonal projections with respect to the norm of V, therefore $\|(I - P_h + P_{1,\ldots,i-q,h})P_{i-q+1,\ldots,i}\|_{\mathscr{L}(V)} \leq 1$. We consider the case

$$\|(I - P_h + P_{1,\ldots,i-q,h})P_{i-q+1,\ldots,i}\|_{\mathscr{L}(V)} < 1,$$

since otherwise the inequality (42) is obviously true. By [53, Theorem 3.6, Cap. I] $\dim(P_h - P_{1,\ldots,i-q,h})E_{i-q+1,\ldots,i} = \dim E_{i-q+1,\ldots,i} = q$.

We choose $\bar{u} \in (P_h - P_{1,\ldots,i-q,h})E_{i-q+1,\ldots,i}$ such that $\|\bar{u}\|_V = 1$ and

$$\lambda(\bar{u}) = \max_{\substack{w \in (P_h - P_{1,\ldots,i-q,h})E_{i-q+1,\ldots,i} \\ w \neq 0}} \lambda(w),$$

where $\lambda(\bar{u})$ is the Rayleigh quotient. Let us consider the following V-orthogonal decomposition of $\bar{u}$:

$$\bar{u} = u + v, \qquad \text{with } u \in E_{1,\ldots,i}, \ v \in (E_{1,\ldots,i})^\perp,$$

then working as in (39) we have

$$\begin{aligned}
\|v\|_V &= \delta(\bar{u}, E_{1,\ldots,i}) \\
&\leq \delta((P_h - P_{1,\ldots,i-q,h})E_{i-q+1,\ldots,i}, E_{i-q+1,\ldots,i}) \\
&= \|(I - P_h + P_{1,\ldots,i-q,h})P_{i-q+1,\ldots,i}\|_{\mathscr{L}(V)}.
\end{aligned}$$

Then the required estimate

$$0 \leq \frac{\lambda_{i,h} - \lambda_i}{\lambda_{i,h}} \leq \|v\|_V$$

follows from the following chain of inequalities working as in the proof of Theorem 4

$$\lambda(u) \leq \lambda_i \leq \lambda_{i,h} \leq \lambda(\bar{u}).$$

The last estimate is a consequence of the definition of $\bar{u}$ and of (29).

To obtain the second line of (42) it is enough to apply [56, Theorem 3.2]. $\qquad\square$

The first consequence of Theorem 6 concerns the case of multiple eigenvalues.

Corollary 5. *Assume that the eigenvalue λ_p with $p > 1$ has multiplicity $m > 1$ so that (35) holds with $p + m - 1 \le N$ and*

$$\min_{j=1,\ldots,p-1} |\lambda_{j,h} - \lambda_p| \ne 0.$$

Then, for $i = p, \ldots, p + m - 1$ we have

$$0 \le \frac{\lambda_{i,h} - \lambda_p}{\lambda_{i,h}} \le \|(I - P_h + P_{1,\ldots,p-1,h})P_{p,\ldots,i}\|^2_{\mathscr{L}(V)}$$

$$\le \left(1 + \max_{j=1,\ldots,p-1} \frac{\lambda_{j,h}^2 \lambda_p^2}{|\lambda_{j,h} - \lambda_p|^2} \|(I-P_h)TP_{1,\ldots,p-1,h}\|^2_{\mathscr{L}(V)}\right) \|(I-P_h)P_{p,\ldots,i}\|^2_{\mathscr{L}(V)},$$

where $P_{1,\ldots,p-1,h}$ is the orthogonal projection with respect to the norm of V onto $E_{1,\ldots,p-1,h} = \mathrm{span}\{u_{1,h}, \ldots, u_{p-1,h}\}$ and $P_{p,\ldots,i}$ is the orthogonal projection onto any $i - p + 1$ dimensional subspace of the eigenspace $E_{p,\ldots,p+m-1}$ corresponding to the eigenvalue λ_p.

Remark 6. We remark that the error estimates for the eigenvalues of Theorems 5 and 6 contain multiplicative constants which approach 1, provided that assumptions (37) and (41) hold true.

Let us consider some particular cases in order to see the strength of Theorem 6 and of Corollary 5.

CASE 1: $\lambda_1 < \lambda_2 = \lambda_3 < \lambda_4$
Let us suppose that λ_2 has multiplicity 2, so we can apply Corollary 5 with $p = m = 2$. Assumption (37) gives

$$\min_{j=1,\ldots,p-1} |\lambda_{j,h} - \lambda_p| = |\lambda_{1,h} - \lambda_2| \approx \lambda_2 - \lambda_1.$$

In Corollary 5 we can take $i = 2$ or $i = 3$. For $i = 2$ we obtain

$$\frac{\lambda_{2,h} - \lambda_2}{\lambda_{2,h}} \le \left(1 + \frac{\lambda_{1,h}^2 \lambda_2^2}{|\lambda_{1,h} - \lambda_2|^2} \|(I - P_h)TP_{1,h}\|^2_{\mathscr{L}(V)}\right) \|(I - P_h)P_2\|^2_{\mathscr{L}(V)}. \quad (43)$$

For $i = 3$ we have

$$\frac{\lambda_{3,h} - \lambda_3}{\lambda_{3,h}} = \frac{\lambda_{3,h} - \lambda_2}{\lambda_{3,h}}$$

$$\le \left(1 + \frac{\lambda_{1,h}^2 \lambda_2^2}{|\lambda_{1,h} - \lambda_2|^2} \|(I - P_h)TP_{1,h}\|^2_{\mathscr{L}(V)}\right) \|(I - P_h)P_{2,3}\|^2_{\mathscr{L}(V)}.$$

$$(44)$$

In the first inequality the error is bounded by the best approximation error for eigenfunction u_2, while in the second one we have the bound in terms of the best approximation error for span$\{u_2, u_3\}$. Therefore, we can separate the rate of convergence according to approximability of each eigenfunction. Notice that if u_2 is less regular than u_3, we have from both the estimates (43) and (44) the same rate of convergence, while in the opposite case the inequality (43) could give the best rate of convergence corresponding to the most regular eigenfunction. Moreover, we also see here the improvement with respect to the result of Theorem 5, which would not give a valid estimate in this case since the denominator tends to zero.

CASE 2:$\lambda_1 < \lambda_2 \approx \lambda_3 < \lambda_4$
This case is similar to the previous one but we have clustered eigenvalues. We obtain results similar to the ones quoted for Case 1. In particular, as in the previous case Theorem 5 would not give a good estimate for $i = 3$ since $|\lambda_3 - \lambda_{2,h}| \approx 0$.

CASE 3: $\lambda_1 < \lambda_2 = \lambda_3 < \lambda_4$
We apply Theorem 6 with $i = 3$. Then we can choose $q = 1, 2, 3$ and obtain the following bounds.
    For $q = 1$ we have

$$\frac{\lambda_{3,h} - \lambda_3}{\lambda_{3,h}} \leq \left(1 + \max_{j=1,2} \frac{\lambda_{j,h}^2 \lambda_3^2}{|\lambda_{j,h} - \lambda_3|^2} \|(I - P_h)TP_{1,2,h}\|_{\mathscr{L}(V)}^2\right) \|(I - P_h)P_3\|_{\mathscr{L}(V)}^2.$$

but this estimate is not optimal since $\min_{j=1,2} |\lambda_{j,h} - \lambda_3| \approx 0$.
    For $q = 2$ we have

$$\frac{\lambda_{3,h} - \lambda_3}{\lambda_{3,h}} \leq \left(1 + \frac{\lambda_{1,h}^2 \lambda_3^2}{|\lambda_{1,h} - \lambda_3|^2} \|(I - P_h)TP_{1,h}\|_{\mathscr{L}(V)}^2\right) \|(I - P_h)P_{2,3}\|_{\mathscr{L}(V)}^2.$$

If $\lambda_3 - \lambda_1$ is large enough this inequality gives a sharp estimate in the case we have u_1 with poor approximability property.
    For $q = 3$ we obtain

$$\frac{\lambda_{3,h} - \lambda_3}{\lambda_{3,h}} \leq \|(I - P_h)P_{1,2,3}\|_{\mathscr{L}(V)}^2,$$

and this estimate recovers the result of Theorem 4 so that we estimate the error in terms of the approximability of the span of all the previous eigenfunctions.

5.3 Numerical Results

In this section we report some numerical results of eigenproblems with multiple eigenvalues, whose associated eigenfunctions can have different regularities.

The first example in this direction is due to Babuška and Osborn [12]. Let us consider the following one-dimensional differential equation: find $\lambda \in \mathbb{R}$ such that there exists $u \neq 0$ with:

$$-\left(\frac{1}{\varphi'(x)}u'(x)\right)' = \lambda\varphi'(x)u(x), \quad x \in (-\pi, \pi)$$

$$u(-\pi) = u(\pi), \quad \left(\frac{1}{\varphi}u'\right)(-\pi) = \left(\frac{1}{\varphi}u'\right)(\pi),$$

where

$$\varphi(x) = \pi^{-\alpha}|x|^{1+\alpha}\,\mathrm{sign}(x), \quad 0 < \alpha < 1.$$

It is easy to check that the continuous eigenvalues and eigenfunctions are given by

$$\lambda_0 = 0, \quad \lambda_{2i-1} = \lambda_{2i} = i^2 \quad \text{for } i = 1, 2, \ldots$$

$$u_0 = 1, \quad u_{2i-1} = \cos(i\varphi(x)), \ u_{2i} = \sin(i\varphi(x)) \quad \text{for } i = 1, 2, \ldots.$$

Due to the definition of φ we have that $\cos(\varphi(x)) \in H^2(-\pi, \pi)$ while $\sin(\varphi(x)) \in H^{1+\alpha}(-\pi, \pi)$. Each eigenvalue has multiplicity 2 and its eigenspace contains a regular eigenfunction approximated optimally by piecewise linear finite elements and a less regular eigenfunction for which the optimal rate of convergence cannot be reached. We refer to the numerical results reported in [12, Sect. 10] showing that for each double eigenvalue the rate of convergence is either 2 or $1 + \alpha$.

The second example does not fit the theory presented so far, however the problem is an important one and the numerical experiments show that the results presented in Sect. 5.2 also hold in this case. It would also be interesting to extend the theory to this situation.

Let $\Omega \subseteq \mathbb{R}^2$ be an open polygon, denote by $\mathbf{n}$ the outward normal vector to its boundary $\partial\Omega$ and by $\mathbf{t}$ the counterclockwise oriented tangent vector. We consider the following eigenproblem which describes the vibration frequencies of a fluid in a cavity, hence it can be considered as the simplest problem in fluid-structure interaction (see e.g. [19, 24, 34]):

$$\begin{cases} -\nabla\,\mathrm{div}\,\mathbf{u} = \lambda\mathbf{u} & \text{in } \Omega \\ \mathrm{rot}\,\mathbf{u} = 0 & \text{in } \Omega \\ \mathbf{u}\cdot\mathbf{n} = 0 & \text{on } \partial\Omega. \end{cases} \tag{45}$$

By standard orthogonalities in $\mathbb{R}^2$ between the operators ∇ and $\mathbf{rot}$, (45) can be transformed into

$$\begin{cases} -\mathbf{rot}\,\mathrm{rot}\,\mathbf{u} = \lambda\mathbf{u} & \text{in } \Omega \\ \mathrm{div}\,\mathbf{u} = 0 & \text{in } \Omega \\ \mathbf{u}\cdot\mathbf{t} = 0 & \text{on } \partial\Omega, \end{cases}$$

which arises in electromagnetic applications (see e.g. [20, 25, 54, 63]).

Here we focus on (45). One can observe that the constraint $\operatorname{rot}\mathbf{u} = 0$ follows automatically from the first equation if $\lambda \neq 0$, hence one could drop the irrotationality constraint and add a zero frequency corresponding to the infinite dimensional null space of the functions which belong to $\nabla H_0^1(\Omega)$. Numerical methods based on this idea have been analyzed, for instance, in [20, 27, 54, 74] for the Maxwell's problem and in [19, 34] for the fluid-structure example. Another approach is based on a penalization strategy (see, e.g., [18, 47, 54, 73]), which we are going to consider in the present paper.

Let s be a positive real number, then the penalized formulation of (45) reads: find $\lambda \in \mathbb{R}$ and $\mathbf{u} \neq 0$ such that:

$$\begin{cases} -\nabla \operatorname{div}\mathbf{u} + \dfrac{1}{s}\operatorname{rot}\operatorname{rot}\mathbf{u} = \lambda\mathbf{u} & \text{in } \Omega \\ \mathbf{u}\cdot\mathbf{n} = 0 & \text{on } \partial\Omega \\ \operatorname{rot}\mathbf{u} = 0 & \text{on } \partial\Omega \end{cases} \qquad (46)$$

Let us introduce the following Hilbert spaces

$$H_0^1(\Omega) = \{v \in H^1(\Omega) : v = 0 \text{ on } \partial\Omega\}$$

$$\mathbf{H}_0(\operatorname{div};\Omega) = \{\mathbf{v} \in L^2(\Omega)^2 : \operatorname{div}\mathbf{v} \in L^2(\Omega),\ \mathbf{v}\cdot\mathbf{n} = 0\}$$

$$\mathbf{H}(\operatorname{rot};\Omega) = \{\mathbf{v} \in L^2(\Omega)^2 : \operatorname{rot}\mathbf{v} \in L^2(\Omega)\}$$

The variational formulation of (46) reads: given $s \in \mathbb{R}$ with $s > 0$, find $\lambda \in \mathbb{R}$ and $\mathbf{u} \in \mathbf{H}_0(\operatorname{div};\Omega) \cap \mathbf{H}(\operatorname{rot};\Omega)$ with $\mathbf{u} \neq 0$ such that

$$(\operatorname{div}\mathbf{u}, \operatorname{div}\mathbf{v}) + \frac{1}{s}(\operatorname{rot}\mathbf{u}, \operatorname{rot}\mathbf{v}) = \lambda(\mathbf{u}, \mathbf{v}) \quad \forall\mathbf{v} \in \mathbf{H}_0(\operatorname{div};\Omega) \cap \mathbf{H}(\operatorname{rot};\Omega). \quad (47)$$

It is well-known that if Ω is convex the space $\mathbf{H}_0(\operatorname{div};\Omega) \cap \mathbf{H}(\operatorname{rot};\Omega)$ is equal to $H^1(\Omega)^2 \cap \mathbf{H}_0(\operatorname{div};\Omega)$. But this equivalence fails if Ω is a nonconvex polygon, as it has been shown in [37]. On the other hand, we observe that it is not possible to construct a piecewise polynomial function which is contained in $\mathbf{H}_0(\operatorname{div};\Omega) \cap \mathbf{H}(\operatorname{rot};\Omega)$ but not in $H^1(\Omega)^2$. For this reason we introduce a mixed formulation of (47) by setting $sp = \operatorname{rot}\mathbf{u}$, thus we obtain the following problem: given $s > 0$, find $\lambda \in \mathbb{R}$ and $\mathbf{u} \in \mathbf{H}_0(\operatorname{div};\Omega)$ with $\mathbf{u} \neq 0$ such that for some $p \in H_0^1(\Omega)$

$$\begin{aligned} (\operatorname{div}\mathbf{u}, \operatorname{div}\mathbf{v}) + (\operatorname{rot}p, \mathbf{v}) &= \lambda(\mathbf{u}, \mathbf{v}) & \forall\mathbf{v} \in \mathbf{H}_0(\operatorname{div};\Omega) \\ (\operatorname{rot}q, \mathbf{u}) - s(p, q) &= 0 & \forall q \in H_0^1(\Omega) \end{aligned} \qquad (48)$$

Notice that if we take $s = 0$ in (48), then the second equation implies that $\operatorname{rot}\mathbf{u} = 0$, so that p is the Lagrange multiplier associated to the irrotational constraint (see [54] for the analogous situation in the case of Maxwell eigenproblem).

Given a regular family $\{\mathcal{T}_h\}$ of triangulations of the domain Ω, we consider the following finite element spaces

$$\mathbf{V}_h = \{\mathbf{v} \in \mathbf{H}_0(\text{div}; \Omega) : \mathbf{v}_{|K} \in RT_0(K) \ \forall K \in \mathcal{T}_h\}$$
$$Q_h = \{q \in H_0^1(\Omega) : q_{|K} \in P_1(K) \ \forall K \in \mathcal{T}_h\} \tag{49}$$

where $P_k(K)$ is the set of polynomials of degree less than or equal to k on K and $RT_0(K) = P_0(K)^2 + P_0(K)(x, y)^t$ is the space of lowest order Raviart-Thomas elements (see [65]). Then the discrete counterpart of (48) reads: given $s > 0$, find $\lambda_h \in \mathbb{R}$ and $\mathbf{u}_h \in \mathbf{V}_h$ with $\mathbf{u}_h \neq 0$ such that for some $p_h \in Q_h$

$$(\text{div } \mathbf{u}_h, \text{div } \mathbf{v}) + (\mathbf{rot} \, p_h, \mathbf{v}) = \lambda_h(\mathbf{u}_h, \mathbf{v}) \quad \forall \mathbf{v} \in \mathbf{V}_h$$
$$(\mathbf{rot} \, q, \mathbf{u}_h) - s(p_h, q) = 0 \qquad\qquad \forall q \in Q_h. \tag{50}$$

Problem (48) with $s = 0$ is an eigenproblem in mixed form (see Sect. 4) and the term which is added when $s > 0$ contributes to its stability (since it has the right negative sign). We refer to [10] for the analysis in a more general framework.

Here we use this problem as an example of problems with multiple eigenvalues associated to eigenfunctions of different regularity. We observe that, thanks to the Helmholtz decomposition, the eigensolutions of problem (48) split into two families. The first one is given by the eigensolutions $(\lambda^n, \mathbf{u}^n)$ such that $\mathbf{rot} \, \mathbf{u}^n = 0$ and $\mathbf{u}^n = \nabla\varphi$ with (λ^n, φ) the eigensolution of the following Laplace equation with Neumann boundary conditions

$$-\Delta\varphi = \lambda^n\varphi \quad \text{in } \Omega$$
$$\frac{\partial\varphi}{\partial\mathbf{n}} = 0 \qquad \text{on } \partial\Omega.$$

The second family $(\lambda^d, \mathbf{u}^d)$ satisfies $\text{div } \mathbf{u}^d = 0$ so that $\mathbf{u}^d = -\mathbf{rot} \, \psi$ with (λ^d, ψ) the eigensolution of the following Laplace equation with Dirichlet boundary conditions

$$-\frac{1}{s}\Delta\psi = \lambda^d\psi \quad \text{in } \Omega$$
$$\psi = 0 \qquad \text{on } \partial\Omega.$$

As a consequence of this carachterization, the Neumann eigenvalues λ^n do not depend on s, while the Dirichlet ones λ^d grow linearly with $\frac{1}{s}$.

We consider the L-shaped domain Ω reported in Fig. 9. Since Ω is not convex, there are eigenfunctions which are not in $H^2(\Omega)$. We use as reference values for the eigenvalues of the Neumann problem the solution published in [39], which are computed by a Galerkin approximation with a geometrical refined mesh near the

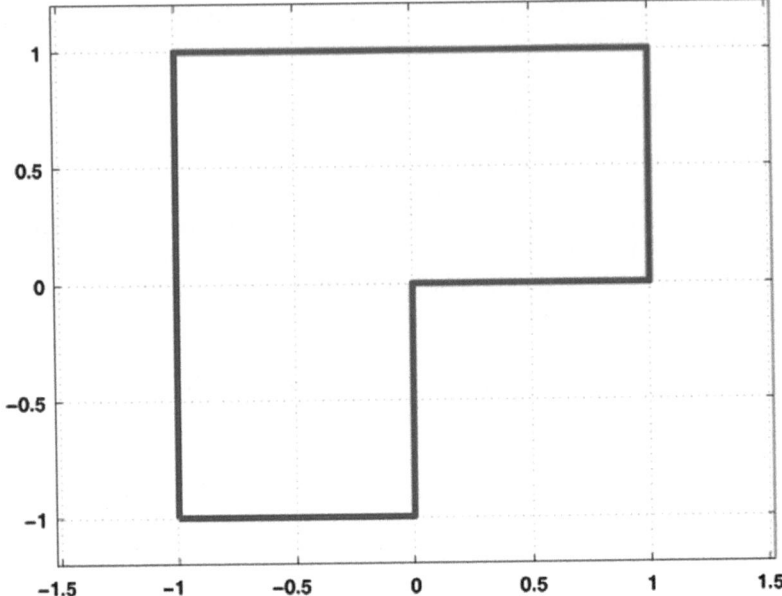

Fig. 9 The L-shaped domain

corner (10 layers and ratio 4) and polynomials of "high" degree (degree 10). The first 5 eigenvalues with 11 correct digits are:

$$
\begin{aligned}
\lambda_1^n &= 0.147562182408E + 01 \\
\lambda_2^n &= 0.353403136678E + 01 \\
\lambda_3^n &= 0.986960440109E + 01 \\
\lambda_4^n &= 0.986960440109E + 01 \\
\lambda_5^n &= 0.113894793979E + 02.
\end{aligned}
\tag{51}
$$

The first eigenvector has a strong singularity; notice, in particular, that $\lambda_3^n = \lambda_4^n = \pi^2$.

We computed the solution of problem (50) using the package FreeFem++ [50]: an open source software for solving partial differential problems by the finite element method. The results of this computation have been taken from [69]. In Listing 1.7 we report the FreeFem++ code which computes eigenvalues and eigenfunctions of problem (48).

In the first computation we set $s = 0$ so that we obtain the first five eigenvalues associated to the irrotational eigenfunction, that is the Neumann eigenvalues λ_i^n for $i = 1, \ldots, 5$. In Table 7 we report the computed eigenvalues together with the rate of convergence estimated by using the exact values quoted in (51). The integer n represents the number of subdivision of the interval $[0, 1]$, the total number of elements is $NE = 6n^2$. Figure 10 reports the mesh corresponding to the value $n = 8$.

Table 7 Eigenvalues computed using the code listed in Listing 1.7 for $s = 0$

Exact	Computed (rate)				
	$n = 2$	$n = 4$	$n = 8$	$n = 16$	$n = 32$
1.48	1.3248	1.4176 (1.4)	1.4531 (1.4)	1.4668 (1.4)	1.4722 (1.4)
3.53	3.4976	3.5217 (1.6)	3.5305 (1.8)	3.5331 (1.9)	3.5338 (1.9)
9.87	9.0496	9.6577 (2.0)	9.8161 (2.0)	9.8562 (2.0)	9.8662 (2.0)
9.87	9.3021	9.7420 (2.0)	9.8385 (2.0)	9.8619 (2.0)	9.8677 (2.0)
11.39	10.7739	11.2193 (1.9)	11.3448 (1.9)	11.3781 (2.0)	11.3866 (2.0)
DOF	65	225	833	3201	12545
$NE = 6n^2$	24	96	384	1563	6144

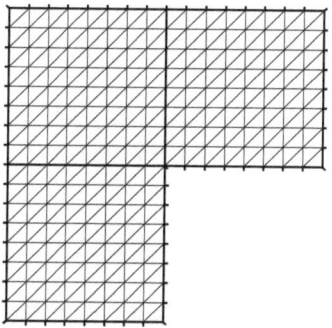

Fig. 10 The mesh in the L-shaped domain Ω for $n = 8$, $NE = 384$

We see in this first example that the discrete eigenvalues stay below the corresponding continuous one. We remark that the technique used to prove the error estimates in Sect. 5.2 cannot be applied to the approximation of problem (48). Indeed the estimates in Theorems 4–6 are based on (30), that is the Ritz approximations of the eigenvalues bound the continuous ones by above. In general, no relation of this type can be deduced for the approximation of the eigenvalues of the problem in mixed form.

The second computation is performed using the value $s = 13.376875077383353$, so that there is a regular Dirichlet eigenvalue which equals the first Neumann eigenvalue λ_1^n that is not regular. This value for s has been computed by solving the equation $2\pi^2/s = \lambda_1^n$: since there is a Dirichlet eigenvalue $\lambda_1^d = 2\pi^2$ (for $s = 1$) associated to a smooth eigenfunction, it follows that this choice for s shifts this eigenvalue in such a way that its value is superimposed to λ_1^n.

Table 8 reports the first 5 computed eigenvalues together with the rate of convergence for those eigenvalues converging to λ_1^n.

We see that in this case the rate of convergence is different from one eigenvalue to the other and depends on the regularity of the associated eigenfunction as predicted by Theorem 6, even if the problem does not fit within the theory presented in Sect. 5.2. Moreover, we notice that $\lambda_{3,h}$, as $n \geq 4$, converges to λ_1^n from below, while $\lambda_{4,h}$ converges from above to the same value.

Table 8 Eigenvalues computed using the code listed in Listing 1.7 for $s = 13.376875077383353$

Exact	Computed (rate)					
	$n = 2$	$n = 4$	$n = 8$	$n = 16$	$n = 32$	$n = 64$
	0.9867	0.7905	0.7413	0.7273	0.7229	0.7214
	1.3248	1.2669	1.1688	1.1443	1.1381	1.1366
1.48	1.6462	1.4176 (1.6)	1.4531 (1.4)	1.4668 (1.4)	1.4722 (1.3)	1.4743 (1.3)
1.48	2.3922	1.7059 (1.9)	1.5327 (2.0)	1.4899 (2.0)	1.4792 (2.0)	1.4765 (2.0)
	3.4976	2.7079	2.3317	2.2381	2.2147	2.2088
DOF	65	225	833	3201	12545	49665
$NE = 6n^2$	24	96	384	1563	6144	24576

Listing 1.7 FreeFem++ code for problem (50)

```
int n=32; // numbers of subdivision of [0,1]
/*
    Construction of the mesh
    The domain is subdivided into three squares
    The mesh for each square is constructed separately
*/
mesh Th1=square(n,n);
mesh Th2=square(n,n,[x-1,y]);
mesh Th3=square(n,n,[x-1,y-1]);
verbosity=3;
plot(Th1,Th2,Th3);
int[int] r1=[4,0], r2=[2,0], r3=[1,0];
int[int] r4=[3,0], r5=[1,5], r6=[2,6];
Th1=change(Th1,label=r1);
Th2=change(Th2,label=r2);
Th2=change(Th2,label=r3);
Th3=change(Th3,label=r4);
Th3=change(Th3,label=r5);
Th3=change(Th3,label=r6);
// the three meshes are glued together
mesh Th=Th1+Th2+Th3;
plot(Th,wait=1);
//
/* s is the penalization parameter */
//real s=0;
real s=13.376875077383353;
real sigma=0; // shift parameter
// finite element spaces
fespace VQh(Th,[RT0,P1]);
VQh [u1,u2,p],[v1,v2,q];
//
```

```
/* Definition of the partial differential equation */
varf op([u1,u2,p],[v1,v2,q])=
    int2d(Th)( (dx(u1)+dy(u2)) * (dx(v1)+dy(v2))
                -dx(p)*v2 + dy(p)*v1
                +dx(q)*u2 - dy(q)*u1
                + s*p*q - sigma*u1*v1 - sigma*u2*v2)
    +on(1,2,3,4,5,6,u1=0,u2=0,p=0);
/* Construction of the matrices */
matrix OP=op(VQh,VQh,solver=sparsesolver);
varf m([u1,u2,p],[v1,v2,q])=int2d(Th)( u1*v1 + u2*v2 );
matrix M=m(VQh,VQh,solver=GMRES);
/* Number of computed eigenvalues */
real nev=5;
real[int] ev(nev);
VQh[int] [eV,eW,ep](nev);
int k;
int nold=cout.precision(20);
/* computation of eigenvalues and eigenvectors */
k=EigenValue(OP,M,sym=true,nev=nev,value=ev,
             sigma=sigma,tol=1e-10,vector=eV);
/* plot of the eigenvectors */
for(int i=0;i<k;++i)
{
cout << "⎵Eigenvalue⎵⎵" <<i << "⎵=⎵" << ev[i] <<endl;
plot([eV[i],eW[i]],ep[i], cmm="⎵Eigenvalue⎵" + ev[i],
     wait=1,value=true);
}
```

6 A Posteriori Error Analysis

In this section we present an introduction to the subject of a posteriori error estimation and adaptive mesh refinement for the approximation of the eigenvalue problem. The main purpose of this section is to underline the difficulties which arise in the a posteriori error analysis of eigenvalue problems. In this sense, the presentation is by no means exhaustive and focuses on error estimators of the residual type. We first recall the bibliography, which is now very extensive for eigenvalue problem as well as for source problems. Then in Sects. 6.3, 6.4 we highlight the difficulties which arise in dealing with eigenvalue problems, both in standard and mixed form, and we show why the techniques used to develop an a posteriori error analysis for the source problem cannot be trivially extended to the eigenvalue problem.

6.1 Some Introductory Remarks

In the last twenty years there has been a great deal of research work on a posteriori error estimation and adaptive refinement for the finite element approximation of PDEs that arise from physical and engineering applications. The aim is to obtain a numerical solution within a prescribed tolerance using a minimal amount of work.

The a priori error estimates provided by the standard error analysis for the finite element method yield information only on the asymptotic error behavior, and strongly depend on the regularity of the solution. In particular in the presence of local singularities, such as re-entrant corners and interior or boundary layers, the overall accuracy of the numerical solution deteriorates. A first remedy is to refine the mesh; nevertheless the uniform refinement could lead to an excessive computational effort, when actually it is enough to refine only the elements in the neighborhood of the singularities. The question then is how to detect the elements which have to be refined and how to obtain a good balance between the refined and the un-refined regions such that the overall accuracy is optimal.

Another issue is to be able to judge the quality of the numerical solution, namely to obtain reliable estimates of the accuracy of the computed solution in order to decide whether a prescribed tolerance has been achieved or not.

In this context the need of an error estimator which can be computed locally from the numerical solution and the data of the problem appears clear. The error estimator should yield reliable upper and lower bounds for the actual error. Indeed global upper bounds are sufficient to ensure that the numerical solution achieves a prescribed tolerance. Local lower bounds however are essential to guarantee that the error is not overestimated and that its local distribution is correctly resolved. Finally, the calculation of the a posteriori error estimate should be far less expensive than the computation of the numerical solution.

The bibliography on a posteriori error estimators for the finite element method is now very extensive (see, in particular the books by Verfürth [71] and Ainsworth and Oden [1], and the references therein).

For classical finite element approximations of eigenvalue problems, Babuška and Rheinboldt in [13] first introduced an a posteriori error estimator for a one dimensional problem; Verfürth in [70] derived an upper bound of suboptimal order in the approximation of the eigenvalue and only a global lower bound by considering the eigenvalue problem as a parameter-dependent nonlinear equation and using general results for the Galerkin approximation of such type of problems. Alonso et al. in [5] presented an error estimator of the residual type for piecewise linear finite element approximation of structure vibration problems. Then Larson in [60] obtained optimal order estimates assuming the H^2-regularity of the eigenfunctions, which however excludes domain with re-entrant corners or discontinuous coefficients; Heuveline and Rannacher in [51] obtained results based on a general analysis for nonlinear equations without requiring the H^2-regularity of the problem. In [41] Durán et al. developed a simple analysis for a residual type error estimator for the linear finite element approximation of a second order elliptic problem. Finally, more recent results can be found in [9, 49, 72].

Concerning the mixed approximation of eigenvalue problems, Durán *et al.* in [40] analyzed an error estimator for the approximation of the eigensolutions of a second-order elliptic problem by using the equivalence between the mixed finite element method of Raviart-Thomas of the lowest order and the non-conforming piecewise linear approximation of Crouzeix and Raviart. Then Alonso *et al.* in [6] adapted the techniques used in [40] to derive an error estimator for the lowest order Raviart-Thomas approximation of the acoustic vibration problem. Moreover in [3, 4] this error estimator has been used together with the one presented in [5] to deal with structural-acoustic vibration problems on matching and non matching grids, respectively. A posteriori error estimators for eigenvalue problems in mixed form have been studied in [44, 45] without using the equivalence with a non-conforming approximation. Finally, the Stokes eigenvalue problem has been considered in [61].

Eventually, the convergence of adaptive methods for the finite element approximation of eigenvalue problems has been studied quite recently in [32, 38, 42, 43, 48, 62].

6.2   A Posteriori Error Estimators

In this section we recall the properties of the error estimators and we describe a simple mesh-refinement algorithm.

Let η denote the error estimator, which is usually given by the sum of local error indicators

$$\eta^2 = \sum_{K \in \mathcal{T}_h} \eta_K^2.$$

The error estimator has to satisfy the following properties:

Reliability. It should bound from above the global error e_h in a suitable norm $\| \ \|$:

$$\|e_h\| \leq C\eta + \text{h.o.t.},$$

where h.o.t. denotes higher order terms.

Efficiency. It should provide local lower error estimates, in order to point out which elements should be effectively refined:

$$\eta_K \leq C \|e_h\|_{V,K^*} + \text{h.o.t.},$$

where K^* is the union of K and few neighboring elements.

Low computational cost . The computation of η_K should be inexpensive in comparison with the overall computation of the discrete solution.

Here and thereafter, C denotes a generic constant, not necessarily always the same, but always independent of the mesh size.

Having an a posteriori error estimator, an adaptive mesh-refinement algorithm, which applies a usual procedure from [71], reads as follows:

1. Start with an initial coarse mesh $\mathcal{T}_0$. Set $k = 0$.
2. Solve the discrete problem on $\mathcal{T}_k$.
3. For each element $K \in \mathcal{T}_k$ compute the local error indicator η_K.
4. Evaluate stopping criterion and decide to finish or go to next step.
5. Decide which elements have to be refined and construct the new mesh.
6. Define resulting mesh as $\mathcal{T}_{k+1}$, replace k by $k + 1$ and go to step 2.

Heuristic arguments show that for a linear finite element discretization of the Poisson problem the optimal mesh among all partitions with a given number of elements is the one which equilibrates the error, i.e. the error in each element is almost the same (see [13, 14, 66]).

Based on this strategy, there are two main possibilities to decide which elements have to be refined (see [71]):

1. Let K be an element in $\mathcal{T}_h$ and let τ be an element obtained by subdividing K. It is quite reasonable to assume that the errors in K and τ behave like $C h_K^r$ and $C h_\tau^r$ with unknown constants C and r. Computing error estimators η_K and η_τ, one can roughly calculate C and r and hence approximately predict the error in an element τ obtained by subdividing K.
2. For each element $K \in \mathcal{T}_h$ compute the local error indicator η_K. Then refine each element K' whose error indicator $\eta_{K'}$ satisfies

$$\eta_{K'} \geq \gamma \max\{\eta_K : K \in \mathcal{T}_h\},$$

where $\gamma \in (0, 1)$ is a prescribed threshold (very often $\gamma = 0.5$).

The first possibility clearly is more sophisticated. However, in practice the second possibility, which is cheaper, gives satisfactory results.

Another issue concerns the shape regularity of the mesh: it is mandatory to preserve it during the refinement process. For triangular meshes there exist essentially three different strategies which preserve the shape regularity:

1. *Regular refinement*: divide triangles into four by joining the midpoints of edges.
2. *Longest edge bisection*: bisect triangles by joining the midpoint of the longest edge with the vertex opposite to this edge.
3. *Marked edge bisection*: bisect triangles by joining the midpoint of a marked edge with the vertex opposite to this edge.

Finally, in order to obtain an admissible triangulation, i.e. without hanging nodes, the refinement process has to obey some additional rules (see, in particular, [15–17, 59, 67, 68]).

6.3   Standard Finite Element Approximation

The aim of the present section is to highlight the difficulties which arise in the a posteriori error analysis moving from the source problem to the eigenvalue problem. For the sake of simplicity, we consider the Laplace problem as a model:

$$\text{given } f \in L^2(\Omega), \text{ find } u \text{ such that}$$

$$\begin{cases} -\Delta u = f \text{ in } \Omega \\ u = 0 \text{ on } \partial\Omega, \end{cases} \tag{52}$$

where $\Omega \subset \mathbb{R}^d$ $(d = 2, 3)$ is a polygonal or polyhedral domain.

Let us first consider the standard variational formulation of problem (52), which reads

$$\text{given } f \in L^2(\Omega), \text{ find } u \in V \text{ such that}$$

$$(\nabla u, \nabla v) = (f, v) \quad \forall v \in V,$$

where as usual $(\cdot, \cdot)$ denotes the L^2-inner product and $V = H_0^1(\Omega)$.

Let $\{\mathcal{T}_h\}$ denote a shape-regular family (i.e., satisfying the minimum angle condition, see [35]) of triangulations of Ω. As usual we require that any two elements in $\mathcal{T}_h$ share at most a common face, edge or a common vertex, and we denote by h the maximum diameter of the elements K in $\mathcal{T}_h$. Let $\mathcal{E}$ be the set of interior edges (faces in three dimensions) of the mesh and $\mathcal{E}_K \subset \mathcal{E}$ be the subset of the edges (faces in three dimensions) of the element K.

We denote by $V_h \subset V$ a finite element subspace of V (we can take for example the space consisting of continuous piecewise linear functions vanishing on the boundary of Ω). The discrete problem reads:

$$\text{given } f \in L^2(\Omega), \text{ find } u_h \in V_h \text{ such that}$$

$$(\nabla u_h, \nabla v_h) = (f, v_h) \quad \forall v_h \in V_h.$$

In the following we shall denote by $a(u, v)$ the bilinear form $(\nabla u, \nabla v)$ and by $e_h = u - u_h$ the error which belongs to the space V and satisfies the residual equation

$$a(e_h, v) = (f, v) - a(u_h, v) \quad \forall v \in V. \tag{53}$$

Moreover, the standard Galerkin orthogonality property holds

$$a(e_h, v_h) = 0 \quad \forall v \in V_h.$$

The first step in the error analysis is to write the residual equation (53) as a sum of local contribution and to apply integration by parts on each element. Moreover, observing that the traces of functions in V match along the interface between two elements, we obtain the following expression

$$a(e_h, v) = \sum_{K \in \mathcal{T}_h} \left\{ (f + \Delta u_h, v)_K - \frac{1}{2} \sum_{e \in \mathcal{E}_K} \int_e \left[\!\!\left[\frac{\partial u_h}{\partial n} \right]\!\!\right]_e v \right\} \qquad \forall v \in V,$$

where $(\cdot, \cdot)_K$ and $[\![\cdot]\!]_e$ denote respectively the L^2-inner product restricted to the element K and the jump across the edge (face in three dimensions) e, which is defined as follows

$$\left[\!\!\left[\frac{\partial u_h}{\partial n} \right]\!\!\right]_e = \nabla u_h|_{K_e^1} \cdot \boldsymbol{n}_{K_e^1} + \nabla u_h|_{K_e^2} \cdot \boldsymbol{n}_{K_e^2},$$

K_e^1 and K_e^2 being the two elements in $\mathcal{T}_h$ sharing the edge (face in three dimensions) e and $\boldsymbol{n}_{K_e^1}$ and $\boldsymbol{n}_{K_e^2}$ the unit outward normals on ∂K_e^1 and ∂K_e^2, respectively.

Given any $v \in V$, let $v^I \in V_h$ be such that

$$\|v - v^I\|_{0,K} \le C h_K |v|_{1,\tilde{K}}$$

and

$$\|v - v^I\|_{0,e} \le C h_e^{1/2} |v|_{1,\tilde{K}},$$

where $\tilde{K}$ is the union of all the elements sharing a vertex with K and h_K and h_e denote the diameter of K and the diameter of edge (face in three dimensions) e, respectively. We can take, for example, the well-known Clément interpolant (see [36]).

Then, thanks to the Galerkin orthogonality, we have

$$a(e_h, v) = \sum_{K \in \mathcal{T}_h} \left\{ (f + \Delta u_h, v - v^I)_K - \frac{1}{2} \sum_{e \in \mathcal{E}_K} \int_e \left[\!\!\left[\frac{\partial u_h}{\partial n} \right]\!\!\right]_e (v - v^I) \right\} \qquad \forall v \in V.$$

Using the Cauchy-Schwarz inequality and the properties of v^I, we obtain

$$a(e_h, v) \le C \|v\|_1 \left\{ \sum_{K \in \mathcal{T}_h} \left(h_K^2 \|r\|_{0,K}^2 + \frac{1}{2} \sum_{e \in \mathcal{E}_K} h_e \|R\|_{0,e}^2 \right) \right\}^{1/2},$$

where r is the interior residual

$$r|_K = f + \Delta u_h \qquad \forall K \in \mathcal{T}_h$$

and R is the boundary residual

$$R|_e = \left[\!\!\left[\frac{\partial u_h}{\partial n} \right]\!\!\right]_e \quad \forall e \in \mathcal{E}_K.$$

Hence, we can estimate the energy norm, or equivalently the H^1-norm of the error, as follows:

$$|e_h|_1 \le C \left\{ \sum_{K \in \mathcal{T}_h} \left(h_K^2 \|r\|_{0,K}^2 + \frac{1}{2} \sum_{e \in \mathcal{E}_K} h_e \|R\|_{0,e}^2 \right) \right\}^{1/2}.$$

Apart from the constant C, all of the quantities in the right-hand side can be computed explicitly from the data of the problem and the finite element approximation. This suggests to define the local error indicator associated with the element K by

$$\eta_K^2 = h_K^2 \|r\|_{0,K}^2 + \frac{1}{2} \sum_{e \in \mathcal{E}_K} h_e \|R\|_{0,e}^2$$

and the corresponding error estimator by

$$\eta = \left(\sum_{K \in \mathcal{T}_h} \eta_K^2 \right)^{1/2}.$$

As usual in residual-type error indicators, η_K consist of the L^2-norm of the volumetric and edge (face in three dimensions) residual, suitably weighted.

We have then proved the reliability of the error estimator for the source problem.

Theorem 7. *There exists a constant C, depending only on the regularity of the mesh, such that*

$$|e_h|_1 \le C\eta.$$

Let us now consider the eigenvalue problem corresponding to problem (52), which reads

find $\lambda \in \mathbb{R}$ such that there exists u, with $u \ne 0$:

$$\begin{cases} -\Delta u = \lambda u & \text{in } \Omega \\ u = 0 & \text{on } \partial\Omega. \end{cases} \tag{54}$$

The classical variational formulation of (54) is given by

> find $\lambda \in \mathbb{R}$ such that there exists $u \in V$, with $u \neq 0$:
> $$\begin{cases} a(u,v) = \lambda(u,v) & \forall v \in V \\ \|u\|_0 = 1, \end{cases}$$

and its discretization by

> find $\lambda_h \in \mathbb{R}$ such that there exists $u_h \in V_h$, with $u_h \neq 0$:
> $$\begin{cases} a(u_h, v_h) = \lambda_h(u_h, v_h) & \forall v_h \in V_h \\ \|u_h\|_0 = 1. \end{cases}$$

Let (λ, u) be an eigensolution of the continuous problem with λ simple, and (λ_h, u_h) be the corresponding discrete eigensolution. We denote by $e_h = u - u_h$ the error in the approximation of the eigenfunction, which satisfies the following residual equation:

$$a(e_h, v) - (\lambda u - \lambda_h u_h, v) = -a(u_h, v) + \lambda_h(u_h, v) \quad \forall v \in V. \tag{55}$$

First of all we observe that the Galerkin orthogonality property does not hold anymore, indeed

$$a(e_h, v_h) = (\lambda u - \lambda_h u_h, v_h) \quad \forall v_h \in V_h.$$

Therefore, the analysis developed for the source problem cannot be extended in a straightfoward fashion.

Trying to generalize the analysis developed for the source problem to the eigenvalue problem, one has to deal with terms such as

$$(\lambda u - \lambda_h u_h, e_h)$$

and, in order to prove the efficiency of the estimator, as

$$\|\lambda u - \lambda_h u_h\|_{0,K},$$

which appear in the error estimate since the Galerkin orthogonality does not hold. Duran *et al.* in [41] proved that these terms are of higher order than the error. Due to the normalization on the eigenfunction, the first term can be treated in the following way

$$(\lambda u - \lambda_h u_h, e_h) = (\lambda + \lambda_h)(1 - (u, u_h)) = \frac{\lambda + \lambda_h}{2} \|e_h\|_0^2.$$

Then, proceeding as for the source problem rewriting the residual equation (55) as a sum over the elements and applying integration by part yields

$$a(e_h, v) = a(e_h, v - v^I) + a(e_h, v^I)$$

$$= \sum_{K \in \mathscr{T}_h} \left\{ (\Delta u_h + \lambda_h u_h, v - v^I)_K - \frac{1}{2} \sum_{e \in \mathscr{E}_K} \int_e \left[\!\!\left[\frac{\partial u_h}{\partial n} \right]\!\!\right]_e (v - v^I) \right\}$$

$$+ (\lambda u - \lambda_h u_h, v - v^I) + (\lambda u - \lambda_h u_h, v^I)$$

$$= \sum_{K \in \mathscr{T}_h} \left\{ (\Delta u_h + \lambda_h u_h, v - v^I)_K - \frac{1}{2} \sum_{e \in \mathscr{E}_K} \int_e \left[\!\!\left[\frac{\partial u_h}{\partial n} \right]\!\!\right]_e (v - v^I) \right\}$$

$$+ (\lambda u - \lambda_h u_h, v) \qquad \forall v \in V.$$

Taking $v = e_h$ in the above equation, we obtain

$$\|e_h\|_a^2 \le C \left\{ \sum_{K \in \mathscr{T}_h} \left(h_K^2 \|r\|_{0,K}^2 + \frac{1}{2} \sum_{e \in \mathscr{E}_K} h_e \|R\|_{0,e}^2 \right) \right\}^{1/2} \|e_h\|_a + \frac{\lambda + \lambda_h}{2} \|e_h\|_0^2,$$

where for the eigenvalue problem the volumetric and edge (face in three dimensions) residuals are given respectively by

$$r|_K = \Delta u_h + \lambda_h u_h \quad \forall K \in \mathscr{T}_H$$

and

$$R|_e = \left[\!\!\left[\frac{\partial u_h}{\partial n} \right]\!\!\right]_e \quad \forall e \in \mathscr{E}_h.$$

Hence, the local error indicator associated with the element K is defined by

$$\eta_K^2 = h_K^2 \|r\|_{0,K}^2 + \frac{1}{2} \sum_{e \in \mathscr{E}_K} h_e \|R\|_{0,e}^2$$

and the error estimator by

$$\eta = \left(\sum_{K \in \mathscr{T}_h} \eta_K^2 \right)^{1/2}.$$

Then the following theorem, which gives an upper error estimate, holds true.

Theorem 8. *There exists a constant C, depending only on the regularity of the mesh, such that*

$$|e_h|_1 \le C\eta + \left(\frac{\lambda + \lambda_h}{2} \right)^{1/2} \|e_h\|_0.$$

Remark 7. By the a priori error estimates (see [12, 21]), $\|e_h\|_0$ is of higher order than $|e_h|_1$ and thus the estimator provides an upper bound of the error in the energy norm up to a multiplicative constant and a higher order term.

Remark 8. Thanks to the normalization of the eigenfunctions, the following result holds true:

$$|e_h|_1^2 = \lambda + \lambda_h - 2\lambda(u, u_h) = \lambda_h - \lambda + \lambda\|e_h\|_0^2.$$

Therefore, if the error estimator is reliable, then we automatically have that it bounds from above, up to a multiplicative constant dependent on λ and higher order terms, the square root of the error in the approximation of the eigenvalues as well.

The efficiency of the error indicator has been proved using a technique introduced for the source problem firstly by Verfürth in [70]. The key idea is to use *interior* bubble functions, supported on a single element, and *edge* (*face* in three dimensions) bubble functions, supported on a pair of neighboring elements.

The following lemma provides a local upper estimate for the volumetric residual r.

Lemma 3. *There exists a constant C, depending only on the regularity of the mesh, such that*

$$h_K\|r\|_{0,K} \le C \left(|e|_{1,K} + h_K\|\lambda u - \lambda_h u_h\|_{0,K}\right).$$

Proof. Let b_K denote the standard cubic bubble function on the element K. Choosing $v = rb_K$ in the residual equation (55), we get

$$a(e_h, rb_k) = (\lambda u - \lambda_h u_h, rb_k) + (r, rb_K).$$

Thanks to the property of bubble functions, it holds

$$\|r\|_{0,K}^2 \le C \int_K r^2 b_K = C \left[a(e_h, rb_K) - (\lambda u - \lambda_h u_h, rb_K)\right]$$

and hence, since $|rb_K|_{1,K} \le C h_K^{-1}\|r\|_{0,K}$, we get the result. $\qquad\square$

We now estimate from above the edge (face in three dimensions) residual R.

Lemma 4. *There exists a constant C, depending only on the regularity of K_e^1 and K_e^2, such that*

$$h_e^{1/2}\|R\|_{0,e} \le C \left(|e_h|_{1,K_e^1 \cup K_e^2} + h_e\|\lambda u - \lambda_h u_h\|_{0,K_e^1 \cup K_e^2}\right).$$

Proof. Let b_e denote the edge (face in three dimensions) bubble function relative to edge e. Writing the residual equation (55) as a sum of elemental contributions and integrating by part yields

$$a(e_h, v) - (\lambda u - \lambda_h u_h, v) = \sum_{K \in \mathscr{T}_h} \left(\int_K rv - \frac{1}{2}\sum_{e \in \mathscr{E}_K}\int_e Rv\right).$$

Taking $v = Rb_e$ in the above equation, we get

$$\int_e R^2 b_e = (r, Rb_e) - a(e_h, Rb_e) + (\lambda u - \lambda_h u_h, Rb_e).$$

$\square$

We conclude the proof using Lemma 3 together with the fact that

$$\|R\|^2_{0,e} \leq C \int_e R^2 b_e \quad \text{and} \quad h_K^{-1/2} \|Rb_e\|_{0,K} + h_K^{1/2} |Rb_e|_{1,K} \leq C \|Rb_e\|_{0,e}.$$

As an immediate consequence of the previous lemmas, we get the following theorem.

Theorem 9. *Let K^* be the union of K and the neighboring elements K' sharing an edge (face in three dimensions) with K. There exists a positive constant, depending only on the regularity of the mesh, such that*

$$\eta_K \leq C \left(|e_h|_{1,K^*} + h_K \|\lambda u - \lambda_h u_h\|_{0,K^*} \right).$$

Remark 9. The term $h_K \|\lambda u - \lambda_h u_h\|_{0,K^*}$ in the previous theorem is a higher order term. Indeed, for each element $K \in \mathcal{T}_h$, it holds

$$h_K \|\lambda u - \lambda_h u_h\|_{0,K^*} \leq \lambda h_K \|u - u_h\|_{0,K^*} + |\lambda - \lambda_h| h_K \|u_h\|_{0,K^*}.$$

By the a priori error estimates (see [12, 21]) both terms in the above equation are higher order than the local error $|e_h|_{1,K^*}$.

In [41] it has also been proved that, for linear elements, the volumetric part of the error indicator is dominated, up to higher order terms, by the edge (face in three dimensions) residuals. This result was known for the source problem, see [33]. The simpler error indicator is defined as

$$\tilde{\eta}_K = \left(\frac{1}{2} \sum_{e \in \mathcal{E}_K} h_e \|R\|^2_{0,e} \right)^{1/2}$$

and the corresponding global error estimator as

$$\tilde{\eta} = \left(\sum_{K \in \mathcal{T}_h} \tilde{\eta}_k^2 \right)^{1/2}.$$

The following theorem states the reliability and efficiency of the new error estimator.

Theorem 10. *There exists a constant* C, *depending only on the regularity of the mesh, such that*

$$|e_h|_1 \leq C \left[\tilde{\eta} + \left(\frac{\lambda + \lambda_h}{2} \right)^{1/2} \|e_h\|_0 + \lambda_h^{3/2} h^2 \right]$$

and

$$\tilde{\eta}_K \leq C \left(|e_h|_{1,K^*} + h_K \|\lambda u - \lambda_h u_h\|_{0,K^*} \right).$$

The additional term $\lambda_h^{3/2} h^2$ in the first estimate of the above theorem is of higher order (see [41, Remark 4.1])

6.4 Mixed Finite Element Approximation

Error estimators for mixed approximations of the Poisson equation have been introduced in [2, 31] starting from the Helmholtz decomposition of the error and using the Galerkin orthogonality which holds for the second equation of the problem.

Changing to the eigenvalue problem, difficulties similar to the ones underlined for the classical approximation and due to the lack of the Galerkin orthogonality arise.

The first a posteriori error estimators for mixed approximation of eigenvalue problems have been obtained in a very particular situation using the equivalence between the mixed method of Raviart-Thomas of the lowest order and the non-conforming Crouzeix-Raviart approximation of the classical formulation (see [40] and [6] for an application to fluid-structure interactions).

An a posteriori error analysis for mixed approximations of eigenvalue problems has been developed in [44, 45] without resorting to the equivalence with a non-conforming discretization. Moreover, an a posteriori error analysis for the finite element approximation of the Stokes eigenvalue problem has been presented in [61].

6.4.1 A Posteriori Error Analysis for Brezzi–Douglas–Marini Finite Elements

In this section we present the a posteriori error analysis developed in [44, 45] for the Brezzi–Douglas–Marini (BDM) approximation of an eigenvalue problem which arises from the displacement formulation to compute the vibration modes of an acoustic fluid contained within a rigid cavity. We define an error estimator of the residual type and prove that, under some regularity conditions on the continuous eigensolution, it is equivalent to the $H(\text{div})$-norm of the error up to higher order terms. The constants involved in this equivalence depend on the corresponding eigenvalue, but are independent of the mesh size. Moreover, the square root of the error in the approximation of the eigenvalue is also bounded by a constant times the estimator.

We consider an eigenvalue problem which has already been introduced in Sect. 5.3. For the ease of the reader, we present the problem again.

$$\text{find } \lambda \in \mathbb{R} \text{ such that there exists } \boldsymbol{u} \neq \boldsymbol{0}$$

$$\begin{cases} -\nabla \operatorname{div} \boldsymbol{u} = \lambda \boldsymbol{u} \text{ in } \Omega \\ \operatorname{rot} \boldsymbol{u} = 0 \quad \text{in } \Omega \\ \boldsymbol{u} \cdot \boldsymbol{n} = 0 \quad \text{on } \partial\Omega, \end{cases} \tag{56}$$

where $\Omega \subset \mathbb{R}^2$ is a simply connected polygonal domain, $\partial\Omega$ its boundary, and $\boldsymbol{n}$ its outward normal unit vector.

This problem has been studied by many authors concerning fluid-structure interaction (see [19, 73]). Moreover, since in two dimensions the divergence and rotational operators are isomorphic, it is equivalent to Maxwell's eigenproblem for a cavity resonator with dielectric constant ε and magnetic permeability μ constant and equal to 1 (see [25, 52]).

A variational formulation of (56) reads:

$$\text{find } \lambda \in \mathbb{R} \text{ such that there exists } \boldsymbol{u} \in H_0(\operatorname{div}, \Omega), \text{ with } \boldsymbol{u} \neq \boldsymbol{0}:$$

$$\begin{cases} (\operatorname{div} \boldsymbol{u}, \operatorname{div} \boldsymbol{v}) = \lambda(\boldsymbol{u}, \boldsymbol{v}) & \forall \boldsymbol{v} \in H_0(\operatorname{div}, \Omega) \\ (\boldsymbol{u}, \operatorname{rot} q) = 0 & \forall q \in H_0^1(\Omega), \end{cases} \tag{57}$$

where $H_0(\operatorname{div}, \Omega) = \{\boldsymbol{v} \in L^2(\Omega)^2 : \operatorname{div} \boldsymbol{v} \in L^2(\Omega) \text{ and } \boldsymbol{v} \cdot \boldsymbol{n} = 0 \text{ on } \partial\Omega\}$ is endowed with the norm $\|\boldsymbol{v}\|_{\operatorname{div}}^2 = \|\boldsymbol{v}\|_0^2 + \|\operatorname{div} \boldsymbol{v}\|_0^2$. It is well known that problem (57) admits a countable set of real and positive eigenvalues, which can be ordered in an increasing divergent sequence. Moreover the eigenfunctions satisfy $\boldsymbol{u} \in H^s(\operatorname{div}, \Omega) = \{\boldsymbol{v} \in H^s(\Omega)^2 : \operatorname{div} \boldsymbol{v} \in H^s(\Omega)\}$, for some $s > 1/2$ depending on Ω ($s = 1$ when Ω is convex) (see [7]).

Let $\{\mathcal{T}_h\}$ be a regular family of triangulations of Ω, where as usual h denotes the maximum diameter of the elements K in $\mathcal{T}_h$. The Brezzi-Douglas-Marini spaces are defined for $k \geq 1$ by

$$BDM_k = \{\boldsymbol{v} \in H(\operatorname{div}, \Omega) : \boldsymbol{v}|_K \in P_k(K)^2 \ \forall K \in \mathcal{T}_h\},$$

where $P_k(K)$ denotes the space of polynomial of degree at most k on K (see [29]).

Setting $V_h = BDM_k \cap H_0(\operatorname{div}, \Omega)$, and denoting by Q_h the subspace of $H_0^1(\Omega)$ consisting of continuous piecewise polynomial of degree at most $k + 1$, the discrete problem is then given by:

$$\text{find } \lambda_h \in \mathbb{R} \text{ such that there exists } \boldsymbol{u}_h \in V_h, \text{ with } \boldsymbol{u}_h \neq \boldsymbol{0}:$$

$$\begin{cases} (\operatorname{div} \boldsymbol{u}_h, \operatorname{div} \boldsymbol{v}) = \lambda_h(\boldsymbol{u}_h, \boldsymbol{v}) & \forall \boldsymbol{v} \in V_h \\ (\boldsymbol{u}_h, \operatorname{rot} q) = 0 & \forall q \in Q_h. \end{cases} \tag{58}$$

Let $(\lambda, \boldsymbol{u})$ be an eigensolution of (57) such that λ is a simple eigenvalue and $\|\boldsymbol{u}\|_0 = 1$. It follows from the abstract theory (see [12, 21]) and known a priori estimates that, for h small enough (depending on λ), there exists $(\lambda_h, \boldsymbol{u}_h)$ eigenpair of (58) with $\|\boldsymbol{u}_h\|_0 = 1$ such that

$$\|\boldsymbol{u} - \boldsymbol{u}_h\|_{\mathrm{div}} = O(h^t) \tag{59}$$

$$\|\boldsymbol{u} - \boldsymbol{u}_h\|_0 = O(h^r) \tag{60}$$

$$|\lambda - \lambda_h| = O(h^{2t}), \tag{61}$$

where $t = \min\{s, k\}$, and $r = \min\{s, k + 1\}$.

Since the problem we are dealing with consists of two equations, it is reasonable to expect that the error estimator will be given by the sum of two terms, related one to the residual of the first equation and the other to the residual of the second equation.

Let $\mathscr{E}$ be the set of the interior edges of the mesh and $\mathscr{E}_K \subset \mathscr{E}$ be the subset of edges of K. We denote by $\boldsymbol{e}_h = \boldsymbol{u} - \boldsymbol{u}_h$ the error in the approximation of the eigenfunctions.

For any $K \in \mathscr{T}_h$ we define two local error indicators by

$$\eta_{1,K}^2 = h_K^2 \|\boldsymbol{\nabla} \operatorname{div} \boldsymbol{u}_h + \lambda_h \boldsymbol{u}_h\|_{0,K}^2 + \frac{1}{2} \sum_{e \in \mathscr{E}_K} h_e \|[\![\operatorname{div} \boldsymbol{u}_h]\!]_e\|_{0,e}^2,$$

$$\eta_{2,K}^2 = h_K^2 \|\operatorname{rot} \boldsymbol{u}_h\|_{0,K}^2 + \frac{1}{2} \sum_{e \in \mathscr{E}_K} h_e \|[\![\boldsymbol{u}_h \cdot \boldsymbol{t}]\!]_e\|_{0,e}^2,$$

and the corresponding error estimators by

$$\eta_1 = \left(\sum_{K \in \mathscr{T}_h} \eta_{1,K}^2 \right)^{1/2},$$

$$\eta_2 = \left(\sum_{K \in \mathscr{T}_h} \eta_{2,K}^2 \right)^{1/2}.$$

The jump of the tangential component across the edge e is defined as follows

$$[\![\boldsymbol{u}_h \cdot \boldsymbol{t}]\!]_e = \boldsymbol{u}_h|_{K_e^1} \cdot \boldsymbol{t}_{K_e^1} + \boldsymbol{u}_h|_{K_e^2} \cdot \boldsymbol{t}_{K_e^2}$$

where, for each triangle K, $\boldsymbol{t}_K$ denotes the unit tangent vector to ∂K oriented counterclockwise.

The following lemmas provide the residual equations which will be the starting points of our error analysis.

Lemma 5. *For $v \in H_0(\mathrm{div}, \Omega) \cap H^\sigma(\mathrm{div}, \mathscr{T}_h)$, with $\sigma > 0$, there holds*

$$(\mathrm{div}\, e_h, \mathrm{div}\, v) - (\lambda u - \lambda_h u_h, v) = -(\mathrm{div}\, u_h, \mathrm{div}\, v) + \lambda_h(u_h, v)$$

$$= \sum_{K \in \mathscr{T}_h} \left[(r_1, v)_K - \frac{1}{2} \sum_{e \in \mathscr{E}_K} \int_e \llbracket \mathrm{div}\, u_h \rrbracket_e v \cdot n \right], (62)$$

where $r_1|_K = \nabla \mathrm{div}\, u_h + \lambda_h u_h$ is the volumetric residual of the first equation of problem (56).

Lemma 6. *For $q \in H_0^1(\Omega)$ there holds*

$$(e_h, \mathrm{rot}\, q) = \sum_{K \in \mathscr{T}_h} \left[(r_2, q)_K + \frac{1}{2} \sum_{e \in \mathscr{E}_K} \int_e q \llbracket u_h \cdot t \rrbracket_e \right], \tag{63}$$

where $r_2|_K = -\mathrm{rot}\, u_h$ is the volumetric residual of the second equation of problem (56).

Proof. The results easily follow by writing the residual equations as a sum over the elements $K \in \mathscr{T}_h$ and integrating by parts over each element. □

Remark 10. In order to write the boundary term coming from the integration by parts as a sum of integrals over the edges of K, we had to require that $v \in H_0(\mathrm{div}, \Omega) \cap H^\sigma(\mathrm{div}, \mathscr{T}_h)$, for some $\sigma > 0$. In the following we shall use Lemma 5 taking $v = e_h - e_h^I$, where $e_h^I \in V_h$ denotes a suitable interpolant of e_h. Hence, the regularity assumption on v is not restrictive for our purposes.

Then the following propositions hold true.

Proposition 1. *There exists a positive constant C, independent of h, such that*

$$\|e_h\|_0 \leq C \|\mathrm{div}\, e_h\|_0 + C \eta_2. \tag{64}$$

Proof. Since $e_h = u - u_h \in H_0(\mathrm{div}, \Omega)$, by the Helmholtz decomposition we can write

$$e_h = \nabla \alpha + \mathrm{rot}\, \beta,$$

where $\alpha \in H^1(\Omega)/\mathbb{R}$ and $\beta \in H_0^1(\Omega)$ are the solutions of the Laplace problem with homogeneous boundary condition and datum $-\mathrm{div}\, e_h$ and $\mathrm{rot}\, e_h$, respectively.

Using the stability of the Laplace problem and the Galerkin orthogonality property of e_h, we have that

$$\|e_h\|_0^2 \leq C \|e_h\|_0 \|\mathrm{div}\, e_h\|_0 + (e_h, \mathrm{rot}\, (\beta - \beta^I)),$$

where β^I denotes the Clément interpolant of β (see [36]).

Applying Lemma 6 with $q = \beta - \beta^I$ and taking into account the properties of Clément interpolation operator, we get

$$\|e_h\|_0^2 \leq C \|e_h\|_0 \| \operatorname{div} e_h\|_0 + \|\beta\|_1 \sum_{K \in \mathscr{T}_h} \left[h_K \|r_2\|_{0,K} + \frac{1}{2} \sum_{e \in \mathscr{E}_K} h_e^{\frac{1}{2}} \|[\![u_h \cdot t]\!]_e\|_{0,e} \right].$$

We conclude the proof using the Helmholtz decomposition and Cauchy-Schwarz inequality.   □

Proposition 2. *There holds*

$$\| \operatorname{div} e_h\|_0^2 \leq C_\lambda \left[\eta_1^2 + \eta_2^2 + (\lambda u - \lambda_h u_h, e_h) \right], \tag{65}$$

where C_λ is a positive constant dependent on λ.

Proof. Let $e_h^I \in V_h$ be a suitable interpolant of e_h, then

$$\begin{aligned}
\| \operatorname{div} e_h\|_0^2 &= (\operatorname{div} e_h, \operatorname{div}(e_h - e_h^I)) + (\operatorname{div} e_h, \operatorname{div} e_h^I) \\
&= (\lambda u - \lambda_h u_h, e_h) - (\operatorname{div} u_h, \operatorname{div}(e_h - e_h^I)) + \lambda_h(u_h, e_h - e_h^I),
\end{aligned}$$

where the last equality follows from the residual equation (62).

The following decomposition of $H_0(\operatorname{div}, \Omega)$ plays a key role in the proof (see [64] for the details of the proof):

Proposition 3. *For any $v \in H_0(\operatorname{div}, \Omega)$ there exists $z \in H_0^1(\Omega)^2$ and $\varphi \in H^1(\Omega)$ such that*

$$v = z + \operatorname{rot} \varphi$$

and the following estimates hold:

$$\|z\|_1 \leq C \| \operatorname{div} v\|_0 \text{ and } \|\varphi\|_1 \leq C \|v\|_0.$$

Then we write the error as $e_h = z + \operatorname{rot} \varphi$ and define $e_h^I = z^I + \operatorname{rot} \varphi^I$, where z^I and φ^I denote the Clément interpolant of z and φ, respectively. Note that with this definition, $e_h^I \in V_h$ and $e_h - e_h^I = (z - z^I) + \operatorname{rot}(\varphi - \varphi^I)$. It follows from Lemmas 5 and 6 that

$$\begin{aligned}
&- (\operatorname{div} u_h, \operatorname{div}(e_h - e_h^I)) + \lambda_h(u_h, e_h - e_h^I) \\
&= -(\operatorname{div} u_h, \operatorname{div}(z - z^I)) + \lambda_h(u_h, z - z^I) + \lambda_h(u_h, \operatorname{rot}(\varphi - \varphi^I)) \\
&\leq \sum_{K \in \mathscr{T}_h} \left[\|r_1\|_{0,K} h_K h_K^{-1} \|z - z^I\|_{0,K} + \frac{1}{2} \sum_{e \in \mathscr{E}_K} \|[\![\operatorname{div} u_h]\!]_e\|_{0,e} h_e^{\frac{1}{2}} h_e^{-\frac{1}{2}} \|z - z^I\|_{0,e} \right]
\end{aligned}$$

$$+\lambda_h \sum_{K \in \mathscr{T}_h} \left[\|r_2\|_{0,K} h_K h_K^{-1} \|\varphi - \varphi^I\|_{0,K} + \frac{1}{2} \sum_{e \in \mathscr{E}_K} \|[\![u_h \cdot t]\!]_e\|_{0,e} h_e^{\frac{1}{2}} h_e^{-\frac{1}{2}} \|\varphi - \varphi^I\|_{0,e} \right].$$

Using the properties of Clément interpolation operator and the estimates in Proposition 3, we have

$$\| \operatorname{div} e_h\|_0^2 \le C \| \operatorname{div} e_h\|_0 \left[\sum_{K \in \mathscr{T}_h} \left(h_K^2 \|r_1\|_{0,K}^2 + \frac{1}{2} \sum_{e \in \mathscr{E}_K} h_e \|[\![\operatorname{div} u_h]\!]_e\|_{0,e}^2 \right) \right]^{\frac{1}{2}}$$

$$+ C\lambda_h \|e_h\|_0 \left[\sum_{K \in \mathscr{T}_h} (h_K^2 \|r_2\|_{0,K}^2 + \frac{1}{2} \sum_{e \in \mathscr{E}_K} h_e \|[\![u_h \cdot t]\!]_e\|_{0,e}^2) \right]^{\frac{1}{2}}$$

$$+ (\lambda u - \lambda_h u_h, e_h).$$

Since $\lambda_h \to \lambda$ we can bound λ_h by a constant depending on λ. We complete the proof using two times the arithmetic-geometric mean inequality $ab \le \dfrac{\varepsilon}{2} a^2 + \dfrac{b^2}{2\varepsilon}$. $\square$

Remark 11. Since $\|u\|_0 = \|u_h\|_0 = 1$, the last term in (65) can be written as

$$(\lambda u - \lambda_h u_h, e_h) = (\lambda + \lambda_h) [1 - (u, u_h)] = \frac{\lambda + \lambda_h}{2} \|e_h\|_0^2 \tag{66}$$

and hence if the continuous eigensolution is smooth enough (i.e. $u \in H^\sigma(\Omega, \operatorname{div})$ for some $\sigma > k$), then by the a priori estimate (60) it turns out to be of higher order than $\| \operatorname{div} e_h\|_0^2$.

As a consequence of the previous results, we can state the following theorem.

Theorem 11. *Let us assume that $u \in H^\sigma(\operatorname{div}, \Omega)$, for some $\sigma > k$. Then there exists a constant C_λ, depending on λ and on the regularity of the mesh, such that*

$$\|e_h\|_{\operatorname{div}} \le C_\lambda (\eta_1 + \eta_2) + \text{h.o.t.} \tag{67}$$

Remark 12. Thanks to the normalization of the eigenfunctions, the following result holds true:

$$\| \operatorname{div} e_h\|_0^2 = \lambda + \lambda_h - 2\lambda(u, u_h) = \lambda_h - \lambda + \lambda \|e_h\|_0^2.$$

Therefore, if the error estimator is reliable, then we automatically have that it bounds from above, up to a multiplicative constant dependent on λ, the square root of the error in the approximation of the eigenvalues as well.

We split the proof of the efficiency of the error indicators into two steps.

First of all we use the following lemma to prove that the error indicator $\eta_{2,K}$ is bounded above by the L^2-norm of the error in the neighborhood of the element K.

Lemma 7. *Let $K \in \mathcal{T}_h$. Given $q_K \in L^2(K)$, $p_{e,K} \in L^2(e)$, $e \subset \partial K$, there exists a unique $\psi_K \in P_{k+3}(K)$ such that*

$$\begin{cases} (\psi_K, r)_K = (q_K, r)_K & \forall r \in P_k(K) \\ \int_e \psi_K s = \int_e p_{e,K} s & \forall s \in P_{k+1}(e) \\ \psi_K = 0 & \text{at the vertices of } K \end{cases} \tag{68}$$

and

$$\|\psi_K\|_{0,K} \leq C \left(\|q_K\|_{0,K} + \sum_{e \subset \partial K} h_e^{\frac{1}{2}} \|p_{e,K}\|_{0,e} \right), \tag{69}$$

with C constant depending only on the regularity of K.

Then the following result holds.

Proposition 4. *There exists a constant C, depending only on the regularity of the element K, such that*

$$h_K^2 \|r_2\|_{0,K}^2 \leq C \|e_h\|_{0,K}^2.$$

Proof. We apply Lemma 7 with $q_K = r_2$, $p_{e,K} = 0 \; \forall e \subset \partial K$, then $\psi_K \in H_0^1(K)$. Let $\psi \in H_0^1(\Omega)$ denote the zero extension of ψ_K. Taking $q = \psi$ in the residual equation (63) we get

$$\|r_2\|_{0,K}^2 = (r_2, \psi_K)_K = (e_h, \mathbf{rot}\ \psi) \leq \|e_h\|_{0,K} \|\mathbf{rot}\ \psi_K\|_{0,K}$$

$$\leq C \|e_h\|_{0,K} h_K^{-1} \|\psi_K\|_{0,K},$$

where the last bound follows from an inverse inequality. We complete the proof using (69). $\qquad\square$

For any interior edge $\bar{e} \in \mathcal{E}$ let $K_{\bar{e}}^1$ and $K_{\bar{e}}^2$ denote the two elements of $\mathcal{T}_h$ sharing $\bar{e}$.

Proposition 5. *Let $\bar{e} \in \mathcal{E}$. There exists a constant C, depending only on the regularity of $K_{\bar{e}}^1$ and $K_{\bar{e}}^2$, such that*

$$\frac{1}{2} h_{\bar{e}} \|[\![\mathbf{u}_h \cdot \mathbf{t}]\!]_{\bar{e}}\|_{0,\bar{e}}^2 \leq C \|e_h\|_{0,K_{\bar{e}}^1 \cup K_{\bar{e}}^2}^2.$$

Proof. For $i = 1, 2$, we apply Lemma 7 with $q_{K_{\bar{e}}^i} = 0$, $p_{\bar{e},K_{\bar{e}}^i} = [\![\mathbf{u}_h \cdot \mathbf{t}]\!]$, and $p_{e,K_{\bar{e}}^i} = 0$ if $e \neq \bar{e}$. Let $\psi \in H_0^1(K_{\bar{e}}^1 \cup K_{\bar{e}}^2) \cap H_0^1(\Omega)$ defined by

$$\psi = \begin{cases} \psi_{K_{\bar{e}}^1} & \text{in } K_{\bar{e}}^1 \\ \psi_{K_{\bar{e}}^2} & \text{in } K_{\bar{e}}^2 \\ 0 & \text{in } \Omega \setminus (K_{\bar{e}}^1 \cup K_{\bar{e}}^2). \end{cases}$$

Taking $q = \psi$ in (63) we get

$$\frac{1}{2} \| [\![u_h \cdot t]\!]_{\bar{e}} \|_{0,\bar{e}}^2 = (e_h, \mathbf{rot}\,\psi)_{K_{\bar{e}}^1 \cup K_{\bar{e}}^2} \leq C \| e_h \|_{0,K_{\bar{e}}^1 \cup K_{\bar{e}}^2} h_{\bar{e}}^{-\frac{1}{2}} \| [\![u_h \cdot t]\!]_{\bar{e}} \|_{0,\bar{e}},$$

where the last bound follows from an inverse inequality and (69). $\qquad\square$

We can therefore state the following theorem.

Theorem 12. *There exists a constant C, depending only on the regularity of the mesh, such that*

$$\eta_2 \leq C \| e_h \|_0.$$

Moreover, the following local estimate holds

$$\eta_{2,K} \leq C \| e_h \|_{0,K^*},$$

with C constant depending only on the regularity of the elements K of K^*.

Now we prove that the error indicator $\eta_{1,K}$ is bounded above, up to higher order terms, by the L^2-norm of the divergence of the error in the neighborhood of the element K. This, together with the previous result, yields the efficiency of the error indicator $\eta_{1,K} + \eta_{2,K}$.

We shall use the following lemma, which generalizes Lemma 7 to vector-valued functions.

Lemma 8. *Let $K \in \mathcal{T}_h$. Given $q_K \in [L^2(K)]^2$, $p_{e,K} \in [L^2(e)]^2$, $e \subset \partial K$, there exists a unique $\psi_K \in [P_{k+3}(K)]^2$ such that*

$$\begin{cases} (\psi_K, r)_K = (q_K, r)_K & \forall r \in [P_k(K)]^2 \\ \int_e \psi_K \cdot s = \int_e p_{e,K} \cdot s & \forall s \in [P_{k+1}(e)]^2 \\ \psi_K = 0 & \text{at the vertices of } K \end{cases} \tag{70}$$

$$\| \psi_K \|_{0,K} \leq C \Big(\| q_K \|_{0,K} + \sum_{e \subset \partial K} h_e^{\frac{1}{2}} \| p_{e,K} \|_{0,e} \Big), \tag{71}$$

with C constant depending only on the regularity of the element K.

Then the following propositions hold.

Proposition 6. *There exists a constant C, depending only on the regularity of K, such that*

$$h_K \| r_1 \|_{0,K} \leq C \big(\| \operatorname{div} e_h \|_{0,K} + h_K \| \lambda u - \lambda_h u_h \|_{0,K} \big).$$

Proof. We apply Lemma 8 with $\boldsymbol{q}_K = \boldsymbol{r}_1$, $\boldsymbol{p}_{e,K} = 0 \ \forall e \subset \partial K$, then $\psi_K \in [H_0^1(K)]^2$. Let $\psi \in [H_0^1(\Omega)]^2$ denote the zero extension of ψ_K. Then taking $v = \psi$ in (62), we get

$$\|\boldsymbol{r}_1\|_{0,K}^2 = (\boldsymbol{r}_1, \psi_K)_K = (\operatorname{div} \boldsymbol{e}_h, \operatorname{div} \psi_K)_K - (\lambda \boldsymbol{u} - \lambda_h \boldsymbol{u}_h, \psi_K)_K$$

$$\leq \|\operatorname{div} \boldsymbol{e}_h\|_{0,K} \|\operatorname{div} \psi_K\|_{0,K} + \|\lambda \boldsymbol{u} - \lambda_h \boldsymbol{u}_h\|_{0,K} \|\boldsymbol{r}_1\|_{0,K}$$

$$\leq C h_K^{-1} \|\operatorname{div} \boldsymbol{e}_h\|_{0,K} \|\boldsymbol{r}_1\|_{0,K} + \|\lambda \boldsymbol{u} - \lambda_h \boldsymbol{u}_h\|_{0,K} \|\boldsymbol{r}_1\|_{0,K},$$

where the last bound follows from an inverse inequality and from (71). $\qquad\square$

Proposition 7. *Let $\bar{e} \in \mathscr{E}$. There exists a constant C, depending only on the regularity of $K_{\bar{e}}^1$ and $K_{\bar{e}}^2$, such that*

$$\frac{1}{2} h_{\bar{e}}^{\frac{1}{2}} \|\llbracket \operatorname{div} \boldsymbol{u}_h \rrbracket_{\bar{e}}\|_{0,\bar{e}} \leq C(\|\operatorname{div} \boldsymbol{e}_h\|_{0,K_{\bar{e}}^1 \cup K_{\bar{e}}^2} + h_{\bar{e}} \|\lambda \boldsymbol{u} - \lambda_h \boldsymbol{u}_h\|_{0,K_{\bar{e}}^1 \cup K_{\bar{e}}^2}).$$

Proof. For $i = 1,2$, we apply Lemma 8 with $\boldsymbol{q}_{K_{\bar{e}}^i} = \boldsymbol{0}$, $\boldsymbol{p}_{\bar{e},K_{\bar{e}}^i} = (0, -\llbracket \operatorname{div} \boldsymbol{u}_h \rrbracket_{\bar{e}})$, and $\boldsymbol{p}_{e,K_{\bar{e}}^i} = \boldsymbol{0}$ if $e \neq \bar{e}$. Let $\psi \in [H_0^1(K_{\bar{e}}^1 \cup K_{\bar{e}}^2)]^2 \cap [H_0^1(\Omega)]^2$ be defined by

$$\psi = \begin{cases} \psi_{K_{\bar{e}}^1} & \text{in } K_{\bar{e}}^1 \\ \psi_{K_{\bar{e}}^2} & \text{in } K_{\bar{e}}^2 \\ \boldsymbol{0} & \text{in } \Omega \backslash (K_{\bar{e}}^1 \cup K_{\bar{e}}^2). \end{cases}$$

Taking $v = \psi$ in the residual equation (62) we get

$$\frac{1}{2} \|\llbracket \operatorname{div} \boldsymbol{u}_h \rrbracket_{\bar{e}}\|_{0,\bar{e}}^2 = (\operatorname{div} \boldsymbol{e}_h, \operatorname{div} \psi)_{K_{\bar{e}}^1 \cup K_{\bar{e}}^2} - (\lambda \boldsymbol{u} - \lambda_h \boldsymbol{u}_h, \psi)_{K_{\bar{e}}^1 \cup K_{\bar{e}}^2}$$

$$\leq C(\|\operatorname{div} \boldsymbol{e}_h\|_{0,K_{\bar{e}}^1 \cup K_{\bar{e}}^2} h_{\bar{e}}^{-\frac{1}{2}} \|\llbracket \operatorname{div} \boldsymbol{u}_h \rrbracket_{\bar{e}}\|_{0,\bar{e}}$$

$$+ \|\lambda \boldsymbol{u} - \lambda_h \boldsymbol{u}_h\|_{0,K_{\bar{e}}^1 \cup K_{\bar{e}}^2} h_{\bar{e}}^{\frac{1}{2}} \|\llbracket \operatorname{div} \boldsymbol{u}_h \rrbracket_{\bar{e}}\|_{0,\bar{e}}),$$

where the last bound follows from an inverse inequality and from (71). $\qquad\square$

As a consequence of Propositions 6 and 7 the following theorem holds.

Theorem 13. *There exists a constant C, depending only on the regularity of the elements of K^*, such that*

$$\eta_{1,K} \leq C (\|\operatorname{div} \boldsymbol{e}_h\|_{0,K^*} + h_K \|\lambda \boldsymbol{u} - \lambda_h \boldsymbol{u}_h\|_{0,K^*}).$$

Remark 13. The term $h_K \| \lambda u - \lambda_h u_h \|_{0,K^*}$ in the previous theorem is a higher order term. Indeed, for each element $K \in \mathscr{T}_h$

$$h_K \| \lambda u - \lambda_h u_h \|_{0,K^*} \leq | \lambda - \lambda_h | h_K \| u_h \|_{0,K^*} + \lambda h_K \| e_h \|_{0,K^*} \leq C h^{2t+1}$$
$$+ \lambda h_K \| e_h \|_{0,K^*},$$

where the last bound follows from the a priori estimate (61). Note that the right hand side is asymptotically negligible with respect to the local error $\| \operatorname{div} e_h \|_{0,K^*}$.

Putting together the results of Theorems 12 and 13, we have that the error indicator $\eta_{1,K} + \eta_{2,K}$ is bounded above by the local error up to a multiplicative constant and higher order terms, namely,

$$\eta_{1,K} + \eta_{2,K} \leq C \| e_h \|_{\operatorname{div},K^*} + O(h^{2t+1}) + O(h_K) \| e_h \|_{0,K^*}.$$

We summarize all the result we presented in the following theorem.

Theorem 14. *There exists a constant C, depending only on the regularity of the mesh, such that*

$$\eta_{1,K} + \eta_{2,K} \leq C \| e_h \|_{\operatorname{div},K^*} + \text{h.o.t.} \tag{72}$$

Moreover if $u \in H^\sigma(\operatorname{div}, \Omega)$ for some $\sigma > k$, then there exist two constants $C_{1,\lambda}$ and $C_{2,\lambda}$, depending on λ and on the regularity of the mesh, such that

$$\| e_h \|_{\operatorname{div}} \leq C_{1,\lambda}(\eta_1 + \eta_2) + \text{h.o.t.} \tag{73}$$

$$| \lambda - \lambda_h |^{\frac{1}{2}} \leq C_{2,\lambda}(\eta_1 + \eta_2) + \text{h.o.t.} \tag{74}$$

6.4.2 A Posteriori Error Analysis for Raviart-Thomas Finite Elements

In this section we develop an a posteriori error analysis for the Raviart-Thomas (RT) approximation of the Laplace eigenproblem with Neumann boundary condition.

It is known (see [25]) that the mixed formulation of the Laplace eigenproblem with Neumann boundary condition is equivalent to the eigenvalue problem considered in the previous section. Moreover, if we consider Raviart-Thomas or Brezzi-Douglas-Marini finite elements, then the corresponding discrete problems are equivalent as well. Nevertheless, the a posteriori error analysis developed for Brezzi-Douglas-Marini approximation does not hold for Raviart-Thomas finite elements. This is due to the fact that, contrary to BDM elements, RT elements provide an approximation of the same order in L^2 and $H(\operatorname{div})$ and thus the term $(\lambda u - \lambda_h u_h, e_h)$, which appears in the analysis, is not a higher order term (see Remark 11).

We prove that, if a superconvergence result holds true, then the error estimator η_2 previously introduced in Sect. 6.4.1 is equivalent to the L^2-norm of the error up to higher order terms.

In order to prove this result, we also prove that the *grad*-part of the L^2-norm of the error is negligible. A similar result is known to hold for the source problem, provided the solution is smooth enough. Nevertheless, the proof given for the source problem cannot be extended in a straightforward fashion to the eigenvalue problem. Indeed the proof strongly relies on the Galerkin orthogonality, which holds for the source problem but not for the eigenvalue problem.

The mixed formulation of the Laplace eigenproblem with Neumann boundary condition reads:

find $\lambda \in \mathbb{R}$ such that there exist $(\sigma, \varphi) \in H_0(\mathrm{div}, \Omega) \times L_0^2(\Omega)$, with $\varphi \neq 0$:
$$\begin{cases} (\sigma, \tau) + (\mathrm{div}\,\tau, \varphi) = 0 & \forall\,\tau \in H_0(\mathrm{div}, \Omega) \\ (\mathrm{div}\,\sigma, \psi) = -\lambda(\varphi, \psi) & \forall\,\psi \in L_0^2(\Omega) \end{cases}$$
(75)

and its discretization by means of Raviart-Thomas finite elements is given by

find $\lambda_h \in \mathbb{R}$ such that there exist $(\sigma_h, \varphi_h) \in \Sigma_h \times \Phi_h$, with $\varphi_h \neq 0$:
$$\begin{cases} (\sigma_h, \tau) + (\mathrm{div}\,\tau, \varphi_h) = 0 & \forall\,\tau \in \Sigma_h \\ (\mathrm{div}\,\sigma_h, \psi) = -\lambda_h(\varphi_h, \psi) & \forall\,\psi \in \Phi_h, \end{cases}$$
(76)

where

$$\Sigma_h = RT_k \cap H_0(\mathrm{div}, \Omega),$$

and

$$\Phi_h = \{\psi \in L_0^2(\Omega) : \psi|_K \in P_k \ \forall\, K \in \mathcal{T}_h\}.$$

The Raviart-Thomas spaces are defined for $k \geq 0$ as follows:

$$RT_k = \{\tau \in H(\mathrm{div}, \Omega) : \tau|_K \in P_k(K)^2 + P_k(K)(x, y)^t \ \forall K \in \mathcal{T}_h\}.$$

Due to regularity results (see [7]), there exists a constant $s > 1/2$ (depending on Ω), such that (σ, φ) belongs to the space $H^s(\Omega)^2 \times H^{1+s}(\Omega)$. Furthermore, the following estimate holds true:

$$\|\sigma\|_s + \|\mathrm{div}\,\sigma\|_{1+s} \leq C \|\sigma\|_0,$$
(77)

where C is a constant depending on the eigenvalue λ. In (77), s is at least one if Ω is convex, while s is at least $\pi/\omega - \varepsilon$ for any $\varepsilon > 0$ for a non convex domain, $\omega < 2\pi$ being the maximum interior angle of Ω.

Let (λ, σ) be an eigensolution of (75) such that λ is a simple eigenvalue and $\|\sigma\|_0 = 1$. From the abstract theory (see [12, 21]) and known a priori estimates it follows that, for h small enough (depending on λ), there exists (λ_h, σ_h) eigenpair of (76) with $\|\sigma_h\|_0 = 1$ such that

$$\|\sigma - \sigma_h\|_{\mathrm{div}} = O(h^t) \tag{78}$$

$$\|\sigma - \sigma_h\|_0 = O(h^t) \tag{79}$$

$$|\lambda - \lambda_h| = O(h^{2t}), \tag{80}$$

where $t = \min\{s, k+1\}$.

In the following we denote by $e_h = \sigma - \sigma_h$ the error in the approximation of the eigenfunction and by P_h the L^2-projection on Φ_h.

The main result of this section is stated in the following theorem.

Theorem 15. *If* $\|P_h\varphi - \varphi_h\|_0$ *is of higher order than* $\|e_h\|_0$, *then there exist a constant* C, *depending on the regularity of the mesh, such that*

$$\|e_h\|_0 \leq C\eta_2 + \text{h.o.t.},$$

where "h.o.t." denotes higher order terms.

The error indicator η_2 in the above theorem is the one introduced in Sect. 6.4.1 and it is given by

$$\eta_{2,K}^2 = h_K^2 \|\operatorname{rot}\sigma_h\|_{0,K}^2 + \frac{1}{2}\sum_{e\in\mathscr{E}_K} h_e \|[\![\sigma_h \cdot t]\!]_e\|_{0,e}^2.$$

As for Brezzi-Douglas-Marini approximation, the a posteriori error analysis starts from the Helmholtz decomposition of the error

$$e_h = \nabla\alpha + \operatorname{rot}\beta.$$

The rot-part of the L^2-norm of the error is then bounded above by the error indicator η_2, as it has been done in Sect. 6.4.1. But, contrary to what has been done for Brezzi-Douglas-Marini approximation, we prove that if a superconvergence property holds, then the $grad$-part of the L^2-norm of the error $(e_h, \nabla\alpha)$ is of higher order than $\|e_h\|_0^2$. It is known that a similar result holds true for the source problem, provided the solution is smooth enough. Indeed, thanks to the Helmholtz decomposition, it holds

$$\|\sigma^s - \sigma_h^s\|_0^2 = (\sigma^s - \sigma_h^s, \nabla a) + (\sigma^s - \sigma_h^s, \operatorname{rot}b),$$

where σ^s and σ_h^s denote the solution of the continuous and discrete source problem, respectively. Moreover, due to the Galerkin orthogonality property, we have

$$(\sigma^s - \sigma_h^s, \nabla a) = -(\operatorname{div}(\sigma^s - \sigma_h^s), a) = -(\operatorname{div}(\sigma^s - \sigma_h^s), a - P_h a). \tag{81}$$

Therefore, if σ^s is smooth enough, then from the standard a priori error analysis for mixed problems it follows that

$$(\sigma^s - \sigma_h^s, \nabla a) \le C h^{k+2} \|\sigma^s - \sigma_h^s\|_0, \tag{82}$$

and hence it turns out to be of higher order than $\|\sigma^s - \sigma_h^s\|_0^2$.

We observe that the previous proof cannot be generalized in a straightforward way to the eigenvalue problem. In fact in this case, the Galerkin orthogonality does not hold and hence we are not allowed to subtract $P_h a$ in the right hand side of (81). In what follows we shall generalize equation (82) to the eigenvalue problem and we will use it to prove Theorem 15.

We start proving the following theorem.

Theorem 16. *There exist a constant C, depending on the regularity of the mesh, such that*

$$\|e_h\|_0 \le C(\eta_2 + |\lambda - \lambda_h| + \|\varphi - P_h\varphi\|_{(H^1/\mathbb{R})^*} + \|P_h\varphi - \varphi_h\|_0) \tag{83}$$

Proof. By the Helmholtz decomposition, we write the error as

$$e_h = \nabla\alpha + \mathbf{rot}\,\beta,$$

where $\alpha \in H^1(\Omega)/\mathbb{R}$ and $\beta \in H_0^1(\Omega)$ are the solutions of the Laplace problem with homogeneous Neumann and Dirichlet boundary condition and datum $-\operatorname{div} e_h$ and $\operatorname{rot} e_h$, respectively. Then the L^2-norm of the error is given by

$$\|e_h\|_0^2 = (e_h, \nabla\alpha) + (e_h, \mathbf{rot}\,\beta).$$

Arguing as in the proof of Proposition 1, we have that

$$(e_h, \mathbf{rot}\,\beta) \le C\,\eta_2\|e_h\|_0.$$

Hence, it remains to prove that

$$(e_h, \nabla\alpha) \le C(|\lambda - \lambda_h| + \|\varphi - P_h\varphi\|_{(H^1/\mathbb{R})^*} + \|P_h\varphi - \varphi_h\|_0)\|e_h\|_0.$$

Integrating $(e_h, \nabla\alpha)$ by parts and using the second equation of problems (75) and (76), we get

$$(e_h, \nabla\alpha) = -(\operatorname{div}(\sigma - \sigma_h), \alpha) = -(\lambda\varphi - \lambda_h\varphi_h, \alpha).$$

We now estimate $(\lambda\varphi - \lambda_h\varphi_h, \alpha)$ in the following way:

$$(\lambda\varphi - \lambda_h\varphi_h, \alpha) \le \|\lambda\varphi - \lambda_h\varphi_h\|_{(H^1/\mathbb{R})^*}\|\alpha\|_{H^1/\mathbb{R}} \le C\|\lambda\varphi - \lambda_h\varphi_h\|_{(H^1/\mathbb{R})^*}\|e_h\|_0.$$

Finally, we bound $\|\lambda \varphi - \lambda_h \varphi_h\|_{(H^1/\mathbb{R})^*}$ as follows:

$$\|\lambda \varphi - \lambda_h \varphi_h\|_{(H^1/\mathbb{R})^*} \leq |\lambda - \lambda_h| \|\varphi\|_0 + |\lambda_h| \|\varphi - \varphi_h\|_{(H^1/\mathbb{R})^*}.$$

We conclude the proof by adding and subtracting $P_h \varphi$ in the last term of the above equation, and a applying triangular inequality. □

The following corollary, which states the reliability of the error estimator η_2, holds true.

Corollary 6. *If $\|P_h \varphi - \varphi_h\|_0$ is of higher order than $\|e_h\|_0$, then there exists a constant C, depending on the regularity of the mesh, such that*

$$\|e_h\|_0 \leq C \eta_2 + \text{h.o.t.}.$$

Proof. The terms $|\lambda - \lambda_h|$ and $\|\varphi - P_h \varphi\|_{(H^1/\mathbb{R})^*}$ in (83) are of higher order than $\|e_h\|_0$. Indeed, thanks to the a priori error estimate (80),

$$|\lambda - \lambda_h| = O(h^{2t}),$$

while, taking into account the properties of the L^2-projection P_h together with the a priori estimate (77), the second term is bounded in this way

$$\|\varphi - P_h \varphi\|_{(H^1/\mathbb{R})^*} = \sup_{\psi \in H^1/\mathbb{R}} \frac{(\varphi - P_h \varphi, \psi - P_h \psi)}{\|\psi\|_{H^1/\mathbb{R}}} \leq Chh^{\min\{k,s\}+1}. \qquad (84)$$

□

The superconvergence result required by Corollary 6 has been proved for the lowest order Raviart-Thomas elements in [46]. Moreover, numerical evidence of the superconvergence property for Brezzi-Douglas-Marini space of lowest order has also been shown.

In order to prove that the local error indicator $\eta_{2,K}$ is bounded above by the L^2-norm of the error in the neighborhood of the element K, we can argue in the same way as it has been done in Sect. 6.4.1 for Brezzi-Douglas-Marini approximation. Indeed, the following propositions holds true for Raviart-Thomas finite elements as well.

Proposition 8. *There exists a constant C, depending only on the regularity of the element K, such that*

$$h_K^2 \|\text{rot}\, \sigma_h\|_{0,K}^2 \leq C \|e_h\|_{0,K}^2.$$

Proposition 9. *Let $\bar{e} \in \mathscr{E}$. There exists a constant C, depending only on the regularity of $K_{\bar{e}}^1$ and $K_{\bar{e}}^2$, such that*

$$\frac{1}{2} h_{\bar{e}} \|[\![\sigma_h \cdot t]\!]_e\|_{0,\bar{e}}^2 \leq C \|e_h\|_{0, K_{\bar{e}}^1 \cup K_{\bar{e}}^2}^2,$$

where $K_{\bar{e}}^1$ and $K_{\bar{e}}^2$ denote the two elements in $\mathscr{T}_h$ sharing $\bar{e}$.

We omit the details of the proof, which are as in Sect. 6.4.1. Collecting the results of the above propositions, we can state the following theorem.

Theorem 17. *For each element $K \in \mathcal{T}_h$ there exists a constant C, depending only on the regularity of the elements in K^*, such that*

$$\eta_{2,K} \leq C \|e_h\|_{0,K^*}.$$

Moreover the following global estimate holds

$$\eta_2 \leq C \|e_h\|_0,$$

with C constant depending only on the regularity of the mesh.

The following theorem summarizes the results we proved.

Theorem 18. *For each element $K \in \mathcal{T}_h$ there exists a constant C, depending only on the regularity of the elements in K^*, such that*

$$\eta_{2,K} \leq C \|e_h\|_{0,K^*}.$$

Moreover if $\|P_h\varphi - \varphi\|_0$ is h.o.t., then there exist a constant C, depending on the regularity of the mesh, such that

$$\|e_h\|_0 \leq C\eta_2 + \text{h.o.t.},$$

where "h.o.t." denotes higher order terms.

6.5 Numerical Results

In this section we present the results of some preliminary numerical computation which confirm the good behavior of the error indicators introduced in Sects. 6.4.1 and 6.4.2 when used to mark the elements to be refined.

The numerical tests concern the problem of a rigid L-shaped cavity, as shown in Fig. 11. Since the domain has a re-entrant corner, eigenfunctions with singularities are expected.

Fig. 11 L-shaped domain

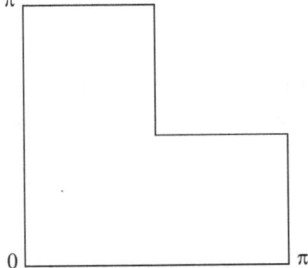

Fig. 12 Initial triangulation

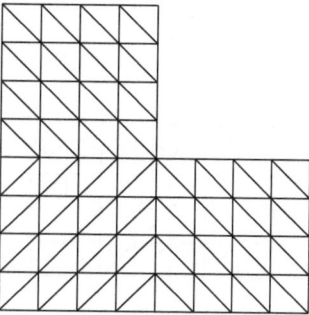

We present the results obtained with meshes generated by the adaptive method described in Sect. 6.2. We use the longest edge bisection refinement and the refinement strategy 2 with $\gamma = 0.5$. The process starts with a coarse uniform triangulation $\mathcal{T}_0$ shown in Fig. 12.

The tests have been performed taking as approximation spaces the lowest order Brezzi-Douglas-Marini (BDM_1) and Raviart-Thomas elements (RT_0), respectively. In this particular case, the error indicators introduced in Sects. 6.4.1 and 6.4.2 reduce to

$$\eta_K = \eta_{1,K} + \eta_{2,K}$$

with

$$\eta_{1,K}^2 = h_K^2 \lambda_h^2 \|\boldsymbol{\sigma}_h\|_{0,K}^2 + \frac{1}{2} \sum_{e \in \mathscr{E}_K} h_e \|[\![\operatorname{div} \boldsymbol{\sigma}_h]\!]_e\|_{0,e}^2$$

and

$$\eta_{2,K}^2 = h_K^2 \|\operatorname{rot} \boldsymbol{\sigma}_h\|_{0,K}^2 + \frac{1}{2} \sum_{e \in \mathscr{E}_K} h_e \|[\![\boldsymbol{\sigma}_h \cdot \boldsymbol{t}]\!]_e\|_{0,e}^2$$

for the first order Brezzi-Douglas-Marini finite element, while

$$\eta_K^2 = \frac{1}{2} \sum_{e \in \mathscr{E}_K} h_e \|[\![\boldsymbol{\sigma}_h \cdot \boldsymbol{t}]\!]_e\|_{0,e}^2$$

for the lowest order Raviart-Thomas elements.

The numerical computations concern the first eigensolution, which is the one which presents local singularities in the neighborhood of the re-entrant corner.

Figure 13 shows the triangulations obtained after four, six and seven steps of the refinement process for BDM approximation, whereas Fig. 14 uses the RT approximation. As can be seen, in both cases the error indicators correctly detect the elements which have to be refined, namely the ones in the neighborhood regions of the singularities.

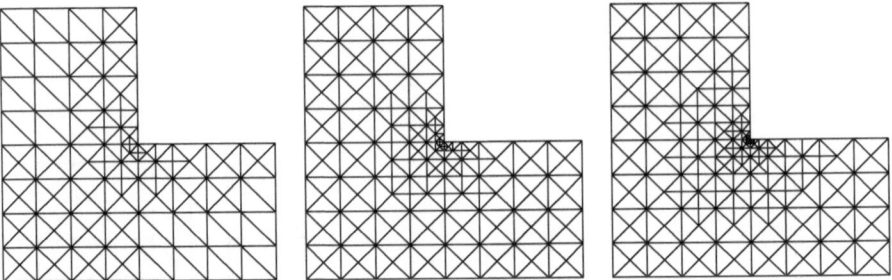

Fig. 13 Fourth, sixth, and seventh refinement steps (BDM_1)

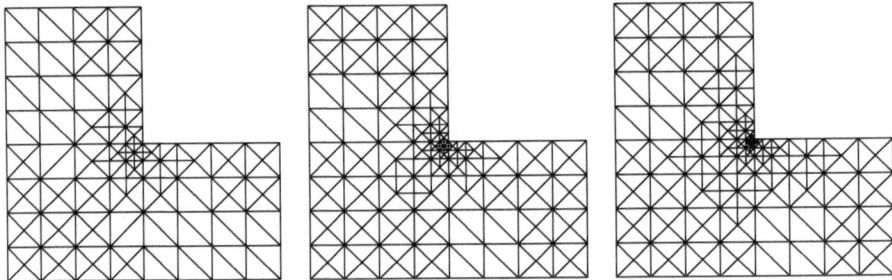

Fig. 14 Fourth, sixth, and seventh refinement steps (RT_0)

References

1. M. Ainsworth and J. T. Oden. *A posteriori error estimation in finite element analysis*. Pure and Applied Mathematics (New York). Wiley-Interscience [John Wiley & Sons], New York, 2000.
2. A. Alonso. Error estimators for a mixed method. *Numer. Math.*, 74(4):385–395, 1996.
3. A. Alonso, A. Dello Russo, C. Otero-Souto, C. Padra, and R. Rodríguez. An adaptive finite element scheme to solve fluid-structure vibration problems on non-matching grids. *Comput. Vis. Sci.*, 4(2):67–78, 2001. Second AMIF International Conference (Il Ciocco, 2000).
4. A. Alonso, A. Dello Russo, C. Padra, and R. Rodríguez. A posteriori error estimates and a local refinement strategy for a finite element method to solve structural-acoustic vibration problems. *Adv. Comput. Math.*, 15(1-4):25–59 (2002), 2001. A posteriori error estimation and adaptive computational methods.
5. A. Alonso, A. Dello Russo, and V. Vampa. A posteriori error estimates in finite element solution of structure vibration problems with applications to acoustical fluid-structure analysis. *Comput. Mech.*, 23(3):231–239, 1999.
6. A. Alonso, A. Dello Russo, and V. Vampa. A posteriori error estimates in finite element acoustic analysis. *J. Comput. Appl. Math.*, 117(2):105–119, 2000.
7. C. Amrouche, C. Bernardi, M. Dauge, and V. Girault. Vector potential in three-dimensional nonsmooth domains. *Math Methods Appl. Sci.*, 21(9):823–864, 1998.
8. P. M. Anselone. *Collectively compact operator approximation theory and applications to integral equations*. Prentice-Hall Inc., Englewood Cliffs, N. J., 1971. With an appendix by Joel Davis, Prentice-Hall Series in Automatic Computation.

9. M. G. Armentano and C. Padra. A posteriori error estimates for the Steklov eigenvalue problem. *Appl. Numer. Math.*, 58(5):593–601, 2008.
10. D. N. Arnold, R. S. Falk, and R. Winther. Finite element exterior calculus: from Hodge theory to numerical stability. *Bull. Amer. Math. Soc. (N.S.)*, 47(2):281–354, 2010.
11. I. Babuška and R. Narasimhan. The Babuška-Brezzi condition and the patch test: an example. *Comput. Methods Appl. Mech. Engrg.*, 140(1-2):183–199, 1997.
12. I. Babuška and J. Osborn. Eigenvalue problems. In *Handbook of numerical analysis, Vol. II*, Handb. Numer. Anal., II, pages 641–787. North-Holland, Amsterdam, 1991.
13. I. Babuška and W. C. Rheinboldt. A-posteriori error estimates for the finite element method. *Int. J. Numer. Meth. Eng.*, 12:1597–1615, 1978.
14. I. Babuška and W. C. Rheinboldt. Error estimates for adaptive finite element computations. *SIAM J. Numer. Anal.*, 15(4):736–754, 1978.
15. R. E. Bank. The efficient implementation of local mesh refinement algorithms. In *Adaptive computational methods for partial differential equations (College Park, Md., 1983)*, pages 74–81. SIAM, Philadelphia, PA, 1983.
16. R. E. Bank. *PLTMG: a software package for solving elliptic partial differential equations.* Software, Environments, and Tools. Society for Industrial and Applied Mathematics (SIAM), Philadelphia, PA, 1998. Users' guide 8.0.
17. R. E. Bank, A. H. Sherman, and A. Weiser. Refinement algorithms and data structures for regular local mesh refinement. In *Scientific computing (Montreal, Que., 1982)*, IMACS Trans. Sci. Comput., I, pages 3–17. IMACS, New Brunswick, NJ, 1983.
18. K.-J. Bathe, C. Nitikitpaiboon, and X. Wang. A mixed displacement-based finite element formulation for acoustic fluid-structure interaction. *Comput. & Structures*, 56(2-3):225–237, 1995.
19. A. Bermúdez, R. Durán, M. A. Muschietti, R. Rodríguez, and J. Solomin. Finite element vibration analysis of fluid-solid systems without spurious modes. *SIAM J. Numer. Anal.*, 32(4):1280–1295, 1995.
20. A. Bermúdez and D. G. Pedreira. Mathematical analysis of a finite element method without spurious solutions for computation of dielectric waveguides. *Numer. Math.*, 61(1):39–57, 1992.
21. D. Boffi. Finite element approximation of eigenvalue problems. *Acta Numer.*, 19:1–120, 2010.
22. D. Boffi, F. Brezzi, and L. Gastaldi. On the convergence of eigenvalues for mixed formulations. *Ann. Scuola Norm. Sup. Pisa Cl. Sci. (4)*, 25(1-2):131–154 (1998), 1997. Dedicated to Ennio De Giorgi.
23. D. Boffi, F. Brezzi, and L. Gastaldi. On the problem of spurious eigenvalues in the approximation of linear elliptic problems in mixed form. *Math. Comp.*, 69(229):121–140, 2000.
24. D. Boffi, C. Chinosi, and L. Gastaldi. Approximation of the grad div operator in nonconvex domains. *CMES Comput. Model. Eng. Sci.*, 1(2):31–43, 2000.
25. D. Boffi, P. Fernandes, L. Gastaldi, and I. Perugia. Computational models of electromagnetic resonators: analysis of edge element approximation. *SIAM J. Numer. Anal.*, 36(4):1264–1290 (electronic), 1999.
26. D. Boffi and C. Lovadina. Remarks on augmented Lagrangian formulations for mixed finite element schemes. *Boll. Un. Mat. Ital. A (7)*, 11(1):41–55, 1997.
27. A. Bossavit. Solving Maxwell equations in a closed cavity and the question of spurious modes. *IEEE Trans. on Magnetics*, 26:702–705, 1990.
28. J. H. Bramble, J. E. Pasciak, and A. V. Knyazev. A subspace preconditioning algorithm for eigenvector/eigenvalue computation. *Adv. Comput. Math.*, 6(2):159–189 (1997), 1996.
29. F. Brezzi and M. Fortin. *Mixed and hybrid finite element methods*, volume 15 of *Springer Series in Computational Mathematics*. Springer-Verlag, New York, 1991.
30. C. Canuto, M. Y. Hussaini, A. Quarteroni, and T. A. Zang. *Spectral methods*. Scientific Computation. Springer-Verlag, Berlin, 2006. Fundamentals in single domains.
31. C. Carstensen. A posteriori error estimate for the mixed finite element method. *Math. Comp.*, 66(218):465–476, 1997.
32. C. Carstensen and J. Gedicke. An oscillation-free adaptive FEM for symmetric eigenvalue problems. Technical report, DFG Research Center MATHEON, Berlin, 2008.

33. C. Carstensen and R. Verfürth. Edge residuals dominate a posteriori error estimates for low order finite element methods. *SIAM J. Numer. Anal.*, 36(5):1571–1587, 1999.
34. H. Chen and R. Taylor. Vibration analysis of fluid-solid systems using a finite element displacement formulation. *Int. J. Numer. Methods Eng.*, 29:683–698, 1990.
35. P. G. Ciarlet. *The finite element method for elliptic problems.* North-Holland Publishing Co., Amsterdam, 1978. Studies in Mathematics and its Applications, Vol. 4.
36. P. Clément. Approximation by finite element functions using local regularization. *Rev. Française Automat. Informat. Recherche Opérationnelle Sér. Rouge Anal. Numér.*, 9(R-2):77–84, 1975.
37. M. Costabel and M. Dauge. Singularities of electromagnetic fields in polyhedral domains. *Arch. Ration. Mech. Anal.*, 151(3):221–276, 2000.
38. X. Dai, J. Xu, and A. Zhou. Convergence and optimal complexity of adaptive finite element eigenvalue computations. *Numer. Math.*, 110(3):313–355, 2008.
39. M. Dauge. Benchmark computations for Maxwell equations for the approximation of highly singular solutions. http://perso.univ-rennes1.fr/monique.dauge/benchmax.html.
40. R. Durán, L. Gastaldi, and C. Padra. A posteriori error estimators for mixed approximations of eigenvalue problems. *Math. Models Methods Appl. Sci.*, 9(8):1165–1178, 1999.
41. R. Durán, C. Padra, and R. Rodríguez. A posteriori error estimates for the finite element approximation of eigenvalue problems. *Math. Models Methods Appl. Sci.*, 13(8):1219–1229, 2003.
42. E M. Garau and P. Morin. Convergence and quasi-optimality of adaptive FEM for Steklov eigenvalue problems. *Cuad. Mat. Mec.*, page 30, 2009.
43. E. M. Garau, P. Morin, and C. Zuppa. Convergence of adaptive finite element methods for eigenvalue problems. *Math. Models Methods Appl. Sci.*, 19(5):721–747, 2009.
44. F. Gardini. A posteriori error estimates for an eigenvalue problem arising from fluid-structure interaction. *Istit. Lombardo Accad. Sci. Lett. Rend. A*, 138:17–34 (2005), 2004.
45. F. Gardini. *A posteriori error estimates for eigenvalue problems in mixed form.* PhD thesis, Università degli Studi di Pavia, 2005.
46. F. Gardini. Mixed approximation of eigenvalue problems: a superconvergence result. *M2AN Math. Model. Numer. Anal.*, 43(5):853–865, 2009.
47. L. Gastaldi. Mixed finite element methods in fluid structure systems. *Numer. Math.*, 74(2):153–176, 1996.
48. S. Giani and I. G. Graham. A convergent adaptive method for elliptic eigenvalue problems. *SIAM J. Numer. Anal.*, 47(2):1067–1091, 2009.
49. L. Grubišić and J. S. Ovall. On estimators for eigenvalue/eigenvector approximations. *Math. Comp.*, 78(266):739–770, 2009.
50. F. Hecht. *FreeFem++, Third Edition, Version 3.12*, 2011. http://www.freefem.org/ff++/.
51. V. Heuveline and R. Rannacher. A posteriori error control for finite approximations of elliptic eigenvalue problems. *Adv. Comput. Math.*, 15(1-4):107–138 (2002), 2001.
52. R. Hiptmair. Finite elements in computational electromagnetism. *Acta Numerica*, 11:237–339, 2002.
53. T. Kato. *Perturbation theory for linear operators.* Springer-Verlag, New York, 1976.
54. F. Kikuchi. Mixed and penalty formulations for finite element analysis of an eigenvalue problem in electromagnetism. *Comput. Methods Appl. Mech. Engrg.*, 64:509–521, 1987.
55. A. V. Knyazev. Sharp a priori error estimates of the rayleigh-ritz method without assumptions of fixed sign or compactness. *Math. Notes*, 38:998–1002, 1986.
56. A. V. Knyazev. New estimates for Ritz vectors. *Math. Comp.*, 66(219):985–995, 1997.
57. A. V. Knyazev and J. E. Osborn. New a priori FEM error estimates for eigenvalues. *SIAM J. Numer. Anal.*, 43(6):2647–2667 (electronic), 2006.
58. W. G. Kolata. Approximation in variationally posed eigenvalue problems. *Numer. Math.*, 29(2):159–171, 1978.
59. I. Kossaczký. A recursive approach to local mesh refinement in two and three dimensions. *J. Comput. Appl. Math.*, 55(3):275–288, 1994.

60. M. G. Larson. A posteriori and a priori error analysis for finite element approximations of self-adjoint elliptic eigenvalue problems. *SIAM J. Numer. Anal.*, 38(2):608–625 (electronic), 2000.

61. C. Lovadina, M. Lyly, and R. Stenberg. A posteriori estimates for the Stokes eigenvalue problem. *Numer. Methods Partial Differential Equations*, 25(1):244–257, 2009.

62. D. Mao, L. Shen, and A. Zhou. Adaptive finite element algorithms for eigenvalue problems based on local averaging type a posteriori error estimates. *Adv. Comput. Math.*, 25(1-3):135–160, 2006.

63. P. Monk. *Finite element methods for Maxwell's equations*. Numerical Mathematics and Scientific Computation. Oxford University Press, New York, 2003.

64. J. E. Pasciak and J. Zhao. Overlapping Schwarz methods in $H(\text{curl})$ on polyhedral domains. *J. Numer. Math.*, 10(3):221–234, 2002.

65. P.-A. Raviart and J.-M. Thomas. A mixed finite element method for second order elliptic problems. In I. Galligani and E. Magenes, editors, *Mathematical Aspects of the Finite Element Method*, volume 606 of *Lecture Notes in Math.*, pages 292–315, New York, 1977. Springer-Verlag.

66. W. C. Rheinboldt. On a theory of mesh-refinement processes. *SIAM J. Numer. Anal.*, 17(6):766–778, 1980.

67. M.-C. Rivara. Algorithms for refining triangular grids suitable for adaptive and multigrid techniques. *Internat. J. Numer. Methods Engrg.*, 20(4):745–756, 1984.

68. M.-C. Rivara. Design and data structure of fully adaptive, multigrid, finite-element software. *ACM Trans. Math. Software*, 10(3):242–264, 1984.

69. G. Valle. Problemi agli autovalori per equazioni alle derivate parziali: autovalori multipli. Tesi di laurea, Università degli Studi di Pavia, 2011.

70. R. Verfürth. A posteriori error estimates for nonlinear problems. *Math. Comp.*, 62(206):445–475, 1994.

71. R. Verfürth. *A Review of a posteriori error estimation and adaptive mesh-refinement techniques*. Wiley and Teubner, 1996.

72. T. F. Walsh, G. M. Reese, and U. L. Hetmaniuk. Explicit a posteriori error estimates for eigenvalue analysis of heterogeneous elastic structures. *Comput. Methods Appl. Mech. Engrg.*, 196(37-40):3614–3623, 2007.

73. X. Wang and K.-J. Bathe. On mixed elements for acoustic fluid-structure interactions. *Math. Models Methods Appl. Sci.*, 7(3):329–343, 1997.

74. J. Webb. Edge elements and what they can do for you. *IEEE Trans. on Magnetics*, 29:1460–1465, 1993.

C^0 Interior Penalty Methods

Susanne C. Brenner

Abstract C^0 interior penalty methods are discontinuous Galerkin methods for fourth order problems. In this article we discuss various aspects of such methods including a priori error analysis, a posteriori error analysis and fast solution techniques.

1 Introduction

There are three classical approaches in finite element methods for fourth order problems. The first one uses conforming finite element methods [6, 43, 97]. The advantage of this approach is that convergence is automatically guaranteed. The disadvantage is that it requires C^1 finite elements, which can be quite complicated, especially in three dimensions. The second approach uses nonconforming finite element methods [12,74,83,84]. The advantage of this approach is that nonconforming finite elements are simpler. The disadvantage is that such elements do not come in a natural hierarchy. The existing nonconforming elements only use low order polynomials and hence are not efficient for capturing smooth solutions. It is also nontrivial to design nonconforming finite element methods that converge, especially in three dimensions. The third approach uses a mixed formulation [42, 45, 61, 71]. The advantage of this approach is that it only requires C^0 finite elements. The disadvantage is that it replaces the symmetric positive definite continuous problem by a saddle point problem, and it is not easy to choose finite element pairs that satisfy the Ladyshenskaya-Babuška-Brezzi stability condition [8,41]. Furthermore it is nontrivial to design a good mixed formulation for complicated problems.

S.C. Brenner (✉)
Department of Mathematics and Center for Computation & Technology, Louisiana State University, Baton Rouge, LA 70803, USA
e-mail: brenner@math.lsu.edu

J. Blowey and M. Jensen (eds.), *Frontiers in Numerical Analysis – Durham 2010*, 79
Lecture Notes in Computational Science and Engineering 85,
DOI 10.1007/978-3-642-23914-4_2, © Springer-Verlag Berlin Heidelberg 2012

C^0 interior penalty methods [52] are discontinuous Galerkin methods [7, 81] for fourth order problems that can overcome the shortcomings of the classical approaches. These methods use standard C^0 Lagrange finite elements for second order problems. They are simpler than C^1 elements and they come in a natural hierarchy. The higher order C^0 interior penalty methods can capture smooth solutions efficiently. Unlike mixed methods, they preserve the positive definiteness of the continuous problem and their derivation is straight-forward, which makes them attractive for complicated problems. Moreover, since C^0 interior penalty methods use standard finite element spaces for second order problems, fast solvers for second order problems can be naturally employed as preconditioners for the resulting discrete problems.

Below we will derive C^0 interior penalty methods for several two dimensional model problems, present the error analysis of these methods, and discuss fast solution techniques for the discrete problems.

1.1 Basic Notations and Definitions

For later reference we collect here some basic notations and definitions for finite element methods, differential operators and Sobolev spaces. Detailed information on these concepts can be found for example in [1, 32, 44].

Finite Element Methods

- $\Omega \subset \mathbb{R}^2$ is a bounded polygonal domain.
- $\mathcal{T}_h$ is a triangulation of Ω.
- $V_h \ (\subset H^1(\Omega))$ is a C^0 Lagrange finite element space associated with $\mathcal{T}_h$.
- $\Pi_h : C(\bar{\Omega}) \longrightarrow V_h$ is the nodal interpolation operator for V_h.
- v_T is the restriction of $v \in V_h$ to the element T.
- $\mathcal{E}_h^i$ is the set of the interior edges of $\mathcal{T}_h$.
- $\mathcal{E}_h^b$ is the set of the boundary edges of $\mathcal{T}_h$.
- $\mathcal{E}_h = \mathcal{E}_h^i \cup \mathcal{E}_h^b$ is the set of all the edges of $\mathcal{T}_h$.
- h_T is the diameter of an element $T \in \mathcal{T}_h$ and $h = \max_{T \in \mathcal{T}_h} h_T$.
- $|e|$ is the length of an edge $e \in \mathcal{E}_h$.
- $\mathcal{T}_p$ is the set of the elements in $\mathcal{T}_h$ that share the common vertex p.
- $\mathcal{T}_e$ is the set of the elements in $\mathcal{T}_h$ that share the common edge e.
- $|\mathcal{T}_p|$ (resp. $|\mathcal{T}_e|$) is the number of elements (resp. edges) in $\mathcal{T}_p$ (resp. $\mathcal{T}_e$).

Differential Operators

- The order of $\alpha = (\alpha_1, \alpha_2)$ is $|\alpha| = \alpha_1 + \alpha_2$, and $\partial^\alpha v = \partial^{\alpha_1 + \alpha_2} v / \partial x_1^{\alpha_1} \partial x_2^{\alpha_2}$.
- $\Delta = (\partial^2/\partial x_1^2) + (\partial^2/\partial x_2^2)$ is the Laplacian operator and Δ^2 is the biharmonic operator.
- ∇v is the 2×1 gradient vector of v and $\nabla^2 v$ is the 2×2 Hessian matrix of v.
- The outer normal derivative along $\partial\Omega$ is denoted by $\partial/\partial n$.

Sobolev Spaces

- $H^m(\Omega)$ is the Sobolev space of square integrable functions whose weak derivatives up to order m are also square integrable.
- $|v|_{H^k(\Omega)}$ is the seminorm defined by $|v|^2_{H^k(\Omega)} = \sum\limits_{|\alpha|=k} \|\partial^\alpha v\|^2_{L_2(\Omega)}$.
- $\|v\|_{H^m(\Omega)}$ is the norm of $H^m(\Omega)$ defined by $\|v\|^2_{H^m(\Omega)} = \sum\limits_{0 \leq k \leq m} |v|^2_{H^k(\Omega)}$.
- For a positive number s that is not an integer, $H^s(\Omega)$ is the fractional order Sobolev space. Let $m = \lfloor s \rfloor$ be the largest integer $< s$ and $\tau = s - m$. Then the seminorm $|v|_{H^s(\Omega)}$ and the norm $\|v\|_{H^s(\Omega)}$ are given by

$$|v|^2_{H^s(\Omega)} = \sum\limits_{|\alpha|=m} \int_\Omega \int_\Omega \frac{|(\partial^\alpha v)(x) - (\partial^\alpha v)(y)|^2}{|x-y|^{2+2\tau}} \, dx\, dy,$$

$$\|v\|^2_{H^s(\Omega)} = \|v\|^2_{H^m(\Omega)} + |v|^2_{H^s(\Omega)}.$$

- For $s \geq 0$, $H^s_0(\Omega)$ is the closure in $H^s(\Omega)$ of the space of C^∞ functions with compact supports in Ω, and $H^{-s}(\Omega)$ is the dual space of $H^s_0(\Omega)$.

Piecewise Sobolev Spaces

- $H^s(\Omega, \mathscr{T}_h) = \{v \in L_2(\Omega) : v\big|_T \in H^s(T) \; \forall\, T \in \mathscr{T}_h\}$ is the piecewise Sobolev space with respect to the triangulation $\mathscr{T}_h$.

We will use C with or without a subscript to represent a generic positive constant that can depend on the shape regularity of the triangulation $\mathscr{T}_h$ and the polynomial degree of the finite element space, but not on the mesh parameter h.

Remark 1. The results in this article are also valid for the Q_k tensor product Lagrange finite element spaces.

2 Model Problems

For simplicity we consider the two dimensional biharmonic equation with various boundary conditions as our model problems, but the results discussed in this article can be extended to more complicated problems in three dimensions [52].

Example 1.

$$\Delta^2 u = f \qquad \text{in } \Omega \tag{2.1a}$$

$$u = \frac{\partial u}{\partial n} = 0 \qquad \text{on } \partial\Omega \tag{2.1b}$$

This boundary value problem is related to the bending of clamped Kirchhoff plates and stationary incompressible Stokes equations with the no-slip boundary condition.

Example 2.

$$\Delta^2 u = f \quad \text{in } \Omega \tag{2.2a}$$

$$u = \Delta u = 0 \quad \text{on } \partial\Omega \tag{2.2b}$$

This boundary value problem is related to the bending of simply supported Kirchhoff plates

Example 3.

$$\Delta^2 u = f \quad \text{in } \Omega \tag{2.3a}$$

$$\frac{\partial u}{\partial n} = \frac{\partial \Delta u}{\partial n} = 0 \quad \text{on } \partial\Omega \tag{2.3b}$$

This boundary value problem is related to the Cahn-Hilliard model for phase separation phenomena.

We assume $f \in L_2(\Omega)$ in all the examples. For Example 3 we also assume the solvability condition

$$\int_\Omega f \, dx = 0. \tag{2.4}$$

2.1   Weak Formulations

The weak formulation of the model problems is to find $u \in V$ such that

$$a(u, v) = \int_\Omega f v \, dx \quad \forall v \in V, \tag{2.5}$$

where

$$a(w, v) = \int_\Omega \nabla^2 w : \nabla^2 v \, dx, \tag{2.6}$$

$\nabla^2 w : \nabla^2 v$ is the inner product of the Hessian matrices of w and v, and V is a closed subspace of the Sobolev space $H^2(\Omega)$ chosen as follows:

- $V = H_0^2(\Omega)$ for (2.1)
- $V = H^2(\Omega) \cap H_0^1(\Omega)$ for (2.2)
- $V = \{v \in H^2(\Omega) : \partial v/\partial n = 0 \text{ on } \partial\Omega\}$ for (2.3)

The well-posedness of the weak formulations follow from Poincaré-Friedrichs inequalities [79] for $H^2(\Omega)$. For the model problem (2.3) the solution is unique up to an additive constant under the solvability condition (2.4).

2.2 Elliptic Regularity

If the domain is smooth, the solution u of (2.5) belongs to $H^4(\Omega)$ by classical elliptic regularity results [2]. This is however not the case for a polygonal domain. The weak solutions of the boundary value problem in Example 1 belong in general to $H^{2+\alpha}(\Omega)$ for some $\alpha \in (1/2, 2]$. The value of α depends on the interior angles at the corners of Ω. For a convex Ω, we have $\alpha > 1$ and α is close to 1 if one of the interior angles of Ω is close to π. The weak solutions of the boundary value problems in Example 2 and Example 3 belong in general to $H^{2+\alpha}(\Omega)$ for some $\alpha \in (0, 2]$, and α can be close to 0 if one of the interior angle is close to π. Thus for these two model problems α can be close to 0 even for convex Ω. Details of the elliptic regularity theory for the biharmonic equations can be found in [15,46,57,63,64,77].

We will refer to α as the index of elliptic regularity, and the elliptic regularity estimate

$$|u|_{H^{2+\alpha}(\Omega)} \leq C_\Omega \|f\|_{L_2(\Omega)} \tag{2.7}$$

holds for the solution u of the model problems.

2.3 Ramifications of Elliptic Regularity

The regularity of the solutions of the model problems will play an important role in their error analysis. But there is another important consequence of elliptic regularity for the boundary value problems in Example 2 and Example 3.

The boundary value problem in Example 2 is formally equivalent to the following two second order boundary value problems:

$$-\Delta u = v \quad \text{in } \Omega \qquad\qquad -\Delta v = f \quad \text{in } \Omega$$
$$u = 0 \quad \text{on } \partial\Omega \qquad\qquad v = 0 \quad \text{on } \partial\Omega$$

However, the solution obtained from the second order equations coincides with the solution of the fourth order problem if and only if Ω is convex. Indeed the solution obtained from the second order equations in general does not belong to $H^2(\Omega)$ when Ω is nonconvex.

A similar phenomenon holds for the boundary value problem in Example 3. These observations indicate that mixed finite element methods for the model problems in Example 2 and Example 3 are problematic when Ω is nonconvex.

3 C^0 Interior Penalty Methods

We will only provide details for the derivation of C^0 interior penalty methods for the model problem (2.1). In the following derivation we assume that u is sufficiently smooth, say $u \in H^4(\Omega)$.

Let $V_h \subset H_0^1(\Omega)$ be the P_k $(k \geq 2)$ Lagrange finite element space associated with the triangulation $\mathcal{T}_h$ of Ω whose members vanish on $\partial\Omega$. We begin with the following integration by parts formula:

$$\int_T (\Delta^2 w) v \, dx = \int_{\partial T} \left[\left(\frac{\partial \Delta w}{\partial n} \right) v - \left(\frac{\partial^2 w}{\partial n^2} \right) \left(\frac{\partial v}{\partial n} \right) - \left(\frac{\partial^2 w}{\partial n \partial t} \right) \left(\frac{\partial v}{\partial t} \right) \right] ds \qquad (3.1)$$
$$+ \int_T (\nabla^2 w : \nabla^2 v) \, dx,$$

where T is a triangle, $w \in H^4(T)$, $v \in H^2(T)$, and $\partial/\partial n$ (resp. $\partial/\partial t$) denotes the exterior normal derivative (resp. the counterclockwise tangential derivative). Summing up (3.1) (with $w = u$ and $v \in V_h$) over all the triangles in $\mathcal{T}_h$, we have, after cancelations

$$\sum_{T \in \mathcal{T}_h} \int_T (\Delta^2 u) v \, dx = - \sum_{T \in \mathcal{T}_h} \int_{\partial T} \left(\frac{\partial^2 u}{\partial n^2} \right) \left(\frac{\partial v}{\partial n} \right) ds + \sum_{T \in \mathcal{T}_h} \int_T (\nabla^2 u : \nabla^2 v) \, dx,$$
$$(3.2)$$

and, since $\Delta^2 u = f$,

$$\sum_{T \in \mathcal{T}_h} \int_T (\Delta^2 u) v \, dx = \int_\Omega f v \, dx. \qquad (3.3)$$

We can rewrite the first sum on the right-hand side of (3.2) as a sum over the edges in $\mathcal{E}_h$:

$$- \sum_{T \in \mathcal{T}_h} \int_{\partial T} \left(\frac{\partial^2 u}{\partial n^2} \right) \left(\frac{\partial v}{\partial n} \right) ds = \sum_{e \in \mathcal{E}_h} \int_e \left(\frac{\partial^2 u}{\partial n_e^2} \right) \left[\!\left[\frac{\partial v}{\partial n_e} \right]\!\right] ds, \qquad (3.4)$$

where n_e is a unit vector normal to the edge e and the jump $[\![\partial v/\partial n_e]\!]$ is defined as follows.

For an interior edge e shared by two triangles $T_\pm$ where n_e points from T_- to T_+ (cf. Fig. 1), we define on the edge e

$$\left[\!\left[\frac{\partial v}{\partial n_e} \right]\!\right] = n_e \cdot (\nabla v_+ - \nabla v_-), \qquad (3.5)$$

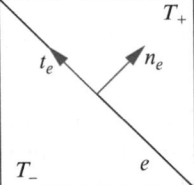

Fig. 1 An interior edge shared by two triangles $T_\pm$ with the unit normal n_e and the unit tangent t_e

$$\frac{\partial^2 u}{\partial n_e^2} = n_e \cdot (\nabla^2 u) n_e, \tag{3.6}$$

where $v_\pm = v|_{T_\pm}$. Note that the definitions of $[\![\partial v/\partial n_e]\!]$ and $\partial^2 u/\partial n_e^2$ are independent of the choice of $T_\pm$, or equivalently, independent of the choice of n_e.

For a boundary edge e which is an edge of the triangle $T \in \mathscr{T}_h$, we take n_e to be the unit normal pointing towards the outside of Ω and define on the edge e

$$\left[\!\!\left[\frac{\partial v}{\partial n_e}\right]\!\!\right] = -n_e \cdot \nabla v_T, \tag{3.7}$$

$$\frac{\partial^2 u}{\partial n_e^2} = n_e \cdot (\nabla^2 u) n_e. \tag{3.8}$$

where $v_T = v|_T$.

Combining (3.2)–(3.4), we find

$$\sum_{T \in \mathscr{T}_h} \int_T (\nabla^2 u : \nabla^2 v)\, dx + \sum_{e \in \mathscr{E}_h} \int_e \left(\frac{\partial^2 u}{\partial n_e^2}\right) \left[\!\!\left[\frac{\partial v}{\partial n_e}\right]\!\!\right] ds = \int_\Omega f v\, dx \quad \forall\, v \in V_h. \tag{3.9}$$

Since $\partial^2 u/\partial n_e^2$ has the same trace from either side of the edge e, we can write

$$\frac{\partial^2 u}{\partial n_e^2} = \left\{\!\!\left\{\frac{\partial^2 u}{\partial n_e^2}\right\}\!\!\right\}, \tag{3.10}$$

where the average of the second order normal derivative from the two sides of e is defined as follows.

Let $w \in H^3(\Omega, \mathscr{T}_h)$ be a piecewise H^3 function. Then we define

$$\left\{\!\!\left\{\frac{\partial^2 w}{\partial n_e^2}\right\}\!\!\right\} = \frac{1}{2}\left(\frac{\partial^2 w_-}{\partial n_e^2} + \frac{\partial^2 w_+}{\partial n_e^2}\right) \tag{3.11}$$

on an interior edge and

$$\left\{\!\!\left\{\frac{\partial^2 w}{\partial n_e^2}\right\}\!\!\right\} = \frac{\partial^2 w_T}{\partial n_e^2} \tag{3.12}$$

on a boundary edge. These definitions are independent of the choice of n_e.

Now we have, by (3.9) and (3.10),

$$\sum_{T \in \mathscr{T}_h} \int_T (\nabla^2 u : \nabla^2 v)\, dx + \sum_{e \in \mathscr{E}_h} \int_e \left\{\!\!\left\{\frac{\partial^2 u}{\partial n_e^2}\right\}\!\!\right\} \left[\!\!\left[\frac{\partial v}{\partial n_e}\right]\!\!\right] ds = \int_\Omega f v\, dx \quad \forall\, v \in V_h. \tag{3.13}$$

Since $[\![\partial u/\partial n_e]\!] = 0$ for all $e \in \mathscr{E}_h$, we can rewrite (3.13) as

$$\sum_{T \in \mathscr{T}_h} \int_T (\nabla^2 u : \nabla^2 v)\, dx + \sum_{e \in \mathscr{E}_h} \int_e \left\{\!\!\left\{ \frac{\partial^2 u}{\partial n_e^2} \right\}\!\!\right\} \left[\!\!\left[\frac{\partial v}{\partial n_e} \right]\!\!\right] ds$$

$$+ \sum_{e \in \mathscr{E}_h} \int_e \left\{\!\!\left\{ \frac{\partial^2 v}{\partial n_e^2} \right\}\!\!\right\} \left[\!\!\left[\frac{\partial u}{\partial n_e} \right]\!\!\right] ds = \int_\Omega f v \, dx \quad \forall v \in V_h, \qquad (3.14)$$

and finally as

$$\sum_{T \in \mathscr{T}_h} \int_T (\nabla^2 u : \nabla^2 v)\, dx + \sum_{e \in \mathscr{E}_h} \int_e \left\{\!\!\left\{ \frac{\partial^2 u}{\partial n_e^2} \right\}\!\!\right\} \left[\!\!\left[\frac{\partial v}{\partial n_e} \right]\!\!\right] ds$$

$$+ \sum_{e \in \mathscr{E}_h} \int_e \left\{\!\!\left\{ \frac{\partial^2 v}{\partial n_e^2} \right\}\!\!\right\} \left[\!\!\left[\frac{\partial u}{\partial n_e} \right]\!\!\right] ds + \sigma \sum_{e \in \mathscr{E}_h} \frac{1}{|e|} \int_e \left[\!\!\left[\frac{\partial u}{\partial n_e} \right]\!\!\right]\left[\!\!\left[\frac{\partial v}{\partial n_e} \right]\!\!\right] ds \qquad (3.15)$$

$$= \int_\Omega f v \, dx \quad \forall v \in V_h,$$

where $\sigma > 0$ is a penalty parameter to be chosen.

In summary, the solution u of (2.1) satisfies the mesh-dependent problem

$$a_h(u, v) = \int_\Omega f v \, dx \qquad \forall v \in V_h, \qquad (3.16)$$

where the bilinear form $a_h(\cdot, \cdot)$ is defined on the piecewise Sobolev space $H^3(\Omega, \mathscr{T}_h)$ by

$$a_h(w, v) = \sum_{T \in \mathscr{T}_h} \int_T (\nabla^2 w : \nabla^2 v)\, dx + \sum_{e \in \mathscr{E}_h} \int_e \left\{\!\!\left\{ \frac{\partial^2 w}{\partial n_e^2} \right\}\!\!\right\} \left[\!\!\left[\frac{\partial v}{\partial n_e} \right]\!\!\right] ds$$

$$+ \sum_{e \in \mathscr{E}_h} \int_e \left\{\!\!\left\{ \frac{\partial^2 v}{\partial n_e^2} \right\}\!\!\right\} \left[\!\!\left[\frac{\partial w}{\partial n_e} \right]\!\!\right] ds + \sigma \sum_{e \in \mathscr{E}_h} \frac{1}{|e|} \int_e \left[\!\!\left[\frac{\partial w}{\partial n_e} \right]\!\!\right]\left[\!\!\left[\frac{\partial v}{\partial n_e} \right]\!\!\right] ds. \qquad (3.17)$$

We can now formulate the C^0 interior penalty method for (2.1) as follows. Find $u_h \in V_h$ such that

$$a_h(u_h, v) = \int_\Omega f v \, dx \qquad \forall v \in V_h. \qquad (3.18)$$

Remark 2. The two (3.9) and (3.15) are equivalent for the solution u of the continuous problem (2.1). But they are *not* equivalent for $u_h \in V_h$. The two additional terms in (3.15) symmetrize and stabilize the discrete problem (3.18).

Similarly we can derive C^0 interior penalty methods for the model problems (2.2) and (2.3) through integration by parts, symmetrization and stabilization. For the model problem (2.2), we take $V_h \subset H_0^1(\Omega)$ to be the P_k Lagrange finite element space as before. But we change the definition of the bilinear form $a_h(\cdot, \cdot)$ to

$$
\begin{aligned}
a_h(w, v) = &\sum_{T \in \mathcal{T}_h} \int_T (\nabla^2 w : \nabla^2 v) \, dx + \sum_{e \in \mathscr{E}_h^i} \int_e \left\{\!\!\left\{ \frac{\partial^2 w}{\partial n_e^2} \right\}\!\!\right\} \left[\!\!\left[\frac{\partial v}{\partial n_e} \right]\!\!\right] ds \\
&+ \sum_{e \in \mathscr{E}_h^i} \int_e \left\{\!\!\left\{ \frac{\partial^2 v}{\partial n_e^2} \right\}\!\!\right\} \left[\!\!\left[\frac{\partial w}{\partial n_e} \right]\!\!\right] ds + \sigma \sum_{e \in \mathscr{E}_h^i} \frac{1}{|e|} \int_e \left[\!\!\left[\frac{\partial w}{\partial n_e} \right]\!\!\right] \left[\!\!\left[\frac{\partial v}{\partial n_e} \right]\!\!\right] ds. \quad (3.19)
\end{aligned}
$$

Note that the only difference between (3.17) and (3.19) is in the set of edges over which the sums take place.

For the model problem (2.3) we keep the same bilinear form defined by (3.17), but we change V_h to be the P_k Lagrange finite element space whose members vanish at a chosen point on $\bar{\Omega}$ (say a corner of Ω).

3.1 Well-Posedness of the Discrete Problems

Again we only discuss the discrete problem for (2.1) in detail. By a standard inverse estimate [32, 44] and the Cauchy-Schwarz inequality we have

$$
\sum_{e \in \mathscr{E}_h} \left| \int_e \left\{\!\!\left\{ \frac{\partial^2 w}{\partial n_e^2} \right\}\!\!\right\} \left[\!\!\left[\frac{\partial v}{\partial n_e} \right]\!\!\right] ds \right|
$$

$$
\leq \left(\sum_{e \in \mathscr{E}_h} |e| \| \{\!\!\{ \partial^2 w / \partial n_e^2 \}\!\!\} \|_{L_2(e)}^2 \right)^{\frac{1}{2}} \left(\sum_{e \in \mathscr{E}_h} |e|^{-1} \| [\!\![\partial v / \partial n_e]\!\!] \|_{L_2(e)}^2 \right)^{\frac{1}{2}}
$$

$$
\leq C \left(\sum_{e \in \mathscr{E}_h} \sum_{T \in \mathcal{T}_e} |w|_{H^2(T)}^2 \right) \left(\sum_{e \in \mathscr{E}_h} |e|^{-1} \| [\!\![\partial v / \partial n_e]\!\!] \|_{L_2(e)}^2 \right)^{\frac{1}{2}} \quad (3.20)
$$

$$
\leq C \left(\sum_{T \in \mathcal{T}_h} |w|_{H^2(T)}^2 \right) \left(\sum_{e \in \mathscr{E}_h} |e|^{-1} \| [\!\![\partial v / \partial n_e]\!\!] \|_{L_2(e)}^2 \right)^{\frac{1}{2}}
$$

which implies

$$
|a_h(w, v)| \leq C \|w\|_h \|v\|_h \qquad \forall\, v, w \in V_h, \quad (3.21)
$$

where the mesh-dependent norm $\| \cdot \|_h$ is defined by

$$
\|v\|_h^2 = \sum_{T \in \mathcal{T}_h} |v|_{H^2(T)}^2 + \sigma \sum_{e \in \mathscr{E}_h} |e|^{-1} \| [\!\![\partial v / \partial n_e]\!\!] \|_{L_2(e)}^2. \quad (3.22)
$$

It also follows from (3.20) and the inequality of geometric and arithmetic means that

$$
a_h(v, v) \geq \sum_{T \in \mathscr{T}_h} |v|^2_{H^2(T)} - C \left(\sum_{T \in \mathscr{T}_h} |v|^2_{H^2(T)} \right)^{\frac{1}{2}} \left(\sum_{e \in \mathscr{E}_h} |e|^{-1} \| [\![\partial v / \partial n_e]\!] \|^2_{L_2(e)} \right)^{\frac{1}{2}}
$$

$$
+ \sigma \sum_{e \in \mathscr{E}_h} |e|^{-1} \| [\![\partial v / \partial n_e]\!] \|^2_{L_2(e)}
$$

$$
\geq \frac{1}{2} \sum_{T \in \mathscr{T}_h} |v|^2_{H^2(T)} + \left(\sigma - \frac{C^2}{2} \right) \sum_{e \in \mathscr{E}_h} |e|^{-1} \| [\![\partial v / \partial n_e]\!] \|^2_{L_2(e)} \qquad (3.23)
$$

$$
\geq \frac{1}{2} \|v\|^2_h \qquad \forall v \in V_h,
$$

provided the penalty parameter σ is sufficiently large. From now on we assume this is the case so that the discrete problem (3.18) is symmetric positive definite and hence uniquely solvable. This also holds for the model problems (2.2) and (2.3). Without loss of generality, we will assume from here on that

$$
\sigma \geq 1. \qquad (3.24)
$$

Note that (3.21) and (3.23) imply

$$
C_1 \|v\|^2_h \leq a_h(v, v) \leq C_2 \|v\|^2_h \qquad \forall v \in V_h. \qquad (3.25)
$$

3.2 Galerkin Orthogonality

It follows from (3.16) and (3.18) that

$$
a_h(u - u_h, v) = 0 \qquad \forall v \in V_h \qquad (3.26)
$$

provided that $u \in H^4(\Omega)$. However, as was noted in Sect. 2.2, this is usually not the case when Ω is a polygon. Nevertheless it is possible to establish (3.26) for the model problems using the singular function representation of u [15, 46, 57]. Details for the model problem (2.1) can be found in [34].

3.3 A Standard Error Analysis

A standard error analysis can be carried out using the Galerkin orthogonality (3.26) for the model problem (2.1). Let the norm $\| \cdot \|_h$ be defined on $H^s(\Omega, \mathscr{T}_h)$ ($s > 5/2$) by

$$\|v\|_h^2 = \sum_{T \in \mathcal{T}_h} |v|_{H^2(T)}^2 + \sigma \sum_{e \in \mathcal{E}_h} |e|^{-1} \| [\![\partial v / \partial n_e]\!] \|_{L_2(e)}^2$$

$$+ \frac{1}{\sigma} \sum_{e \in \mathcal{E}_h} |e| \| \{\!\{ \partial^2 v / \partial n_e^2 \}\!\} \|_{L_2(e)}^2. \tag{3.27}$$

It follows from the Cauchy-Schwarz inequality that

$$|a_h(w, v)| \leq 2 \|w\|_h \|v\|_h \qquad \forall \, v, w \in H^s(\Omega, \mathcal{T}_h), \ s > 5/2. \tag{3.28}$$

Note that we need the stronger norm $\|\!\|\cdot\|\!\|_h$ for the estimate on the infinite dimensional space $H^s(\Omega, \mathcal{T}_h)$ that contains u. On the finite element space V_h the two norms $\|\cdot\|_h$ and $\|\!\|\cdot\|\!\|_h$ are equivalent by (3.20).

Let $v \in V_h$ be arbitrary. It follows from (3.23), (3.26) and (3.28) that

$$\|u - u_h\|_h \leq \|u - v\|_h + \|v - u_h\|_h$$

$$\leq \|u - v\|_h + C \max_{w \in V_h \setminus \{0\}} \frac{a_h(v - u_h, w)}{\|w\|_h}$$

$$= \|u - v\|_h + C \max_{w \in V_h \setminus \{0\}} \frac{a_h(v - u, w)}{\|w\|_h}$$

$$\leq C \|u - v\|_h,$$

and hence

$$\|u - u_h\| \leq C \inf_{v \in V_h} \|u - v\|_h. \tag{3.29}$$

Let $\Pi_h : C(\bar{\Omega}) \longrightarrow V_h$ be the Lagrange nodal interpolation operator. We have the following standard interpolation error estimates [32, 44]:

$$h_T^{-2\gamma} \|\zeta - \Pi_h \zeta\|_{L_2(T)}^2 + h_T^{2(1-\gamma)} |\zeta - \Pi_h \zeta|_{H^1(T)}^2 + h_T^{2(2-\gamma)} |\zeta - \Pi_h \zeta|_{H^2(T)}^2$$

$$\leq C \|\zeta\|_{H^s(T)}^2 \qquad \forall \, T \in \mathcal{T}_h, \ \zeta \in H^s(\Omega), \tag{3.30}$$

where $\gamma = \min(s, k + 1)$. It follows from the trace theorem with scaling and (3.30) that

$$\|u - \Pi_h u\|_h \leq C h^\alpha \|u\|_{H^{2+\alpha}(\Omega)} \leq C h^\alpha \|f\|_{L_2(\Omega)}, \tag{3.31}$$

where $\alpha \in (1/2, 2]$ is the index of elliptic regularity for the model problem (2.1) that appears in (2.7).

3.4 Complications

The standard analysis works for the model problem (2.1) because the solution u always belongs to $H^{2+\alpha}(\Omega)$ for some $\alpha > 1/2$, which guarantees that the norm

$\|u\|_h$ is well-defined. Unfortunately this is not the case for the model problems (3.19) and (2.3) because the elliptic regularity index for these problems can be less than $1/2$. Therefore the standard error analysis becomes problematic for these problems.

We will consider an alternative error analysis in Sect. 5 using instead the norm $\|\cdot\|_h$, which is well-defined for functions in $H^2(\Omega)$. This alternative approach will yield quasi-optimal error estimates (up to data oscillations) using only the weak formulations of the model problems.

3.5 An Alternative Expression for the Discrete Bilinear Form

For the analysis in Sect. 5 it is convenient to use an alternative expression for (3.17). Let $w \in H^4(\Omega, \mathcal{T}_h) \cap H_0^1(\Omega)$ and $v \in H^2(\Omega, \mathcal{T}_h) \cap H_0^1(\Omega)$. It follows from (3.1) that

$$
\sum_{T \in \mathcal{T}_h} \int_T \nabla^2 w : \nabla^2 v \, dx = \sum_{T \in \mathcal{T}_h} \int_T (\Delta^2 w) v \, dx - \sum_{T \in \mathcal{T}_h} \int_{\partial T} \left(\frac{\partial \Delta w}{\partial n}\right) v \, ds
$$

$$
+ \sum_{T \in \mathcal{T}_h} \int_{\partial T} \left(\frac{\partial^2 w}{\partial n^2}\right)\left(\frac{\partial v}{\partial n}\right) ds + \sum_{T \in \mathcal{T}_h} \int_{\partial T} \left(\frac{\partial^2 w}{\partial n \partial t}\right)\left(\frac{\partial v}{\partial t}\right) ds
$$

$$
= \sum_{T \in \mathcal{T}_h} \int_T (\Delta^2 w) v \, dx + \sum_{e \in \mathcal{E}_h^i} \int_e \left[\!\!\left[\frac{\partial \Delta w}{\partial n_e}\right]\!\!\right] v \, ds \qquad (3.32)
$$

$$
- \sum_{e \in \mathcal{E}_h} \int_e \left\{\!\!\left\{\frac{\partial^2 w}{\partial n_e^2}\right\}\!\!\right\} \left[\!\!\left[\frac{\partial v}{\partial n_e}\right]\!\!\right] ds - \sum_{e \in \mathcal{E}_h^i} \int_e \left[\!\!\left[\frac{\partial^2 w}{\partial n_e^2}\right]\!\!\right] \left\{\!\!\left\{\frac{\partial v}{\partial n_e}\right\}\!\!\right\} ds
$$

$$
- \sum_{e \in \mathcal{E}_h^i} \int_e \left[\!\!\left[\frac{\partial^2 w}{\partial n_e \partial t_e}\right]\!\!\right] \frac{\partial v}{\partial t_e} \, ds
$$

where

$$
\left[\!\!\left[\frac{\partial \Delta w}{\partial n_e}\right]\!\!\right] = n_e \cdot [\nabla(\Delta w_+) - \nabla(\Delta w_-)] \qquad \text{for } e \in \mathcal{E}_h^i, \qquad (3.33)
$$

$$
\left[\!\!\left[\frac{\partial^2 w}{\partial n_e^2}\right]\!\!\right] = n_e \cdot [\nabla^2 w_+ - \nabla^2 w_-] n_e \qquad \text{for } e \in \mathcal{E}_h^i, \qquad (3.34)
$$

$$
\left\{\!\!\left\{\frac{\partial v}{\partial n_e}\right\}\!\!\right\} = \frac{n_e}{2} \cdot [\nabla v_- + \nabla v_+] \qquad \text{for } e \in \mathcal{E}_h^i, \qquad (3.35)
$$

$$
\left[\!\!\left[\frac{\partial^2 w}{\partial n_e \partial t_e}\right]\!\!\right] = t_e \cdot [\nabla^2 w_+ - \nabla^2 w_-] n_e \qquad \text{for } e \in \mathcal{E}_h^i. \qquad (3.36)
$$

The convention (as before) is that n_e is a unit normal of e pointing from T_- into T_+ and t_e is the unit tangent of e obtained by rotating n_e by a counterclockwise right angle (cf. Fig. 1).

After substitution and cancelation, we can rewrite (3.17) as

$$
a_h(w,v) = \sum_{T \in \mathscr{T}_h} \int_T (\Delta^2 w) v \, dx + \sum_{e \in \mathscr{E}_h} \int_e \left\{\!\!\left\{ \frac{\partial^2 v}{\partial n_e^2} \right\}\!\!\right\} \left[\!\!\left[\frac{\partial w}{\partial n_e} \right]\!\!\right] ds
$$

$$
+ \sum_{e \in \mathscr{E}_h^i} \int_e \left(\left[\!\!\left[\frac{\partial \Delta w}{\partial n_e} \right]\!\!\right] v - \left[\!\!\left[\frac{\partial^2 w}{\partial n_e^2} \right]\!\!\right] \left\{\!\!\left\{ \frac{\partial v}{\partial n_e} \right\}\!\!\right\} - \left[\!\!\left[\frac{\partial^2 w}{\partial n_e \partial t_e} \right]\!\!\right] \frac{\partial v}{\partial t_e} \right) ds
$$

$$
+ \sigma \sum_{e \in \mathscr{E}_h} \frac{1}{|e|} \int_e \left[\!\!\left[\frac{\partial w}{\partial n_e} \right]\!\!\right] \left[\!\!\left[\frac{\partial v}{\partial n_e} \right]\!\!\right] ds. \tag{3.37}
$$

Alternative bilinear forms for the model problems (2.2) and (2.3) can be obtained similarly.

3.6 Conditioning

C^0 interior penalty methods share the same ill-conditioning with other discretizations for fourth order problems. Using Poincaré-Friedrichs inequalities for piecewise H^2 functions [39], it can be shown [68] that the condition number of the discrete problem resulting from a C^0 interior penalty method grows at the rate of $O(h^{-4})$. Therefore it is important to have efficient solvers for the discrete problem. This will be addressed in Sects. 8 and 9.

4 Enriching Operators

The distance between the C^0 finite element space V_h (where the discrete problem is posed) and the Sobolev space $H^2(\Omega)$ (where the continuous problem is posed) can be measured using an enriching operator $E_h : V_h \longrightarrow H^2(\Omega)$. The idea of using enriching operators to analyze nonconforming methods was first introduced in [23–25] in the context of fast solvers.

4.1 Enriching Operator Based on the Hsieh-Clough-Tocher Macro Element

For concreteness we consider the case where $V_h \subset H_0^1(\Omega)$ is the P_k Lagrange finite element space for the model problem (2.1). Enriching operators can be constructed for the other model problems with similar properties.

The construction of E_h uses the Hsieh-Clough-Tocher macro finite element space $W_h \subset H_0^2(\Omega)$. The degrees of freedom (dofs) of $w \in W_h$ are (i) the values of the derivatives of w up to order 1 at the interior vertices and (ii) the values of the normal derivative of w at the midpoints of the edges in $\mathscr{E}_h^i$.

Let $v \in V_h$. We define $w = E_h v$ by averaging as follows. We take

$$w(p) = v(p) \quad \text{if } p \text{ is an interior vertex.} \tag{4.1}$$

At an interior vertex p of $\mathscr{T}_h$, we assign the first order derivatives of w so that

$$(\nabla w)(p) = \frac{1}{|\mathscr{T}_p|} \sum_{T \in \mathscr{T}_p} \nabla v_T. \tag{4.2}$$

At the midpoint m_e on an edge in $\mathscr{E}_h^i$, we define

$$\frac{\partial w}{\partial n}(m_e) = \frac{1}{2} \sum_{T \in \mathscr{T}_e} \frac{\partial v_T}{\partial n}(m_e). \tag{4.3}$$

Lemma 1. *We have*

$$\sum_{T \in \mathscr{T}_h} \left(h_T^{-4} \| v - E_h v \|_{L_2(T)}^2 + h_T^{-2} |v - E_h v|_{H^1(T)}^2 + |v - E_h v|_{H^2(T)}^2 \right)$$

$$\leq C \sum_{e \in \mathscr{E}_h} \frac{1}{|e|} \| [\![\partial v / \partial n_e]\!] \|_{L_2(e)}^2 \qquad \forall \, v \in V_h. \tag{4.4}$$

Proof. Let $T \in \mathscr{T}_h$ be arbitrary. It follows from (4.1) and scaling that

$$h_T^{-4} \| v - E_h v \|_{L_2(T)}^2 = \sum_{p \in \mathscr{V}_T} |\nabla(v_T - E_h v)(p)|^2 + \sum_{p \in \mathscr{M}_T} \left| \frac{\partial(v_T - E_h v)}{\partial n}(p) \right|^2, \tag{4.5}$$

where $\mathscr{V}_T$ is the set of the three vertices of T and $\mathscr{M}_T$ is the set of the midpoints of the three edges of T.

At a vertex $p \in \mathscr{V}_T$ that is inside Ω, we have, by (4.2),

$$\nabla(v_T - E_h v)(p) = \frac{1}{|\mathscr{T}_p|} \sum_{T' \in \mathscr{T}_p} \left[\nabla v_T(p) - \nabla v_{T'}(p) \right]. \tag{4.6}$$

On the other hand, at a vertex $p \in \mathscr{V}_T$ that is on $\partial\Omega$, we have

$$\nabla(v_T - E_h v)(p) = (\nabla v_T)(p). \tag{4.7}$$

Since the difference between the gradient of v across an interior edge and the gradient of v on a boundary edge can both be bounded in terms of the jump of the normal derivative, it follows from (4.6) and (4.7) that

$$\sum_{p \in \mathscr{V}_T} |\nabla(v_T - E_h v)(p)|^2 \leq C \sum_{p \in \mathscr{V}_T} \sum_{e \in \mathscr{E}_p} \frac{1}{|e|} \| [\![\partial v/\partial n_e]\!] \|_{L_2(e)}^2, \qquad (4.8)$$

where $\mathscr{E}_p$ is the set of the edges in $\mathscr{E}_h$ that share the common endpoint p.
Similarly we have

$$\sum_{p \in \mathscr{M}_T} \left| \frac{\partial(v_T - E_h v)}{\partial n}(p) \right|^2 \leq C \sum_{e \in \mathscr{E}_T} \frac{1}{|e|} \| [\![\partial v/\partial n_e]\!] \|_{L_2(e)}^2, \qquad (4.9)$$

where $\mathscr{E}_T$ is the set of the three edges of T.
It follows from (4.5), (4.8) and (4.9) that

$$\sum_{T \in \mathscr{T}_h} h_T^{-4} \| v_T - E_h v \|_{L_2(T)}^2 \leq C \sum_{e \in \mathscr{E}_h} \frac{1}{|e|} \| [\![\partial v/\partial n_e]\!] \|_{L_2(e)}^2.$$

The rest of the estimates in (4.4) then follow from standard inverse estimates.   □
The following corollary is immediate by standard inverse estimates.

Corollary 1. *We have*

$$|E_h v|_{H^2(\Omega)} \leq C \|v\|_h \qquad \forall \, v \in V_h, \qquad (4.10)$$

$$\sum_{e \in \mathscr{E}_h} \frac{1}{|e|} \| [\![\partial(v - E_h v)/\partial n_e]\!] \|_{L_2(e)}^2 \leq C \|v\|_h^2 \qquad \forall \, v \in V_h, \qquad (4.11)$$

$$\sum_{e \in \mathscr{E}_h^i} \frac{1}{|e|} \| [\![\partial(v - E_h v)/\partial t_e]\!] \|_{L_2(e)}^2 \leq C \|v\|_h^2 \qquad \forall \, v \in V_h, \qquad (4.12)$$

$$\sum_{e \in \mathscr{E}_h^i} \frac{1}{|e|} \| \{\!\{\partial(v - E_h v)/\partial n_e\}\!\} \|_{L_2(e)}^2 \leq C \|v\|_h^2 \qquad \forall \, v \in V_h, \qquad (4.13)$$

$$\sum_{e \in \mathscr{E}_h^i} \frac{1}{|e|^3} \| v - E_h v \|_{L_2(e)}^2 \leq C \|v\|_h^2 \qquad \forall \, v \in V_h. \qquad (4.14)$$

4.2 Enriching Operator Based on C^1 Relatives

Next we consider a different type of enriching operator that will play important
roles in the design and convergence analysis of fast solvers for C^0 interior penalty
methods.

Fig. 2 Left figure: dofs of
the P_2 Lagrange finite
element, Right figure: dofs of
the P_6 Argyris finite element

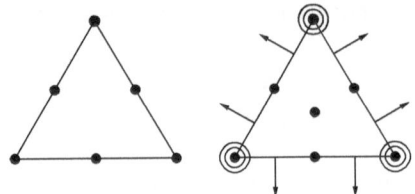

For concreteness we discuss the quadratic case in detail. Let $\tilde{V}_h \subset H^1(\Omega)$ be the P_2 Lagrange finite element space associated with $\mathscr{T}_h$. The construction of $\tilde{E}_h : \tilde{V}_h \longrightarrow H^2(\Omega)$ will use the P_6 Argyris finite element space $\tilde{W}_h \subset H^2(\Omega)$ (cf. [12, 97]). The dofs of the P_2 finite element and the P_6 Argyris finite element are depicted in Fig. 2, where the solid dot represents the value of a shape function, the first (resp. second) circle represent the values of the first (resp. second) order derivatives of a shape function, and the arrow represents the value of the normal derivative of a shape function.

Remark 3. Note that the shape functions and dofs of the P_2 Lagrange finite element are also shape functions and dofs of the P_6 Argyris finite element. We will refer to the Argyris element as a C^1 relative of the C^0 Lagrange element. C^1 relatives for higher order Lagrange finite elements can be found in [34].

The enriching operator $\tilde{E}_h : \tilde{V}_h \longrightarrow \tilde{W}_h$ is defined by averaging as follows. Let p be a vertex, a midpoint or the center of a triangle, we define

$$(\tilde{E}_h v)(p) = v(p). \tag{4.15}$$

At a vertex p, the function $\tilde{E}_h v$ satisfies

$$\left[\nabla(\tilde{E}_h v)\right](p) = \frac{1}{|\mathscr{T}_p|} \sum_{T \in \mathscr{T}_p} (\nabla v_T)(p),$$

$$\left[\nabla^2(\tilde{E}_h v)\right](p) = \frac{1}{|\mathscr{T}_p|} \sum_{T \in \mathscr{T}_p} (\nabla^2 v_T)(p). \tag{4.16}$$

Finally, at a point p of an edge where the normal derivative is a dof for the Argyris element, we define

$$\frac{\partial \tilde{E}_h v}{\partial n}(p) = \frac{1}{|\mathscr{T}_e|} \sum_{T \in \mathscr{T}_e} \frac{\partial v_T}{\partial n}(p).$$

Lemma 2. *We have*

$$\sum_{T \in \mathscr{T}_h} \left(h_T^{-4} \|v - \tilde{E}_h v\|_{L_2(T)}^2 + h_T^{-2} |v - \tilde{E}_h v|_{H^1(T)}^2 + |v - \tilde{E}_h v|_{H^2(T)}^2 \right)$$

$$\leq C \Big(\sum_{T \in \mathscr{T}_h} |v|^2_{H^2(T)} + \sum_{e \in \mathscr{E}_h} \frac{1}{|e|} \| [\![\partial v/\partial n_e]\!] \|^2_{L_2(e)} \Big) \qquad \forall\, v \in \tilde{V}_h. \quad (4.17)$$

Proof. Let $T \in \mathscr{T}_h$ be arbitrary. It follows from (4.15) and scaling that

$$h_T^{-4} \| v - \tilde{E}_h v \|^2_{L_2(T)} = \sum_{p \in \mathscr{V}_T} \big(|\nabla(v_T - \tilde{E}_h v)(p)|^2 + h_T^2 |\nabla^2(v_T - \tilde{E}_h v)(p)|^2 \big)$$

$$+ \sum_{p \in \mathscr{N}_T} \Big| \frac{\partial(v_T - \tilde{E}_h v)}{\partial n}(p) \Big|^2. \quad (4.18)$$

Here $\mathscr{N}_T$ is the set of the six nodes on ∂T where the normal derivative is a dof. The first and third sums on the right-hand side of (4.18) can be estimated as in the proof of Lemma 1, and so we will focus on the second sum.

From (4.16) we have

$$\nabla^2(v_T - \tilde{E}_h v)(p) = \frac{1}{|\mathscr{T}_p|} \sum_{T' \in \mathscr{T}_p} \nabla^2(v_{T'} - v_T)(p). \quad (4.19)$$

It then follows from scaling that

$$\sum_{p \in \mathscr{V}_T} h_T^2 |\nabla^2(v_T - \tilde{E}_h v)(p)|^2 \leq C \sum_{T' \in \mathscr{T}_p} |v_{T'}|^2_{H^2(T)}.$$

The proof of the lemma is completed by following the arguments in the proof of Lemma 1. □

In view of (3.22) and (4.17), the following corollary is immediate.

Corollary 2. *We have*

$$|\tilde{E}_h v|_{H^2(\Omega)} \leq C \|v\|_h \qquad \forall\, v \in \tilde{V}_h. \quad (4.20)$$

By a direct calculation, the nodal interpolation operator $\Pi_h : C(\bar{\Omega}) \longrightarrow \tilde{V}_h$ satisfies the following estimate:

$$\sum_{T \in \mathscr{T}_h} \big(h_T^{-4} \| w - \Pi_h w \|^2_{L_2(T)} + h_T^{-2} |w - \Pi_h w|^2_{H^1(T)} + |\Pi_h w|^2_{H^2(T)} \big)$$

$$\leq C |w|^2_{H^2(\Omega)} \qquad \forall\, w \in \tilde{W}_h. \quad (4.21)$$

Moreover, we have

$$\Pi_h(\tilde{E}_h v) = v \qquad \forall\, v \in \tilde{V}_h \quad (4.22)$$

since $\Pi_h(\tilde{E}_h v)$ and v have identical dofs, a benefit of using a C^1 relative in the construction of the enriching operator.

The following result shows that the composition $\tilde{E}_h \circ \Pi_h$ behaves like a quasi-interpolation operator.

Lemma 3. *We have, for $s \geq 2$,*

$$|\zeta - \tilde{E}_h(\Pi_h\zeta)|_{H^2(\Omega)} \leq C h^{\min(s,3)-2}|\zeta|_{H^s(\Omega)} \qquad \forall\, \zeta \in H^s(\Omega). \qquad (4.23)$$

Proof. Let $T \in \mathcal{T}_h$ be arbitrary and

$$S(T) = \sum_{p \in \mathcal{V}_T} \bigcup_{T' \in \mathcal{T}_p} T'$$

be the collection of triangles in $\mathcal{T}_h$ that shares at least one common vertex with T. It follows from the definition of $\tilde{E}_h$ that

$$|\tilde{E}_h \Pi_h \zeta|_{H^2(T)} \leq C \|\zeta\|_{H^2(S(T))}$$

(cf. (3.30) and the estimates in the proofs of Lemma 1 and Lemma 2) and $\tilde{E}_h(\Pi_h\zeta) = \zeta$ on T if $\zeta \in P_2(S(T))$. Hence we have

$$|\zeta - \tilde{E}_h(\Pi_h\zeta)|_{H^2(T)} \leq C h^{\min(s,3)-2}|\zeta|_{H^s(S(T))} \qquad (4.24)$$

by the Bramble-Hilbert Lemma [18, 51].

The proof is completed by summing up (the square of) (4.24) over all the triangles in $\mathcal{T}_h$. $\qquad\qquad\qquad\qquad\qquad\qquad\qquad\qquad\qquad\qquad\qquad\square$

The corresponding estimate for the enriching operator defined on the P_k Lagrange finite element space reads

$$|\zeta - \tilde{E}_h(\Pi_h\zeta)|_{H^2(\Omega)} \leq C h^{\min(s,k+1)-2}|\zeta|_{H^s(\Omega)} \qquad \forall\, \zeta \in H^s(\Omega). \qquad (4.25)$$

Note that the results for $\tilde{E}_h$ are also valid for P_k Lagrange finite element spaces with boundary conditions, and they can be proved by a slight modification of the estimates for the triangles near the boundary of the domain. Details can be found in [34]. We will use $\tilde{E}_h$ to denote the general enriching operator for Lagrange finite element spaces with or without boundary conditions.

5   Medius Analysis

In this section we present an alternative error analysis for the C^0 interior penalty methods for the model problems that does not require additional information on the regularity of the solution beyond the fact that $u \in H^2(\Omega)$. In particular, it does not rely on the Galerkin orthogonality (3.26) (cf. the discussion in Sect. 3.2). This new approach was first introduced in [58] and it can be applied to many discontinuous finite element methods. The name *medius* analysis indicates that both *a priori* and *a posteriori* techniques are employed in this analysis.

We will present a detailed analysis for the model problem (2.1). Similar results are also valid for the other model problems [28, 30].

5.1 Preliminaries

The first step is similar to the first step in the standard analysis presented in Sect. 3.3. Let $v \in V_h$ be arbitrary. It follows from (3.23) that

$$\|u - u_h\|_h \le \|u - v\|_h + C \max_{w \in V_h \setminus \{0\}} \frac{a_h(v - u_h, w)}{\|w\|_h}. \tag{5.1}$$

But now, instead of using the Galerkin orthogonality, we will bound the numerator of the second term on the right-hand side of (5.1) in terms of $\|u - v\|_h$ and data oscillations.

Let $E_h : V_h \longrightarrow H_0^2(\Omega)$ be the enriching operator from Sect. 4. It follows from (3.18) that

$$a_h(v - u_h, w) = a_h(v, E_h w) + a_h(v, w - E_h w) - \int_\Omega f w \, dx. \tag{5.2}$$

According to (2.5) and (3.17), we have

$$a_h(v, E_h w) = \sum_{T \in \mathcal{T}_h} \int_T \nabla^2 v : \nabla^2 (E_h w) \, dx + \sum_{e \in \mathcal{E}_h} \int_e \left\{\!\!\left\{ \frac{\partial^2 (E_h w)}{\partial n_e^2} \right\}\!\!\right\} \left[\!\!\left[\frac{\partial v}{\partial n_e} \right]\!\!\right] ds$$

$$= \sum_{T \in \mathcal{T}_h} \int_T \nabla^2 (v - u) : \nabla^2 (E_h w) \, dx + \int_\Omega f(E_h w) \, dx \tag{5.3}$$

$$+ \sum_{e \in \mathcal{E}_h} \int_e \left\{\!\!\left\{ \frac{\partial^2 (E_h w)}{\partial n_e^2} \right\}\!\!\right\} \left[\!\!\left[\frac{\partial v}{\partial n_e} \right]\!\!\right] ds.$$

Using the alternative expression (3.37) we can write the second term on the right-hand side of (5.2) as

$$a_h(v, w - E_h w)$$

$$= \sum_{T \in \mathcal{T}_h} \int_T (\Delta^2 v)(w - E_h w) dx + \sum_{e \in \mathcal{E}_h} \int_e \left\{\!\!\left\{ \frac{\partial^2 (w - E_h w)}{\partial n_e^2} \right\}\!\!\right\} \left[\!\!\left[\frac{\partial v}{\partial n_e} \right]\!\!\right] ds$$

$$+ \sum_{e \in \mathscr{E}_h^i} \int_e \left(\left[\!\!\left[\frac{\partial \Delta v}{\partial n_e} \right]\!\!\right] (w - E_h w) - \left[\!\!\left[\frac{\partial^2 v}{\partial n_e^2} \right]\!\!\right] \left\{\!\!\left\{ \frac{\partial (w - E_h w)}{\partial n_e} \right\}\!\!\right\} \right) ds \qquad (5.4)$$

$$- \sum_{e \in \mathscr{E}_h^i} \int_e \left[\!\!\left[\frac{\partial^2 v}{\partial n_e \partial t_e} \right]\!\!\right] \frac{\partial (w - E_h w)}{\partial t_e} ds + \sigma \sum_{e \in \mathscr{E}_h} \frac{1}{|e|} \int_e \left[\!\!\left[\frac{\partial v}{\partial n_e} \right]\!\!\right] \left[\!\!\left[\frac{\partial (w - E_h w)}{\partial n_e} \right]\!\!\right] ds.$$

Combining (5.2)–(5.4) we find

$$a_h(v - u_h, w) = \sum_{T \in \mathscr{T}_h} \int_T \nabla^2 (v - u) : \nabla^2 (E_h w) \, dx + \sum_{e \in \mathscr{E}_h} \int_e \left\{\!\!\left\{ \frac{\partial^2 w}{\partial n_e^2} \right\}\!\!\right\} \left[\!\!\left[\frac{\partial v}{\partial n_e} \right]\!\!\right] ds$$

$$+ \sigma \sum_{e \in \mathscr{E}_h} \frac{1}{|e|} \int_e \left[\!\!\left[\frac{\partial v}{\partial n_e} \right]\!\!\right] \left[\!\!\left[\frac{\partial (w - E_h w)}{\partial n_e} \right]\!\!\right] ds$$

$$- \sum_{e \in \mathscr{E}_h^i} \int_e \left[\!\!\left[\frac{\partial^2 v}{\partial n_e \partial t_e} \right]\!\!\right] \frac{\partial (w - E_h w)}{\partial t_e} ds$$

$$+ \sum_{e \in \mathscr{E}_h^i} \int_e \left(\left[\!\!\left[\frac{\partial \Delta v}{\partial n_e} \right]\!\!\right] (w - E_h w) - \left[\!\!\left[\frac{\partial^2 v}{\partial n_e^2} \right]\!\!\right] \left\{\!\!\left\{ \frac{\partial (w - E_h w)}{\partial n_e} \right\}\!\!\right\} \right) ds$$

$$(5.5)$$

$$- \sum_{T \in \mathscr{T}_h} \int_T (f - \Delta^2 v)(w - E_h w) dx.$$

5.2 First Estimates

Using the estimates for the enriching operator E_h in Sect. 4, the terms on the right-hand side of (5.5) can be estimated as follows:

$$\left| \sum_{T \in \mathscr{T}_h} \int_T \nabla^2 (v - u) : \nabla^2 (E_h w) \, dx \right| \le C \left(\sum_{T \in \mathscr{T}_h} |u - v|_{H^2(T)}^2 \right)^{\frac{1}{2}} \|w\|_h \qquad (5.6)$$

by (4.10);

$$\left| \sum_{e \in \mathscr{E}_h} \int_e \left\{\!\!\left\{ \frac{\partial^2 w}{\partial n_e^2} \right\}\!\!\right\} \left[\!\!\left[\frac{\partial v}{\partial n_e} \right]\!\!\right] ds \right| = \left| \sum_{e \in \mathscr{E}_h} \int_e \left\{\!\!\left\{ \frac{\partial^2 w}{\partial n_e^2} \right\}\!\!\right\} \left[\!\!\left[\frac{\partial (v - u)}{\partial n_e} \right]\!\!\right] ds \right|$$

$$\le \left(\sum_{e \in \mathscr{E}_h} \frac{1}{|e|} \| [\![\partial (u - v)/\partial n_e]\!] \|_{L_2(e)}^2 \right)^{\frac{1}{2}} \left(\sum_{e \in \mathscr{E}_h} |e| \| \{\!\{ \partial^2 w/\partial n_e^2 \}\!\} \|_{L_2(e)}^2 \right)^{\frac{1}{2}}$$

$$\leq C\Big(\sum_{e\in\mathscr{E}_h}\frac{1}{|e|}\|\,[\![\partial(u-v)/\partial n_e]\!]\,\|^2_{L_2(e)}\Big)^{\frac{1}{2}}\Big(\sum_{T\in\mathscr{T}_h}|w|^2_{H^2(T)}\Big)^{\frac{1}{2}} \qquad (5.7)$$

$$\leq C\Big(\sum_{e\in\mathscr{E}_h}\frac{1}{|e|}\|\,[\![\partial(u-v)/\partial n_e]\!]\,\|^2_{L_2(e)}\Big)^{\frac{1}{2}}\|w\|_h$$

by a direct calculation;

$$\Big|\sum_{e\in\mathscr{E}_h}\frac{1}{|e|}\int_e\Big[\!\!\Big[\frac{\partial v}{\partial n_e}\Big]\!\!\Big]\Big[\!\!\Big[\frac{\partial(w-E_hw)}{\partial n_e}\Big]\!\!\Big]ds\Big|$$

$$\leq C\Big(\sum_{e\in\mathscr{E}_h}\frac{1}{|e|}\|\,[\![\partial(u-v)/\partial n_e]\!]\,\|^2_{L_2(e)}\Big)^{\frac{1}{2}}\Big(\sum_{e\in\mathscr{E}_h}\frac{1}{|e|}\|\,[\![\partial(w-E_hw)/\partial n_e]\!]\,\|^2_{L_2(e)}\Big)^{\frac{1}{2}}$$

$$\leq C\Big(\sum_{e\in\mathscr{E}_h}\frac{1}{|e|}\|\,[\![\partial(u-v)/\partial n_e]\!]\,\|^2_{L_2(e)}\Big)^{\frac{1}{2}}\|w\|_h \qquad (5.8)$$

by (4.11);

$$\Big|\sum_{e\in\mathscr{E}^i_h}\int_e\Big[\!\!\Big[\frac{\partial^2 v}{\partial n_e\partial t_e}\Big]\!\!\Big]\frac{\partial(w-E_hw)}{\partial t_e}ds\Big|$$

$$\leq\Big(\sum_{e\in\mathscr{E}_h}|e|\,\|\,[\![\partial^2 v/\partial n_e\partial t_e]\!]\,\|^2_{L_2(e)}\Big)^{\frac{1}{2}}\Big(\sum_{e\in\mathscr{E}^i_h}\frac{1}{|e|}\|\partial(w-E_hw)/\partial t_e\|^2_{L_2(e)}\Big)^{\frac{1}{2}}$$

$$\leq\Big(\sum_{e\in\mathscr{E}^i_h}\frac{1}{|e|}\|\,[\![\partial v/\partial n_e]\!]\,\|^2_{L_2(e)}\Big)^{\frac{1}{2}}\|w\|_h \qquad (5.9)$$

$$\leq C\Big(\sum_{e\in\mathscr{E}^i_h}\frac{1}{|e|}\|\,[\![\partial(u-v)/\partial n_e]\!]\,\|^2_{L_2(e)}\Big)^{\frac{1}{2}}\|w\|_h$$

by (4.12) and a standard inverse estimate;

$$\Big|\sum_{e\in\mathscr{E}^i_h}\int_e\Big[\!\!\Big[\frac{\partial\Delta v}{\partial n_e}\Big]\!\!\Big](w-E_hw)\,ds\Big|$$

$$\leq\Big(\sum_{e\in\mathscr{E}^i_h}|e|^3\|\,[\![\partial\Delta v/\partial n_e]\!]\,\|^2_{L_2(e)}\Big)^{\frac{1}{2}}\Big(\sum_{e\in\mathscr{E}^i_h}\frac{1}{|e|^3}\|w-E_hw\|^2_{L_2(e)}\Big)^{\frac{1}{2}}$$

$$\leq C \Big(\sum_{e \in \mathscr{E}_h^i} |e|^3 \| \, [\![\partial \Delta v / \partial n_e]\!] \, \|_{L_2(e)}^2 \Big)^{\frac{1}{2}} \|w\|_h \tag{5.10}$$

by (4.14);

$$\Big| \sum_{e \in \mathscr{E}_h^i} \int_e \Big[\!\!\Big[\frac{\partial^2 v}{\partial n_e^2} \Big]\!\!\Big] \Big\{\!\!\Big\{ \frac{\partial (w - E_h w)}{\partial n_e} \Big\}\!\!\Big\} \, ds \Big|$$

$$\leq \Big(\sum_{e \in \mathscr{E}_h^i} |e| \| \, [\![\partial^2 v / \partial n_e^2]\!] \, \|_{L_2(e)}^2 \Big)^{\frac{1}{2}} \Big(\sum_{e \in \mathscr{E}_h^i} \frac{1}{|e|} \| \, \{\!\{ \partial (w - E_h w)/\partial n_e \}\!\} \, \|_{L_2(e)}^2 \Big)^{\frac{1}{2}}$$

$$\leq C \Big(\sum_{e \in \mathscr{E}_h^i} |e| \| \, [\![\partial^2 v / \partial n_e^2]\!] \, \|_{L_2(e)}^2 \Big)^{\frac{1}{2}} \|w\|_h \tag{5.11}$$

by (4.13);

$$\Big| \sum_{T \in \mathscr{T}_h} \int_T (f - \Delta^2 v)(w - E_h w) dx \Big|$$

$$\leq \Big(\sum_{T \in \mathscr{T}_h} h_T^4 \| f - \Delta^2 v \|_{L_2(T)}^2 \Big)^{\frac{1}{2}} \Big(\sum_{T \in \mathscr{T}_h} h_T^{-4} \| w - E_h w \|_{L_2(T)}^2 \Big)^{\frac{1}{2}} \tag{5.12}$$

$$\leq \Big(\sum_{T \in \mathscr{T}_h} h_T^4 \| f - \Delta^2 v \|_{L_2(T)}^2 \Big)^{\frac{1}{2}} \|w\|_h$$

by (4.4).

It follows from (5.5)–(5.12) and the Cauchy-Schwarz inequality that

$$a_h(v - u_h, w) \leq C \Big(\sum_{T \in \mathscr{T}_h} |u - v|_{H^2(T)}^2 + \sum_{e \in \mathscr{E}_h} \frac{1}{|e|} \| \, [\![\partial (u - v)/\partial n_e]\!] \, \|_{L_2(e)}^2$$

$$+ \sum_{e \in \mathscr{E}_h^i} |e|^3 \| \, [\![\partial \Delta v / \partial n_e]\!] \, \|_{L_2(e)}^2 + \sum_{e \in \mathscr{E}_h^i} |e| \| \, [\![\partial^2 v / \partial n_e^2]\!] \, \|_{L_2(e)}^2$$

$$+ \sum_{T \in \mathscr{T}_h} h_T^4 \| f - \Delta^2 v \|_{L_2(T)}^2 \Big)^{\frac{1}{2}} \|w\|_h. \tag{5.13}$$

5.3   Local Efficiency Estimates

To complete the *medius* analysis, we need to bound the last three terms inside the bracket on the right-hand side of (5.13) in terms of $\|u - v\|_h$ and data oscillations. We will accomplish this by using bubble function techniques from *a posteriori* error analysis [3, 92, 93].

5.3.1   Data Oscillations

Let $P_j(\Omega, \mathscr{T}_h)$ be the space of piecewise polynomial functions of degree $\leq j$ and $\bar{f} \in P_j(\Omega, \mathscr{T}_h)$ be the L_2 orthogonal projection of f, i.e.,

$$\int_\Omega (f - \bar{f}) v \, dx = 0 \qquad \forall \, v \in P_j(\Omega, \mathscr{T}_h).$$

The oscillation of f (of order j) is defined by

$$\mathrm{Osc}_j(f) = \Big(\sum_{T \in \mathscr{T}_h} h_T^4 \|f - \bar{f}\|_{L_2(T)}^2 \Big)^{\frac{1}{2}}. \tag{5.14}$$

Remark 4. We have flexibility in choosing the order for the data oscillation and we take $\bar{f} = 0$ if $j < 0$.

5.3.2   Estimate for $h_T^4 \|f - \Delta^2 v\|_{L_2(T)}^2$

Let $T \in \mathscr{T}_h$ be arbitrary and $\zeta \in P_6(T)$ be the bubble function that vanishes to the first order on ∂T and equals 1 at the center of T. It follows from scaling that

$$|\zeta|_{H^2(T)} \leq C h_T^{-2} \|\zeta\|_{L_2(T)} \leq C h_T^{-1}. \tag{5.15}$$

Moreover, by the equivalence of norms on finite dimensional spaces, we have

$$C_1 \int_T (\bar{f} - \Delta^2 v)^2 \zeta^2 \, dx \leq \|\bar{f} - \Delta^2 v\|_{L_2(T)}^2 \leq C_2 \int_T (\bar{f} - \Delta^2 v)^2 \zeta \, dx. \tag{5.16}$$

Let $z = (\bar{f} - \Delta^2 v)\zeta$. We can identify z with its trivial extension, which belongs to $H_0^2(\Omega)$. It follows from (2.5), (5.16), integration by parts (involving only polynomials) and a standard inverse estimate that

$$\|\bar{f} - \Delta^2 v\|_{L_2(T)}^2 \leq C \int_T (\bar{f} - \Delta^2 v) z \, dx$$

$$= C \Big[\int_\Omega f z \, dx - \int_T (\Delta^2 v) z \, dx + \int_T (\bar{f} - f) z \, dx \Big]$$

$$= C\left[\int_\Omega \nabla^2 u : \nabla^2 z\, dx - \int_T \nabla^2 v : \nabla^2 z\, dx + \int_T (\bar{f} - f)z\, dx\right]$$

$$= C\left[\int_T (\nabla^2 u - \nabla^2 v) : \nabla^2 z\, dx + \int_T (\bar{f} - f)z\, dx\right]$$

$$\leq C\left[|u - v|_{H^2(T)}|z|_{H^2(T)} + \|f - \bar{f}\|_{L_2(T)}\|z\|_{L_2(T)}\right]$$

$$\leq C\left[h_T^{-2}|u - v|_{H^2(T)} + \|f - \bar{f}\|_{L_2(T)}\right]\|z\|_{L_2(T)}$$

$$\leq C\left[h_T^{-2}|u - v|_{H^2(T)} + \|f - \bar{f}\|_{L_2(T)}\right]\|\bar{f} - \Delta^2 v\|_{L_2(T)},$$

which implies

$$h_T^2\|\bar{f} - \Delta^2 v\|_{L_2(T)} \leq C\left[|u - v|_{H^2(T)} + h_T^2\|f - \bar{f}\|_{L_2(T)}\right],$$

and hence, by the triangle inequality,

$$h_T^4\|f - \Delta^2 v\|_{L_2(T)}^2 \leq C\left[|u - v|_{H^2(T)}^2 + h_T^4\|f - \bar{f}\|_{L_2(T)}^2\right]. \tag{5.17}$$

Summing up (5.17) over all the triangles in $\mathcal{T}_h$ we find

$$\sum_{T \in \mathcal{T}_h} h_T^4\|f - \Delta^2 v\|_{L_2(T)}^2 \leq C\left[\left[\mathrm{Osc}_j(f)\right]^2 + \sum_{T \in \mathcal{T}_h}|u - v|_{H^2(T)}^2\right]. \tag{5.18}$$

5.3.3 Estimate for $|e|\,\|\,[\![\partial^2 v/\partial n_e^2]\!]\,\|_{L_2(e)}^2$

Let $e \in \mathcal{E}_h^i$ be arbitrary and n_e be the unit vector normal to e and pointing from the triangle T_- to T_+ (cf. Fig. 1). We construct a bubble function on $T_- \cup T_+$ as follows.

Let $\beta \in P_{k-2}(\mathbb{R}^2)$ be the polynomial that equals the jump $[\![\partial^2 v/\partial n_e^2]\!]$ on the edge e and which is constant on the lines perpendicular to e. We define $\zeta_1 \in P_{k-1}(T_- \cup T_+)$ to be the polynomial that satisfies

$$\zeta_1 = 0 \quad \text{on } e \quad \text{and} \quad \frac{\partial \zeta_1}{\partial n_e} = \beta. \tag{5.19}$$

It follows from a direct calculation and standard inverse estimates that

$$|e|^{-1}|\zeta_1|_{L_2(T_-\cup T_+)} + \|\zeta_1\|_{L_\infty(T_-\cup T_+)} \leq C|e|^{\frac{1}{2}}\|\,[\![\partial^2 v/\partial n_e^2]\!]\,\|_{L_2(e)}. \tag{5.20}$$

Next we define $\zeta_2 \in P_8(T_- \cup T_+)$ by the following conditions: (i) ζ_2 vanishes to the first order on $(\partial T_- \cup \partial T_+) \setminus e$ (i.e., the boundary of the quadrilateral in Fig. 1), and (ii) ζ_2 equals 1 at the midpoint of e. It follows from scaling that

$$|e|^{-1}|\zeta_2|_{L_2(T_- \cup T_+)} + \|\zeta_2\|_{L_\infty(T_- \cup T_+)} \le C, \tag{5.21}$$

$$\| [[\partial^2 v/\partial n_e^2]] \|_{L_2(e)}^2 \le C \int_e [[\partial^2 v/\partial n_e^2]]^2 \zeta_2\, ds. \tag{5.22}$$

From (5.19) and (5.22) we have

$$\begin{aligned}
\| [[\partial^2 v/\partial n_e^2]] \|_{L_2(e)}^2 &\le C \int_e [[\partial^2 v/\partial n_e^2]] \beta \zeta_2\, ds \\
&= C \int_e [[\partial^2 v/\partial n_e^2]] (\partial \zeta_1/\partial n_e)\zeta_2\, ds \\
&= C \int_e [[\partial^2 v/\partial n_e^2]] [\partial(\zeta_1 \zeta_2)/\partial n_e]ds.
\end{aligned} \tag{5.23}$$

We can identify $\zeta_1 \zeta_2$ with its trivial extension which belongs to $H_0^2(\Omega)$. It then follows from (2.5) and the integration by parts formula (3.1) (involving only polynomials) that

$$\begin{aligned}
\int_e & [[\partial^2 v/\partial n_e^2]] [\partial(\zeta_1 \zeta_2)/\partial n_e]ds \\
&= \sum_{T \in \mathcal{T}_e} \left(-\int_T \nabla^2 v : \nabla^2(\zeta_1 \zeta_2)dx + \int_T (\Delta^2 v)(\zeta_1 \zeta_2)\, dx \right) \\
&= \sum_{T \in \mathcal{T}_e} \int_T \nabla^2(u-v) : \nabla^2(\zeta_1 \zeta_2)dx - \int_\Omega \nabla^2 u : \nabla^2(\zeta_1 \zeta_2)dx \\
&\quad + \sum_{T \in \mathcal{T}_e} \int_T (\Delta^2 v)(\zeta_1 \zeta_2)\, dx \\
&= \sum_{T \in \mathcal{T}_e} \int_T \nabla^2(u-v) : \nabla^2(\zeta_1 \zeta_2)dx - \sum_{T \in \mathcal{T}_e} \int_T (f - \Delta^2 v)(\zeta_1 \zeta_2)dx.
\end{aligned} \tag{5.24}$$

Combining (5.23) and (5.24), we find by the Cauchy-Schwarz inequality and a standard inverse estimate

$$\begin{aligned}
\| [[\partial^2 v/\partial n_e^2]] \|_{L_2(e)}^2 &\le C \Big(\sum_{T \in \mathcal{T}_e} |u - v|_{H^2(T)} |\zeta_1 \zeta_2|_{H^2(T)} \\
&\quad + \sum_{T \in \mathcal{T}_e} \|f - \Delta^2 v\|_{L_2(T)} \|\zeta_1 \zeta_2\|_{L_2(T)} \Big) \\
&\le C \sum_{T \in \mathcal{T}_e} \big[h_T^{-2}|u-v|_{H^2(T)} + \|f - \Delta^2 v\|_{L_2(T)} \big] \|\zeta_1 \zeta_2\|_{L_2(T)}.
\end{aligned} \tag{5.25}$$

From (5.20) and (5.21) we have

$$\|\zeta_1\zeta_2\|_{L_2(T_-\cup T_+)} \le C|e|^{\frac{3}{2}}\|\,[\![\partial^2 v/\partial n_e^2]\!]\,\|_{L_2(e)},$$

which together with (5.25) implies

$$|e|\,\|\,[\![\partial^2 v/\partial n_e^2]\!]\,\|^2_{L_2(e)} \le C\sum_{T\in\mathcal{T}_e}\big[|u-v|^2_{H^2(T)} + h_T^4\|f-\Delta^2 v\|^2_{L_2(T)}\big],$$

and hence, in view of (5.17),

$$|e|\,\|\,[\![\partial^2 v/\partial n_e^2]\!]\,\|^2_{L_2(e)} \le C\sum_{T\in\mathcal{T}_e}\big[|u-v|^2_{H^2(T)} + h_T^4\|f-\bar{f}\|^2_{L_2(T)}\big]. \qquad (5.26)$$

Summing up (5.26) over all the interior edges, we find

$$\sum_{e\in\mathcal{E}_h^i}|e|\,\|\,[\![\partial^2 v/\partial n_e^2]\!]\,\|^2_{L_2(e)} \le C\Big[\big[\mathrm{Osc}_j(f)\big]^2 + \sum_{T\in\mathcal{T}_h}|u-v|^2_{H^2(T)}\Big]. \qquad (5.27)$$

5.3.4 Estimate for $|e|^3\|\,[\![\partial\Delta v/\partial n_e]\!]\,\|^2_{L_2(e)}$

We will follow the convention in Sect. 5.3.3. Let $e \in \mathcal{E}_h^i$ be arbitrary and $\zeta_2 \in P_8(T_-\cup T_+)$ be defined as in Sect. 5.3.3.

Let $\zeta_3 \in P_{k-3}(T_-\cup T_+)$ such that $\zeta_3 = [\![\partial(\Delta v)/\partial n_e]\!]$ on e and ζ_3 is constant on the lines perpendicular to e. By a direct calculation, we have

$$\|\zeta_3\|_{L_2(T_-\cup T_+)} \le C|e|^{\frac{1}{2}}\|\,[\![\partial(\Delta v)/\partial n_e]\!]\,\|_{L_2(e)}. \qquad (5.28)$$

It follows from the equivalence of norms on finite dimensional spaces and scaling that

$$\|\,[\![\partial(\Delta v)/\partial n_e]\!]\,\|^2_{L_2(e)} \le C\int_e [\![\partial(\Delta v)/\partial n_e]\!]^2\,\zeta_2\,ds$$

$$= C\int_e [\![\partial(\Delta v)/\partial n_e]\!]\,(\zeta_2\zeta_3)ds. \qquad (5.29)$$

We can identify $\zeta_2\zeta_3$ with its trivial extension which belongs to $H_0^2(\Omega)$. It then follows from (2.5) and the integration by parts formula (3.1) (involving only polynomials) that

$$\int_e [\![\partial(\Delta v)/\partial n_e]\!]\,(\zeta_2\zeta_3)ds$$

$$= \sum_{T\in\mathcal{T}_e}\Big(\int_T \nabla^2 v : \nabla^2(\zeta_2\zeta_3)dx - \int_T (\Delta^2 v)(\zeta_2\zeta_3)dx\Big)$$

$$+ \int_e \Big[[[\partial^2 v/\partial n_e^2]] \, (\partial(\zeta_2\zeta_3)/\partial n_e) + [[\partial^2 v/\partial n_e \partial t_e]] \, (\partial(\zeta_2\zeta_3)/\partial t_e) \Big] ds$$

$$= \sum_{T \in \mathscr{T}_e} \int_T \nabla^2(v - u) : \nabla^2(\zeta_2\zeta_3) dx + \int_\Omega \nabla^2 u : \nabla^2(\zeta_2\zeta_3) dx \qquad (5.30)$$

$$+ \int_e \Big[[[\partial^2 v/\partial n_e^2]] \, (\partial(\zeta_2\zeta_3)/\partial n_e) + [[\partial^2 v/\partial n_e \partial t_e]] \, (\partial(\zeta_2\zeta_3)/\partial t_e) \Big] ds$$

$$- \sum_{T \in \mathscr{T}_e} \int_T (\Delta^2 v)(\zeta_2\zeta_3) dx$$

$$= \sum_{T \in \mathscr{T}_e} \int_T \nabla^2(v - u) : \nabla^2(\zeta_2\zeta_3) dx + \sum_{T \in \mathscr{T}_e} \int_T (f - \Delta^2 v)(\zeta_2\zeta_3) dx$$

$$+ \int_e \Big[[[\partial^2 v/\partial n_e^2]] \, (\partial(\zeta_2\zeta_3)/\partial n_e) + [[\partial^2 v/\partial n_e \partial t_e]] \, (\partial(\zeta_2\zeta_3)/\partial t_e) \Big] ds.$$

It follows from (5.29), (5.30), the Cauchy-Schwarz inequality and standard inverse estimates that

$$\| [[\partial(\Delta v)/\partial n_e]] \|^2_{L_2(e)}$$

$$\leq C \Big(\sum_{T \in \mathscr{T}_e} \big[|u - v|_{H^2(T)} |\zeta_2\zeta_3|_{H^2(T)} + \| f - \Delta^2 v \|_{L_2(T)} \| \zeta_2\zeta_3 \|_{L_2(T)} \big]$$

$$+ \| [[\partial^2 v/\partial n_e^2]] \|_{L_2(e)} \| \partial(\zeta_2\zeta_3)/\partial n_e \|_{L_2(e)}$$

$$+ \| [[\partial^2 v/\partial n_e \partial t_e]] \|_{L_2(e)} \| \partial(\zeta_2\zeta_3)/\partial t_e \|_{L_2(e)} \Big) \qquad (5.31)$$

$$\leq C \Big(\sum_{T \in \mathscr{T}_e} \big[h_T^{-2} |u - v|_{H^2(T)} + \| f - \Delta^2 v \|_{L_2(T)} \big]$$

$$+ |e|^{-\frac{3}{2}} \| [[\partial^2 v/\partial n_e^2]] \|_{L_2(e)} + |e|^{-\frac{5}{2}} \| [[\partial v/\partial n_e]] \|_{L_2(e)} \Big) \| \zeta_2\zeta_3 \|_{L_2(T)}.$$

From (5.21) and (5.28) we have

$$\| \zeta_2\zeta_3 \|_{L_2(T)} \leq C |e|^{\frac{1}{2}} \| [[\partial(\Delta v)/\partial n_e]] \|_{L_2(e)},$$

which together with (5.31) implies that

$$|e|^3 \| [[\partial(\Delta v)/\partial n_e]] \|^2_{L_2(e)} \leq C \Big(\sum_{T \in \mathscr{T}_e} \big[|u - v|^2_{H^2(T)} + h_T^4 \| f - \Delta^2 v \|^2_{L_2(T)} \big]$$

$$+ |e| \| [[\partial^2 v/\partial n_e^2]] \|^2_{L_2(e)} + \frac{1}{|e|} \| \partial(u - v)/\partial n_e \|^2_{L_2(e)} \Big),$$

and hence, in view of (5.17) and (5.26),

$$|e|^3 \| [\![\partial(\Delta v)/\partial n_e]\!] \|_{L_2(e)}^2 \leq C \Big(\sum_{T \in \mathcal{T}_e} \big[|u - v|_{H^2(T)}^2 + h_T^4 \|f - \bar{f}\|_{L_2(T)}^2 \big]$$

$$+ \frac{1}{|e|} \|\partial(u - v)/\partial n_e\|_{L_2(e)}^2 \Big). \tag{5.32}$$

Summing up (5.32) over all the interior edges, we find

$$\sum_{e \in \mathcal{E}_h^i} |e|^3 \| [\![\partial(\Delta v)/\partial n_e]\!] \|_{L_2(e)}^2 \leq C \Big(\big[\mathrm{Osc}_j(f) \big]^2 + \sum_{T \in \mathcal{T}_h} |u - v|_{H^2(T)}^2$$

$$+ \sum_{e \in \mathcal{E}_h^i} \frac{1}{|e|} \|\partial(u - v)/\partial n_e\|_{L_2(e)}^2 \Big). \tag{5.33}$$

5.4   An Abstract Error Estimate by the Medius Analysis

Putting the estimates (5.13), (5.18), (5.27) and (5.33) together, we arrive at

$$a_h(v - u_h, w) \leq C \Big(\big[\mathrm{Osc}_j(f) \big]^2 + \sum_{T \in \mathcal{T}_h} |u - v|_{H^2(T)}^2$$

$$+ \sum_{e \in \mathcal{E}_h} \frac{1}{|e|} \| [\![\partial(u - v)/\partial n_e]\!] \|_{L_2(e)}^2 \Big)^{\frac{1}{2}} \|w\|_h,$$

which together with (5.1) implies

$$\|u - u_h\|_h \leq C \big[\|u - v\|_h + \mathrm{Osc}_j(f) \big] \qquad \forall v \in V_h.$$

The following error estimate is immediate.

Theorem 1. *We have*

$$\|u - u_h\|_h \leq C \Big[\inf_{v \in V_h} \|u - v\|_h + \mathrm{Osc}_j(f) \Big].$$

Remark 5. Note that throughout this section the integration by parts formula (3.1) has only been applied to polynomials v and w. Therefore there is no need to justify any integration by parts involving u. The only information of u used in the *medius* analysis is that $u \in H_0^2(\Omega)$ satisfies the weak problem (2.5). Thus the *medius* approach puts the analysis of the C^0 interior penalty method on an equal footing with the analysis of conforming finite element methods.

6   A Priori Error Estimates

In this section we derive *a priori* error estimates for the C^0 interior penalty methods, and for C^1 approximate solutions obtained by a post-processing procedure. We also derive error estimates for variants of the C^0 interior penalty methods that can be applied to problems with rough right-hand sides, and error estimates in a lower order Sobolev norm. The techniques developed here are also useful for the convergence analysis of multigrid methods (cf. Sect. 8.4.4).

6.1   Concrete Error Estimates in the Energy Norm

We can use the abstract result in Theorem 1 to derive concrete error estimates for the model problem (2.1).

Theorem 2. *Let $f \in H^\ell(\Omega)$ ($\ell = 0, 1, 2, \ldots$) and the solution u of the model problem (2.1) belong to $H^s(\Omega)$ for $s \in (2, \ell + 4]$. We have the following error estimate*

$$\|u - u_h\|_h \leq C h^{\min(s,k+1)-2} \tag{6.1}$$

for the solution u_h of (3.18), where $k \geq 2$ is the order of the Lagrange finite element space V_h.

Proof. It follows from (3.22) and (3.30) that

$$\|u - \Pi_h u\|_h \leq C h^{\min(s,k+1)-2}. \tag{6.2}$$

Furthermore, by a standard error estimate for the L_2 orthogonal projection into $P_k(\Omega, \mathscr{T}_h)$, we have

$$\mathrm{Osc}_k(f) = \left(\sum_{T \in \mathscr{T}_h} h_T^4 \|f - \bar{f}\|_{L_2(T)}^2 \right)^{\frac{1}{2}} \leq C h^{2+\min(\ell,k+1)}. \tag{6.3}$$

The estimate (6.1) follows from Theorem 1 (with $j = k$), (6.2), (6.3) and the fact that $2 + \min(\ell, k+1) \geq \min(2+\ell, k+3) \geq \min(s-2, k-1) = \min(s, k+1) - 2$. $\qquad\square$

In the case where $u \in H^{\ell+4}(\Omega)$ and $k \geq \ell+3$, the convergence of the C^0 interior penalty method is of the optimal rate of $O(h^{\ell+2})$. In the case where $u \in H^{2+\alpha}(\Omega)$ for $\alpha \in (1/2, 2]$ and $k \geq 1 + \alpha$, the convergence rate is $O(h^\alpha)$.

6.2   Post-Processing

Let $\tilde{E}_h : V_h \longrightarrow H_0^2(\Omega)$ be the enriching operator introduced in Sect. 4.2. The C^1 finite element function $\tilde{E}_h u_h$ obtained from u_h by post-processing provides an approximation of u in the space $H^2(\Omega)$.

Theorem 3. *Under the assumptions of Theorem 2, we have*

$$|u - \tilde{E}_h u_h|_{H^2(\Omega)} \leq C h^{\min(s,k+1)-2}. \tag{6.4}$$

Proof. It follows from Corollary 2, (4.25), Theorem 2 and (6.2) that

$$\begin{aligned}
|u - \tilde{E}_h u_h|_{H^2(\Omega)} &\leq |u - \tilde{E}_h \Pi_h u|_{H^2(\Omega)} + |\tilde{E}_h(\Pi_h u - u_h)|_{H^2(\Omega)} \\
&\leq C \left(h^{\min(s,k+1)-2} + \|\Pi_h u - u_h\|_h \right) \\
&\leq C \left(h^{\min(s,k+1)-2} + \|\Pi_h u - u\|_h + \|u - u_h\|_h \right) \\
&\leq C h^{\min(s,k+1)-2}. \qquad \square
\end{aligned}$$

Therefore the C^0 interior penalty method is also relevant for computing H^2 approximate solutions for the model problems.

6.3 Extended C^0 Interior Penalty Methods

Since $V_h \not\subset H_0^2(\Omega)$, the C^0 interior penalty method is not well-defined when the right-hand side of the model problem (2.1) belongs to $H^{-2}(\Omega) = [H_0^2(\Omega)]'$ while the model problem itself is well-defined for such right-hand sides. Using the enriching operator $\tilde{E}_h$ we can extend the C^0 interior penalty methods to handle this situation as follows:
Find $u_h^* \in V_h$ such that

$$a_h(u_h^*, v) = \langle f, \tilde{E}_h v \rangle \qquad \forall v \in V_h, \tag{6.5}$$

where $\langle \cdot, \cdot \rangle$ is the canonical bilinear form between a (normed) vector space and its dual.

Let $\alpha \in (\frac{1}{2}, 2]$ be the index of elliptic regularity. In addition to (2.7), we also have a regularity estimate [9, 46] of the form

$$\|u\|_{H^{2+\alpha}(\Omega)} \leq C_{\Omega,\alpha} \|f\|_{H^{-2+\alpha}(\Omega)} \qquad \forall f \in H^{-2+\alpha}(\Omega). \tag{6.6}$$

Theorem 4. *We have the error estimate*

$$\|u - u_h^*\|_h \leq C h^{\min(\alpha,k-1)} \|f\|_{H^{-2+\alpha}(\Omega)} \tag{6.7}$$

for the extended C^0 interior penalty method based on the P_k Lagrange finite element space.

Proof. Consider the case where $\alpha \in [1, 2]$ and $k \geq 3$. We have $\min(\alpha, k - 1) = \alpha$, and from (5.1),

$$\|u - u_h^*\|_h \leq \|u - \Pi_h^\dagger u\|_h + C \max_{w \in V_h \setminus \{0\}} \frac{a_h(\Pi_h^\dagger u - u_h^*, w)}{\|w\|_h}, \qquad (6.8)$$

where $\Pi_h^\dagger$ is the nodal interpolation operator for the *cubic* Lagrange finite element space associated with $\mathcal{T}_h$.

Repeating the arguments in Sect. 5.1 but with E_h replaced by $\tilde{E}_h$ and (3.18) replaced by (6.5), we find the following analog of (5.5):

$$a_h(\Pi_h^\dagger u - u_h^*, w) = \sum_{T \in \mathcal{T}_h} \int_T \nabla^2(\Pi_h^\dagger u - u) : \nabla^2(\tilde{E}_h w)\, dx$$

$$+ \sum_{e \in \mathcal{E}_h} \int_e \left\{\!\!\left\{ \frac{\partial^2 w}{\partial n_e^2} \right\}\!\!\right\} \left[\!\!\left[\frac{\partial \Pi_h^\dagger u}{\partial n_e} \right]\!\!\right] ds$$

$$+ \sigma \sum_{e \in \mathcal{E}_h} \frac{1}{|e|} \int_e \left[\!\!\left[\frac{\partial \Pi_h^\dagger u}{\partial n_e} \right]\!\!\right] \left[\!\!\left[\frac{\partial(w - \tilde{E}_h w)}{\partial n_e} \right]\!\!\right] ds$$

$$- \sum_{e \in \mathcal{E}_h^i} \int_e \left[\!\!\left[\frac{\partial^2 \Pi_h^\dagger u}{\partial n_e \partial t_e} \right]\!\!\right] \frac{\partial(w - \tilde{E}_h w)}{\partial t_e}\, ds$$

$$+ \sum_{e \in \mathcal{E}_h^i} \int_e \left[\!\!\left[\frac{\partial(\Delta \Pi_h^\dagger u)}{\partial n_e} \right]\!\!\right] (w - \tilde{E}_h w)\, ds \qquad (6.9)$$

$$- \sum_{e \in \mathcal{E}_h^i} \int_e \left[\!\!\left[\frac{\partial^2 \Pi_h^\dagger u}{\partial n_e^2} \right]\!\!\right] \left\{\!\!\left\{ \frac{\partial(w - \tilde{E}_h w)}{\partial n_e} \right\}\!\!\right\} ds.$$

The first four terms on the right-hand side of (6.9), can be bounded as in Sect. 5.2 by $h^\alpha \|f\|_{H^{-2+\alpha}(\Omega)} \|w\|_h$, using (3.30), (4.17) and (6.6).

Let $e \in \mathcal{E}_h^i$ be the common edge of $T_\pm$. Note that $\left[\!\!\left[\partial^2 \Pi_h^\dagger \zeta / \partial n_e^2 \right]\!\!\right] = 0$ for any polynomial $\zeta \in P_3(T_- \cup T_+)$. Hence it follows from the Bramble-Hilbert Lemma [18, 51] and the trace theorem with scaling that

$$|e| \left\| \left[\!\!\left[\partial^2(\Pi_h^\dagger u)/\partial n_e^2 \right]\!\!\right] \right\|_{L_2(e)}^2 \leq C h^{2\alpha} \sum_{T \in \mathcal{T}_e} |u|_{H^{2+\alpha}(T)}^2 \qquad (6.10)$$

and therefore, in view of (4.17),

$$\left| \sum_{e \in \mathcal{E}_h^i} \int_e \left[\!\!\left[\frac{\partial^2 \Pi_h^\dagger u}{\partial n_e^2} \right]\!\!\right] \left\{\!\!\left\{ \frac{\partial(w - \tilde{E}_h w)}{\partial n_e} \right\}\!\!\right\} ds \right|$$

$$\le \Big(\sum_{e\in\mathscr{E}_h^i} |e| \big\| \big[\!\big[\partial^2 \Pi_h^\dagger u/\partial n_e^2\big]\!\big] \big\|_{L_2(e)}^2\Big)^{\frac{1}{2}} \Big(\sum_{e\in\mathscr{E}_h^i} \frac{1}{|e|} \big\| \{\!\{\partial(w - \tilde{E}_h w)/\partial n_e\}\!\} \big\|_{L_2(e)}^2\Big)^{\frac{1}{2}}$$

$$\le Ch^\alpha \|f\|_{H^{-2+\alpha}(\Omega)} \|w\|_h. \tag{6.11}$$

Similarly, we have

$$\Big| \sum_{e\in\mathscr{E}_h^i} \int_e \Big[\!\Big[\frac{\partial(\Delta \Pi_h^\dagger u)}{\partial n_e}\Big]\!\Big](w - \tilde{E}_h w)ds \Big|$$

$$\le \Big(\sum_{e\in\mathscr{E}_h^i} |e|^3 \big\| \big[\!\big[\partial(\Delta\Pi_h^\dagger u)/\partial n_e\big]\!\big] \big\|_{L_2(e)}^2\Big)^{\frac{1}{2}} \Big(\sum_{e\in\mathscr{E}_h^i} \frac{1}{|e|^3} \|w - \tilde{E}_h w\|_{L_2(e)}^2\Big)^{\frac{1}{2}}$$

$$\le Ch^\alpha \|f\|_{H^{-2+\alpha}(\Omega)} \|w\|_h. \tag{6.12}$$

The estimate (6.7) then follows from (3.30) and (6.8)–(6.12).

The case where $\alpha \in (\frac{1}{2}, 1)$ and $k \ge 3$ and the case where $\alpha \in (\frac{1}{2}, 2]$ and $k = 2$ can be established by similar arguments, where $\Pi_h^\dagger$ becomes the nodal interpolation operator for the *quadratic* Lagrange finite element space.   □

6.4 Error Estimates in a Lower Order Norm

Let $\alpha \in (\frac{1}{2}, 2]$ be the index of elliptic regularity for the model problem (2.1). We can compare $\tilde{E}_h u_h$ and u in the lower order Sobolev norm $\|\cdot\|_{H^{2-\alpha}(\Omega)}$.

Theorem 5. *We have*

$$\|u - \tilde{E}_h u_h\|_{H^{2-\alpha}(\Omega)} \le Ch^{2\min(\alpha, k-1)} \|f\|_{H^{-2+\alpha}(\Omega)}. \tag{6.13}$$

Proof. Since $u - \tilde{E}_h u_h \in H_0^2(\Omega) \subset H_0^{2-\alpha}(\Omega)$, we have, by duality,

$$\|u - \tilde{E}_h u_h\|_{H^{2-\alpha}(\Omega)} = \sup_{\phi\in H^{-2+\alpha}(\Omega)\setminus\{0\}} \frac{\langle \phi, u - \tilde{E}_h u_h\rangle}{\|\phi\|_{H^{-2+\alpha}(\Omega)}}. \tag{6.14}$$

Let $\zeta \in H_0^2(\Omega)$ satisfy

$$a(\zeta, v) = \langle \phi, v\rangle \qquad \forall v \in H_0^2(\Omega), \tag{6.15}$$

and $\zeta_h^* \in V_h$ satisfy

$$a_h(\zeta_h^*, v) = \langle \phi, \tilde{E}_h v\rangle \qquad \forall v \in V_h. \tag{6.16}$$

Then we have, by Theorem 4,

$$\|\zeta - \zeta_h^*\|_h \leq C h^{\min(\alpha, k-1)} \|\phi\|_{H^{-2+\alpha}(\Omega)}. \tag{6.17}$$

Assume that $\alpha \in (\frac{1}{2}, 1)$. It follows from (2.6), (3.17), (3.32), (6.15) and (6.16) that

$$\langle \phi, u - \tilde{E}_h u_h \rangle$$

$$= a(\zeta, u) - a_h(\zeta_h^*, u_h)$$

$$= \sum_{T \in \mathscr{T}_h} \int_T \nabla^2 \zeta : \nabla^2(u - u_h) \, dx + \sum_{T \in \mathscr{T}_h} \int_T \nabla^2(\zeta - \zeta_h^*) : \nabla^2 u_h \, dx$$

$$- \sum_{e \in \mathscr{E}_h} \int_e \left\{\!\!\left\{ \frac{\partial^2 \zeta_h^*}{\partial n_e^2} \right\}\!\!\right\} \left[\!\!\left[\frac{\partial u_h}{\partial n_e} \right]\!\!\right] ds - \sum_{e \in \mathscr{E}_h} \int_e \left\{\!\!\left\{ \frac{\partial^2 u_h}{\partial n_e^2} \right\}\!\!\right\} \left[\!\!\left[\frac{\partial \zeta_h^*}{\partial n_e} \right]\!\!\right] ds$$

$$- \sigma \sum_{e \in \mathscr{E}_h} \frac{1}{|e|} \int_e \left[\!\!\left[\frac{\partial \zeta_h^*}{\partial n_e} \right]\!\!\right] \left[\!\!\left[\frac{\partial u_h}{\partial n_e} \right]\!\!\right] ds$$

$$= \sum_{T \in \mathscr{T}_h} \int_T \nabla^2(\zeta - \Pi_h^\dagger \zeta) : \nabla^2(u - u_h) \, dx$$

$$+ \sum_{T \in \mathscr{T}_h} \int_T \nabla^2(\zeta - \zeta_h^*) : \nabla^2(u_h - \Pi_h^\dagger u) \, dx$$

$$+ \sum_{e \in \mathscr{E}_h} \int_e \left\{\!\!\left\{ \frac{\partial^2(\Pi_h^\dagger \zeta - \zeta_h^*)}{\partial n_e^2} \right\}\!\!\right\} \left[\!\!\left[\frac{\partial u_h}{\partial n_e} \right]\!\!\right] ds - \sum_{e \in \mathscr{E}_h^i} \int_e \left[\!\!\left[\frac{\partial^2 \Pi_h^\dagger \zeta}{\partial n_e^2} \right]\!\!\right] \left\{\!\!\left\{ \frac{\partial(u - u_h)}{\partial n_e} \right\}\!\!\right\} ds$$

$$- \sum_{e \in \mathscr{E}_h^i} \int_e \left[\!\!\left[\frac{\partial^2 \Pi_h^\dagger \zeta}{\partial n_e \partial t_e} \right]\!\!\right] \frac{\partial(u - u_h)}{\partial t_e} ds + \sum_{e \in \mathscr{E}_h} \int_e \left\{\!\!\left\{ \frac{\partial^2(\Pi_h^\dagger u - u_h)}{\partial n_e^2} \right\}\!\!\right\} \left[\!\!\left[\frac{\partial \zeta_h^*}{\partial n_e} \right]\!\!\right] ds$$

$$- \sum_{e \in \mathscr{E}_h^i} \int_e \left[\!\!\left[\frac{\partial^2 \Pi_h^\dagger u}{\partial n_e^2} \right]\!\!\right] \left\{\!\!\left\{ \frac{\partial(\zeta - \zeta_h^*)}{\partial n_e} \right\}\!\!\right\} ds - \sum_{e \in \mathscr{E}_h^i} \int_e \left[\!\!\left[\frac{\partial^2 \Pi_h^\dagger u}{\partial n_e \partial t_e} \right]\!\!\right] \frac{\partial(\zeta - \zeta_h^*)}{\partial t_e} ds$$

$$- \sigma \sum_{e \in \mathscr{E}_h} \frac{1}{|e|} \int_e \left[\!\!\left[\frac{\partial \zeta_h^*}{\partial n_e} \right]\!\!\right] \left[\!\!\left[\frac{\partial u_h}{\partial n_e} \right]\!\!\right] ds, \tag{6.18}$$

where $\Pi_h^\dagger$ is the nodal interpolation operator for the *quadratic* Lagrange finite element space associated with $\mathscr{T}_h$. Combining this relation with (3.30), (6.1), (6.2), (6.10) (which is valid for the nodal interpolation operator for the quadratic finite element) and (6.17), we find

$$\langle \phi, u - \tilde{E}_h u_h \rangle \leq C h^{2\alpha} \|f\|_{H^{-2+\alpha}(\Omega)} \|\phi\|_{H^{-2+\alpha}(\Omega)},$$

which together with (6.14) implies (6.13).

The case where $\alpha \in [1, 2]$ can be similarly established using the interpolation operator for the *cubic* Lagrange finite element space. $\square$

The following result for the solution of the extended C^0 interior penalty method (6.5) is obtained by similar arguments.

Theorem 6. *We have*

$$\|u - \tilde{E}_h u_h^*\|_{H^{2-\alpha}(\Omega)} \le C h^{2\min(\alpha, k-1)} \|f\|_{H^{-2+\alpha}(\Omega)}.$$

7   A Posteriori Error Estimates

In this section we develop a reliable and efficient residual-based *a posteriori* error estimator for the solution u_h of (3.18). We will only discuss the model problem (2.1) in detail. The results in this section generalize the results in [29] to higher order finite elements. Results for the model problem (2.3) can be found in [28].

Note that the results from Sect. 5.3 are useful for proving the efficiency of the error estimator.

7.1   An A Posteriori Error Estimator

Let $T \in \mathcal{T}_h$ be arbitrary. The residual error

$$\eta_T = h_T^2 \|f - \Delta^2 u_h\|_{L_2(T)} \tag{7.1}$$

measures the extent to which u_h fails to satisfy the biharmonic equation.

Let $e \in \mathcal{E}_h$ be arbitrary. The residual

$$\eta_{e,1} = \frac{\sigma}{|e|^{\frac{1}{2}}} \|\, [\![\partial u_h / \partial n_e]\!] \,\|_{L_2(e)} \tag{7.2}$$

measures the extent to which u_h fails to be in $H_0^2(\Omega)$.

Let $e \in \mathcal{E}_h^i$ be arbitrary. The residual

$$\eta_{e,2} = |e|^{\frac{1}{2}} \|\, [\![\partial^2 u_h / \partial n_e^2]\!] \,\|_{L_2(e)} \tag{7.3}$$

measures the extent that u_h fails to be in $H^3(\Omega)$, while the residual

$$\eta_{e,3} = |e|^{\frac{3}{2}} \|\, [\![\partial(\Delta u_h) / \partial n_e]\!] \,\|_{L_2(e)} \tag{7.4}$$

measures the extent that u_h fails to be in $H^4(\Omega)$.

The residual-based error estimator η_h is defined by

$$\eta_h = \left[\sum_{T \in \mathscr{T}_h} \eta_T^2 + \sum_{e \in \mathscr{E}_h} \eta_{e,1}^2 + \sum_{e \in \mathscr{E}_h^i} (\eta_{e,2}^2 + \eta_{e,3}^2) \right]^{\frac{1}{2}}. \qquad (7.5)$$

Remark 6. We can replace the definition of $\eta_{e,3}$ by $|e|^{\frac{3}{2}} \left\| [\![\partial^3 u_h / \partial n_e^3]\!] \right\|_{L_2(e)}$. Note that the error estimator $\eta_{e,3}$ is identically 0 for the quadratic C^0 interior penalty method.

7.2 Reliability

In this section we show that the error estimator η_h provides an upper bound of the true error $\|u - u_h\|_h$.

It is clear from (3.24) and (7.2) that

$$\sigma \sum_{e \in \mathscr{E}_h} \frac{1}{|e|} \left\| [\![\partial (u - u_h)/\partial n_e]\!] \right\|_{L_2(e)}^2 = \sigma \sum_{e \in \mathscr{E}_h} \frac{1}{|e|} \left\| [\![\partial u_h/\partial n_e]\!] \right\|_{L_2(e)}^2 \leq \sum_{e \in \mathscr{E}_h} \eta_{e,1}^2, \qquad (7.6)$$

so we only need to bound $\sum_{T \in \mathscr{T}_h} |u - u_h|_{H^2(T)}^2$.

It follows from (3.24), Lemma 1 and (7.2) that

$$\sum_{T \in \mathscr{T}_h} |u - u_h|_{H^2(T)}^2 \leq 2 \sum_{T \in \mathscr{T}_h} \left(|u - E_h u_h|_{H^2(T)}^2 + |u_h - E_h u_h|_{H^2(T)}^2 \right)$$

$$\leq 2 |u - E_h u|_{H^2(\Omega)}^2 + C \sum_{e \in \mathscr{E}_h} \eta_{e,1}^2, \qquad (7.7)$$

and we have, by duality,

$$|u - E_h u_h|_{H^2(\Omega)} = \sup_{\phi \in H_0^2(\Omega) \setminus \{0\}} \frac{a(u - E_h u_h, \phi)}{|\phi|_{H^2(\Omega)}}. \qquad (7.8)$$

We can use (2.5), (2.6) and (3.18) to rewrite the numerator on the right-hand side of (7.8) as

$$a(u - E_h u_h, \phi) = \int_\Omega \nabla^2 (u - E_h u_h) : \nabla^2 \phi \, dx$$

$$= \sum_{T \in \mathscr{T}_h} \int_T \nabla^2 (u_h - E_h u_h) : \nabla^2 \phi \, dx - \sum_{T \in \mathscr{T}_h} \int_T \nabla^2 u_h : \nabla^2 (\phi - \Pi_h \phi) \, dx$$

$$+ \int_{\Omega} \nabla^2 u : \nabla^2 \phi \, dx - \sum_{T \in \mathscr{T}_h} \int_T \nabla^2 u_h : \nabla^2 (\Pi_h \phi) dx \qquad (7.9)$$

$$= \sum_{T \in \mathscr{T}_h} \int_T \nabla^2 (u_h - E_h u_h) : \nabla^2 \phi \, dx - \sum_{T \in \mathscr{T}_h} \int_T \nabla^2 u_h : \nabla^2 (\phi - \Pi_h \phi) \, dx$$

$$+ a_h(u_h, \Pi_h \phi) - \sum_{T \in \mathscr{T}_h} \int_T \nabla^2 u_h : \nabla^2 (\Pi_h \phi) dx + \int_{\Omega} f(\phi - \Pi_h \phi) dx.$$

We have, by (3.32),

$$\sum_{T \in \mathscr{T}_h} \int_T \nabla^2 u_h : \nabla^2 (\phi - \Pi_h \phi) \, dx$$

$$= \sum_{T \in \mathscr{T}_h} \int_T (\Delta^2 u_h)(\phi - \Pi_h \phi) dx + \sum_{e \in \mathscr{E}_h^i} \int_e \left[\!\!\left[\frac{\partial (\Delta u_h)}{\partial n_e} \right]\!\!\right] (\phi - \Pi_h \phi) ds$$

$$+ \sum_{e \in \mathscr{E}_h} \int_e \left\{\!\!\left\{ \frac{\partial^2 u_h}{\partial n_e^2} \right\}\!\!\right\} \left[\!\!\left[\frac{\partial \Pi_h \phi}{\partial n_e} \right]\!\!\right] ds - \sum_{e \in \mathscr{E}_h^i} \int_e \left[\!\!\left[\frac{\partial^2 u_h}{\partial n_e^2} \right]\!\!\right] \left\{\!\!\left\{ \frac{\partial (\phi - \Pi_h \phi)}{\partial n_e} \right\}\!\!\right\} ds$$

$$- \sum_{e \in \mathscr{E}_h^i} \int_e \left[\!\!\left[\frac{\partial^2 u_h}{\partial n_e \partial t_e} \right]\!\!\right] \frac{\partial (\phi - \Pi_h \phi)}{\partial t_e} ds, \qquad (7.10)$$

and by (3.17),

$$a_h(u_h, \Pi_h \phi) - \sum_{T \in \mathscr{T}_h} \int_T \nabla^2 u_h : \nabla^2 (\Pi_h \phi) dx$$

$$= \sum_{e \in \mathscr{E}_h} \int_e \left\{\!\!\left\{ \frac{\partial^2 u_h}{\partial n_e^2} \right\}\!\!\right\} \left[\!\!\left[\frac{\partial \Pi_h \phi}{\partial n_e} \right]\!\!\right] ds + \sum_{e \in \mathscr{E}_h} \int_e \left\{\!\!\left\{ \frac{\partial^2 \Pi_h \phi}{\partial n_e^2} \right\}\!\!\right\} \left[\!\!\left[\frac{\partial u_h}{\partial n_e} \right]\!\!\right] ds$$

$$+ \sigma \sum_{e \in \mathscr{E}_h} \frac{1}{|e|} \int_e \left[\!\!\left[\frac{\partial u_h}{\partial n_e} \right]\!\!\right] \left[\!\!\left[\frac{\partial \Pi_h \phi}{\partial n_e} \right]\!\!\right] ds. \qquad (7.11)$$

Putting (7.9)–(7.11) together, we find

$$a(u - E_h u, \phi)$$

$$= \sum_{T \in \mathscr{T}_h} \int_T \nabla^2 (u_h - E_h u_h) : \nabla^2 \phi \, dx$$

$$+ \sum_{T \in \mathscr{T}_h} \int_T (f - \Delta^2 u_h)(\phi - \Pi_h \phi) dx$$

$$-\sum_{e\in\mathcal{E}_h^i}\int_e\left[\!\!\left[\frac{\partial(\Delta u_h)}{\partial n_e}\right]\!\!\right](\phi-\Pi_h\phi)ds+\sum_{e\in\mathcal{E}_h^i}\int_e\left[\!\!\left[\frac{\partial^2 u_h}{\partial n_e^2}\right]\!\!\right]\left\{\!\!\left\{\frac{\partial(\phi-\Pi_h\phi)}{\partial n_e}\right\}\!\!\right\}ds$$

$$+\sum_{e\in\mathcal{E}_h^i}\int_e\left[\!\!\left[\frac{\partial^2 u_h}{\partial n_e\partial t_e}\right]\!\!\right]\frac{\partial(\phi-\Pi_h\phi)}{\partial t_e}ds+\sum_{e\in\mathcal{E}_h}\int_e\left\{\!\!\left\{\frac{\partial^2\Pi_h\phi}{\partial n_e^2}\right\}\!\!\right\}\left[\!\!\left[\frac{\partial u_h}{\partial n_e}\right]\!\!\right]ds$$

$$+\sigma\sum_{e\in\mathcal{E}_h}\frac{1}{|e|}\int_e\left[\!\!\left[\frac{\partial u_h}{\partial n_e}\right]\!\!\right]\left[\!\!\left[\frac{\partial\Pi_h\phi}{\partial n_e}\right]\!\!\right]ds. \tag{7.12}$$

The terms on the right-hand side of (7.12) can be estimated as follows:

$$\left|\sum_{T\in\mathcal{T}_h}\int_T\nabla^2(u_h-E_hu_h):\nabla^2\phi\,dx\right|\le C\left(\sum_{e\in\mathcal{E}_h}\eta_{e,1}^2\right)^{\frac{1}{2}}|\phi|_{H^2(\Omega)} \tag{7.13}$$

by (3.24), Lemma 1 and (7.2);

$$\left|\sum_{T\in\mathcal{T}_h}\int_T(f-\Delta^2u_h)(\phi-\Pi_h\phi)dx\right|$$

$$\le\left(\sum_{T\in\mathcal{T}_h}h_T^4\|f-\Delta^2u_h\|_{L_2(T)}^2\right)^{\frac{1}{2}}\left(\sum_{T\in\mathcal{T}_h}h_T^{-4}\|\phi-\Pi_h\phi\|_{L_2(T)}^2\right)^{\frac{1}{2}}$$

$$\le\left(\sum_{T\in\mathcal{T}_h}\eta_T^2\right)^{\frac{1}{2}}|\phi|_{H^2(\Omega)} \tag{7.14}$$

by (3.30) and (7.1);

$$\left|\sum_{e\in\mathcal{E}_h^i}\int_e\left[\!\!\left[\frac{\partial(\Delta u_h)}{\partial n_e}\right]\!\!\right](\phi-\Pi_h\phi)ds\right|$$

$$\le\left(\sum_{e\in\mathcal{E}_h^i}|e|^3\|\,[\![\partial(\Delta u_h)/\partial n_e]\!]\,\|_{L_2(e)}^2\right)^{\frac{1}{2}}\left(\sum_{e\in\mathcal{E}_h^i}|e|^{-3}\|\phi-\Pi_h\phi\|_{L_2(e)}^2\right)^{\frac{1}{2}}$$

$$\le C\left(\sum_{e\in\mathcal{E}_h^i}\eta_{e,3}^2\right)^{\frac{1}{2}}|\phi|_{H^2(\Omega)} \tag{7.15}$$

and

$$\left|\sum_{e\in\mathcal{E}_h^i}\int_e\left[\!\!\left[\frac{\partial^2 u_h}{\partial n_e^2}\right]\!\!\right]\left\{\!\!\left\{\frac{\partial(\phi-\Pi_h\phi)}{\partial n_e}\right\}\!\!\right\}ds\right|$$

$$\leq \Big(\sum_{e \in \mathscr{E}_h^i} |e| \| [[\partial^2 u_h / \partial n_e^2]] \|_{L_2(e)}^2 \Big)^{\frac{1}{2}} \Big(\sum_{e \in \mathscr{E}_h^i} |e|^{-1} \| \{\!\{\partial(\phi - \Pi_h \phi)/\partial n_e\}\!\} \|_{L_2(e)}^2 \Big)^{\frac{1}{2}}$$

$$\leq C \Big(\sum_{e \in \mathscr{E}_h^i} \eta_{e,2}^2 \Big)^{\frac{1}{2}} |\phi|_{H^2(\Omega)} \tag{7.16}$$

by (3.30), (7.3), (7.4) and the trace theorem with scaling;

$$\Big| \sum_{e \in \mathscr{E}_h^i} \int_e \Big[\!\Big[\frac{\partial^2 u_h}{\partial n_e \partial t_e} \Big]\!\Big] \frac{\partial(\phi - \Pi_h \phi)}{\partial t_e} ds \Big|$$

$$\leq \Big(\sum_{e \in \mathscr{E}_h^i} |e| \| [[\partial^2 u_h / \partial n_e \partial t_e]] \|_{L_2(e)}^2 \Big)^{\frac{1}{2}} \Big(\sum_{e \in \mathscr{E}_h^i} |e|^{-1} \| \partial(\phi - \Pi_h \phi)/\partial t_e \|_{L_2(e)}^2 \Big)^{\frac{1}{2}}$$

$$\leq C \Big(\sum_{e \in \mathscr{E}_h^i} \eta_{e,1}^2 \Big)^{\frac{1}{2}} |\phi|_{H^2(\Omega)} \tag{7.17}$$

by (3.24), (3.30), (7.2), the trace theorem with scaling and a standard inverse estimate;

$$\Big| \sum_{e \in \mathscr{E}_h} \int_e \Big\{\!\!\Big\{ \frac{\partial^2 \Pi_h \phi}{\partial n_e} \Big\}\!\!\Big\} \Big[\!\Big[\frac{\partial u_h}{\partial n_e} \Big]\!\Big] ds \Big|$$

$$\leq \Big(\sum_{e \in \mathscr{E}_h} \eta_{e,1}^2 \Big)^{\frac{1}{2}} \Big(\sum_{e \in \mathscr{E}_h} |e| \| \{\!\{\partial^2 \Pi_h \phi / \partial n_e^2\}\!\} \|_{L_2(e)}^2 \Big)^{\frac{1}{2}} \tag{7.18}$$

$$\leq C \Big(\sum_{e \in \mathscr{E}_h} \eta_{e,1}^2 \Big)^{\frac{1}{2}} \Big(\sum_{T \in \mathscr{T}_h} |\Pi_h \phi|_{H^2(T)}^2 \Big)^{\frac{1}{2}} \leq C \Big(\sum_{e \in \mathscr{E}_h} \eta_{e,1}^2 \Big)^{\frac{1}{2}} |\phi|_{H^2(\Omega)}$$

by (3.24), (3.30), (7.2) and scaling;

$$\Big| \sigma \sum_{e \in \mathscr{E}_h} \frac{1}{|e|} \int_e \Big[\!\Big[\frac{\partial u_h}{\partial n_e} \Big]\!\Big] \Big[\!\Big[\frac{\partial \Pi_h \phi}{\partial n_e} \Big]\!\Big] ds \Big|$$

$$\leq \Big(\sum_{e \in \mathscr{E}_h} \eta_{e,1}^2 \Big)^{\frac{1}{2}} \Big(\sum_{e \in \mathscr{E}_h} |e|^{-1} \| [[\partial(\Pi_h \phi - \phi)/\partial n_e]] \|_{L_2(e)}^2 \Big)^{\frac{1}{2}}$$

$$\leq \Big(\sum_{e \in \mathscr{E}_h} \eta_{e,1}^2 \Big)^{\frac{1}{2}} |\phi|_{H^2(\Omega)} \tag{7.19}$$

by (3.30), (7.2) and the trace theorem with scaling.

It follows from (7.5) and (7.12)–(7.19) that

$$a(u - E_h u_h, \phi) \leq C \eta_h |\phi|_{H^2(\Omega)}. \tag{7.20}$$

Combining (7.6)–(7.8) and (7.20) we have the following result.

Theorem 7. *There exists a positive constant C independent of h and σ such that*

$$\|u - u_h\|_h \leq C \eta_h.$$

7.3 Efficiency

The error estimator η_h also provides a lower bound for the true error $|u - u_h|_{H^2(\Omega)}$ up to data oscillation.

Theorem 8. *There exists a positive constant C independent of h and σ such that*

$$\eta_h \leq C \left[\sigma^{\frac{1}{2}} \|u - u_h\|_h + \mathrm{Osc}_{k-3}(f) \right].$$

Proof. Let $\bar{f}$ be the L_2 orthogonal projection of f in $P_{k-3}(\Omega, \mathcal{T}_h)$. It follows from (5.17), (5.26), (5.32) and (7.1)–(7.5) that

$$\eta_h^2 \leq \sum_{T \in \mathcal{T}_h} |u - u_h|_{H^2(T)}^2 + \sigma^2 \sum_{e \in \mathcal{E}_h} \frac{1}{|e|} \| [\![\partial(u - u_h)/\partial n_e]\!] \|_{L_2(e)}^2 + \sum_{T \in \mathcal{T}_h} h_T^4 \| f - \bar{f} \|_{L_2(T)}^2$$

which together with (5.14) completes the proof.   □

Remark 7. Note that $\mathrm{Osc}_{k-3}(f)$ is asymptotically smaller than $\|u - u_h\|_h$. Indeed, the magnitude of the error in $\| \cdot \|_h$ for the C^0 interior penalty method based on the P_k Lagrange finite element space is at best $O(h^{k-1})$, which can only happen if $u \in H^{k+1}(\Omega)$. In this case $f \in H^{k-3}(\Omega)$ and the magnitude of $\mathrm{Osc}_{k-3}(f)$ is $o(h^{k-1})$.

7.4 Adaptive Algorithms

Adaptive algorithms based on the error estimator η_h and the bulk marking criteria of Dörfler [47, 73] can be developed for the C^0 interior penalty methods, and optimal convergence is observed in numerical experiments [28, 29]. We note that while rigorous convergence results for discontinuous Galerkin methods for second order problems have been obtained recently [16, 60, 62], a rigorous convergence analysis for adaptive C^0 interior penalty methods (or other discontinuous Galerkin methods) for fourth order problems remains open.

8 Multigrid Methods

As mentioned in Sect. 3.6, it is crucial to have efficient solvers for the very ill-conditioned discrete problems resulting from C^0 interior penalty methods. In this section we will develop and analyze multigrid methods for the discrete problems. We will focus on the model problem (2.1). But the methodology of the convergence analysis, which is different from the one in [35] (cf. Sect. 8.4.4), can also be applied to the other model problems.

8.1 Set-Up

Let $\mathcal{T}_j$ ($j \geq 0$) be a sequence of triangulations of Ω such that $\mathcal{T}_j$ ($j \geq 1$) is obtained from $\mathcal{T}_{j-1}$ by a uniform subdivision, and $h_j = \max_{T \in \mathcal{T}_j} h_T$ (so that $h_{j-1} = 2h_j$). The set of the edges (resp. interior edges) of $\mathcal{T}_j$ is denoted by $\mathcal{E}_j$ (resp. $\mathcal{E}_j^i$).

Let V_j be the P_k ($k \geq 2$) Lagrange finite element space associated with $\mathcal{T}_j$, and $u_j \in V_j$ be the solution of the model problem (2.1) on level j obtained by the C^0 interior penalty method. We can rewrite the discrete problem (3.18) as

$$A_j u_j = \phi_j, \tag{8.1}$$

where $A_j : V_j \longrightarrow V_j'$ and $\phi_j \in V_j'$ are defined by

$$\langle A_j w, v \rangle = a_j(w, v) \qquad \forall\, v, w \in V_j, \tag{8.2}$$

$$\langle \phi_j, v \rangle = \int_\Omega f v\, dx \qquad \forall\, v \in V_j. \tag{8.3}$$

Here the bilinear form $a_j(\cdot, \cdot)$ is the analog of (3.17) on V_j and $\langle \cdot, \cdot \rangle$ is the canonical bilinear form between a vector space and its dual.

Multigrid algorithms [17, 21, 32, 59, 69, 91] are iterative methods for solving equations of the form

$$A_j z = \psi \tag{8.4}$$

where $\psi \in V_j'$ and $z \in V_j$.

Remark 8. We follow the convention that finite element functions are denoted by Roman letters and functionals on the finite element spaces are denoted by Greek letters.

8.2 Intergrid Transfer Operators and Smoothers

There are two ingredients in the design of multigrid algorithms. First we need intergrid transfer operators to move functions and functionals between grids. Since the finite element spaces are nested, i.e., $V_0 \subset V_1 \subset \cdots$, we can take the coarse-to-fine operator $I_{j-1}^j : V_{j-1} \longrightarrow V_j$ ($j \geq 1$) to be the natural injection and the fine-to-coarse operator $I_j^{j-1} : V_j' \longrightarrow V_{j-1}'$ to be the transpose of I_{j-1}^j, i.e.,

$$\langle I_j^{j-1} \gamma, v \rangle = \langle \gamma, I_{j-1}^j v \rangle \qquad \forall \, \gamma \in V_j', \, v \in V_{j-1}. \tag{8.5}$$

Secondly we need a smoothing scheme to damp out the highly oscillatory part of the error of an approximate solution of (8.4) so that the remaining error can be captured accurately on a coarser grid. The smoothing step is given by

$$z_{\text{new}} = z_{\text{old}} + \lambda_j S_j^{-1} (\psi - A_j z_{\text{old}}), \tag{8.6}$$

where $S_j : V_j \longrightarrow V_j'$ and λ_j is a damping factor chosen so that

$$\text{spectral radius of} \left(\lambda_j S_j^{-1} A_j \right) \leq 1. \tag{8.7}$$

Below we will discuss two choices for S_j that will lead to two smoothing schemes.

8.2.1 A Standard Smoother

Let $\mathcal{N}_j$ be the set of all the nodes for the P_k Lagrange finite element space that are interior to Ω. The operator $S_j : V_j \longrightarrow V_j'$ is defined by

$$\langle S_j w, v \rangle = h_j^2 \sum_{p \in \mathcal{N}_j} w(p) v(p).$$

Note that we have

$$C_1 \|v\|_{L_2(\Omega)}^2 \leq \langle S_j v, v \rangle \leq C_2 \|v\|_{L_2(\Omega)}^2 \qquad \forall \, v \in V_j, \tag{8.8}$$

and the computational cost for the evaluation of $S_j^{-1} \gamma$ ($\gamma \in V_j'$) is of order $O(n_j)$, where $n_j = \dim V_j$. By (8.8) and standard inverse estimates, we can choose

$$\lambda_j = C h_j^4 \tag{8.9}$$

so that (8.7) holds.

8.2.2 A Nonstandard Smoother

Let $L_j : V_j \longrightarrow V_j'$ be the discrete Laplace operator defined by

$$\langle L_j w, v \rangle = \int_\Omega \nabla w \cdot \nabla v \, dx \qquad \forall \, v, w \in V_j.$$

We take $S_j^{-1} : V_j' \longrightarrow V_j$ to be an approximate inverse of L_j obtained by a multigrid Poisson solve such that

$$C_1 |v|^2_{H^1(\Omega)} \le \langle S_j v, v \rangle \le C_2 |v|^2_{H^1(\Omega)} \qquad \forall \, v \in V_j. \tag{8.10}$$

Note that it is easy to implement a multigrid Poisson solve because the finite element spaces for the C^0 interior penalty methods are standard finite element spaces for second order problems, and the computational cost for the evaluation of $S_j^{-1} \gamma$ ($\gamma \in V_j'$) is of order $O(n_j)$. By (8.10) and standard inverse estimates, we can choose

$$\lambda_j = C h_j^2 \tag{8.11}$$

so that (8.7) holds.

8.3 Multigrid Algorithms

We will consider the V-cycle, W-cycle and F-cycle algorithms for (8.4).

8.3.1 V-Cycle Algorithm

The V-cycle algorithm computes an approximate solution $MG_V(j, \psi, z_0, m)$ of (8.4) with initial guess $z_0 \in V_j$ and m pre-smoothing and m post-smoothing steps.

For $j = 0$, we take $MG_V(0, \psi, z_0, m)$ to be $A_0^{-1} \psi$. For $j \ge 1$, we compute $MG_V(j, \psi, z_0, m)$ recursively in three steps.

Pre-smoothing   For $1 \le \ell \le m$, compute z_ℓ recursively by

$$z_\ell = z_{\ell-1} + \lambda_j S_j^{-1}(\psi - A_j z_{\ell-1}). \tag{8.12}$$

Coarse Grid Correction   Compute

$$z_{m+1} = z_m + I_{j-1}^j MG_V(j-1, \rho_{j-1}, 0, m), \tag{8.13}$$

where $\rho_{j-1} = I_j^{j-1}(\psi - A_j z_m) \in V_{j-1}'$ is the transferred residual of z_m.

Post-smoothing   For $m + 2 \le \ell \le 2m + 1$, compute z_ℓ recursively by

$$z_\ell = z_{\ell-1} + \lambda_j S_j^{-1}(\psi - A_j z_{\ell-1}). \tag{8.14}$$

Final Output

$$MG_V(j, \psi, z_0, m) = z_{2m+1} \qquad (8.15)$$

Remark 9. $MG_V(j-1, \rho_{j-1}, 0, m)$ is the approximate solution of the coarse grid residual equation

$$A_{j-1}e_{j-1} = I_j^{j-1}(\psi - A_j z_m) = I_j^{j-1}A_j(z - z_m) \qquad (8.16)$$

obtained by the $(j-1)$-st level V-cycle algorithm with initial guess 0.

8.3.2 W-Cycle Algorithm

The W-cycle algorithm computes an approximate solution $MG_W(j, \psi, z_0, m)$ of (8.4) with initial guess $z_0 \in V_j$ and m pre-smoothing and m post-smoothing steps.

The only difference between the V-cycle algorithm and the W-cycle algorithm is in the coarse grid correction step, where the coarse grid algorithm is applied twice to the coarse grid residual equation. More precisely, we have

$$
\begin{aligned}
z_{m+\frac{1}{2}} &= MG_W(j-1, \rho_{j-1}, 0, m), \\
z_{m+1} &= z_m + MG_W(j-1, \rho_{j-1}, z_{m+\frac{1}{2}}, m).
\end{aligned}
\qquad (8.17)
$$

8.3.3 F-Cycle Algorithm

The F-cycle algorithm computes an approximate solution $MG_F(j, \psi, z_0, m)$ of (8.4) with initial guess $z_0 \in V_j$ and m pre-smoothing and m post-smoothing steps.

The only difference between the V-cycle algorithm and the F-cycle algorithm is again in the coarse grid correction step, where the coarse grid algorithm is applied once followed by one application of the coarse grid V-cycle algorithm. More precisely, we have

$$
\begin{aligned}
z_{m+\frac{1}{2}} &= MG_F(j-1, \rho_{j-1}, 0, m), \\
z_{m+1} &= z_m + MG_V(j-1, \rho_{j-1}, z_{m+\frac{1}{2}}, m).
\end{aligned}
\qquad (8.18)
$$

8.4 Convergence Analysis

Throughout this section we assume that the index of elliptic regularity α and the order k of the Lagrange finite element space satisfy the relation

$$\alpha \leq k - 1. \qquad (8.19)$$

Furthermore we assume that $\alpha \notin \{\frac{1}{4}, \frac{1}{2}, \frac{3}{4}\}$ (cf. Lemma 7).

From (8.4) and (8.6) we see that

$$z - z_{\text{new}} = z - \left[z_{\text{old}} + \lambda_j (A_j z - A_j z_{\text{old}})\right] = (Id_j - \lambda_j S_j^{-1} A_j)(z - z_{\text{old}}), \quad (8.20)$$

where Id_j is the identity operator on V_j. Therefore the error reduction operator for one smoothing step is given by

$$R_j = Id_j - \lambda_j S_j^{-1} A_j. \tag{8.21}$$

We define $P_j^{j-1} : V_j \longrightarrow V_{j-1}$ ($j \geq 1$) to be the transpose of I_{j-1}^j with respect to the bilinear forms for the C^0 interior penalty methods, i.e.,

$$a_{j-1}(P_j^{j-1} w, v) = a_j(w, I_{j-1}^j v) \qquad \forall v \in V_{j-1}, \ w \in V_j. \tag{8.22}$$

Remark 10. In the case of a conforming method with nested finite element spaces, the operator P_j^{j-1} is just the restriction to V_j of the Ritz projection operator that projects $H_0^2(\Omega)$ onto V_{j-1}.

Lemma 4. *The following relation holds:*

$$P_j^{j-1} = A_{j-1}^{-1} I_j^{j-1} A_j. \tag{8.23}$$

Proof. Using (8.2), (8.5) and (8.22) we find

$$\langle A_{j-1} P_j^{j-1} w, v \rangle = a_{j-1}(P_j^{j-1} w, v)$$
$$= a_j(w, I_{j-1}^j v) = \langle A_j w, I_{j-1}^j v \rangle = \langle I_j^{j-1} A_j w, v \rangle$$

for any $w \in V_j$ and $v \in V_{j-1}$, which implies (8.23).   □

It follows from (8.16) and Lemma 4 that the solution of the coarse grid residual equation is given by

$$e_{j-1} = A_{j-1}^{-1} I_j^{j-1} A_j (z - z_m) = P_j^{j-1}(z - z_m). \tag{8.24}$$

8.4.1 Recurrence Relations for Error Propagation Operators

Let $\mathbb{E}_j^{\text{v}} : V_j \longrightarrow V_j$ be the error reduction operator for the V-cycle algorithm applied to (8.4), i.e.

$$\mathbb{E}_j^{\text{v}}(z - z_0) = z - MG_V(j, \psi, z_0, m). \tag{8.25}$$

From (8.12), (8.14) and (8.20), we have

$$z - z_m = R_j^m(z - z_0), \tag{8.26}$$

$$z - z_{2m+1} = R_j^m(z - z_{m+1}). \tag{8.27}$$

We can connect $z - z_{m+1}$ to $z - z_m$ through the coarse grid correction step. Since $P_j^{j-1}(z - z_m)$ is the exact solution of (8.16) (cf. (8.24)), it follows from (8.13) and (8.25) (applied to the coarse grid equation (8.16)) that

$$z - z_{m+1} = (z - z_m) - I_{j-1}^j MG_V(j-1, \rho_{j-1}, 0, m)$$

$$= (z - z_m) - I_{j-1}^j \left[P_j^{j-1}(z - z_m) - \mathbb{E}_{j-1}^V P_j^{j-1}(z - z_m) \right] \tag{8.28}$$

$$= \left[(Id_j - I_{j-1}^j P_j^{j-1}) + I_{j-1}^j \mathbb{E}_{j-1}^V P_j^{j-1} \right](z - z_m).$$

Combining (8.25)–(8.28), we find the recurrence relation for the V-cycle error propagation operator:

$$\mathbb{E}_j^V = R_j^m \left[(Id_j - I_{j-1}^j P_j^{j-1}) + I_{j-1}^j \mathbb{E}_{j-1}^V P_j^{j-1} \right] R_j^m. \tag{8.29}$$

Similarly we have the following recurrence relations for the error propagation operators $\mathbb{E}_j^W$ and $\mathbb{E}_j^F$ for the W-cycle and F-cycle algorithms:

$$\mathbb{E}_j^W = R_j^m \left[(Id_j - I_{j-1}^j P_j^{j-1}) + I_{j-1}^j (\mathbb{E}_{j-1}^W)^2 P_j^{j-1} \right] R_j^m, \tag{8.30}$$

$$\mathbb{E}_j^F = R_j^m \left[(Id_j - I_{j-1}^j P_j^{j-1}) + I_{j-1}^j \mathbb{E}_{j-1}^V \mathbb{E}_{j-1}^F P_j^{j-1} \right] R_j^m. \tag{8.31}$$

Remark 11. Note that $\mathbb{E}_j^V z = z - MG_V(j, \psi, 0, m)$ by (8.25). Since $MG_V(j, \psi, 0, m)$ is linear in $\psi = A_j z$, we can write $MG_V(j, \psi, 0, m) = B_j A_j z$ where $B_j : V_j' \longrightarrow V_j$ is linear, and hence we have

$$\mathbb{E}_j^V = Id_j - B_j A_j. \tag{8.32}$$

Using (8.4), (8.32) we can rewrite (8.25) as

$$MG_V(j, \psi, z_0, m) = z - \mathbb{E}_j^V(z - z_0)$$

$$= z_0 + (Id_j - \mathbb{E}_j^V)(z - z_0) = z_0 + B_j(\psi - A_j z_0). \tag{8.33}$$

Therefore the V-cycle algorithm and its error propagation operator can be expressed alternatively by (8.33) and (8.32). Similar alternative expressions for the W-cycle and F-cycle algorithms can also be derived [17, 21].

It is clear from (8.29), (8.30) and (8.31) that we need to understand the effects of R_j^m (smoothing property) and $Id_j - I_{j-1}^j P_j^{j-1}$ (approximation property). We will measure these effects by certain mesh-dependent norms.

8.4.2 Mesh-Dependent Norms and Smoothing Properties

Observe that the operator $S_j^{-1} A_j : V_j \longrightarrow V_j$ is symmetric positive definite with respect to the inner product $\langle S_j \cdot, \cdot \rangle$. Therefore for any $t \in \mathbb{R}$ we can define the mesh-dependent norm

$$\|\|v\|\|_{t,j} = \left[\langle S_j (S_j^{-1} A_j)^t v, v \rangle \right]^{\frac{1}{2}} \qquad \forall v \in V_j. \tag{8.34}$$

It is clear that

$$\|\|v\|\|_{1,j} = \langle A_j v, v \rangle^{\frac{1}{2}} = \sqrt{a_j(v,v)} \qquad \forall v \in V_j. \tag{8.35}$$

For the standard smoother defined in Sect. 8.2.1 we have, by (8.8),

$$C_1 \|v\|_{L_2(\Omega)}^2 \le \|\|v\|\|_{0,j}^2 = \langle S_j v, v \rangle \le C_2 \|v\|_{L_2(\Omega)}^2 \qquad \forall v \in V_j; \tag{8.36}$$

while for the nonstandard smoother defined in Sect. 8.2.2 we have, by (8.10),

$$C_1 \|v\|_{H^1(\Omega)}^2 \le \|\|v\|\|_{0,j}^2 = \langle S_j v, v \rangle \le C_2 \|v\|_{H^1(\Omega)}^2 \qquad \forall v \in V_j. \tag{8.37}$$

The following *generalized Cauchy-Schwarz inequality* can be easily derived from (8.2), (8.35) and the standard Cauchy-Schwarz inequality:

$$\|\|v\|\|_{1+s,j} = \max_{w \in V_j \setminus \{0\}} \frac{a_h(v,w)}{\|\|w\|\|_{1-s,j}}. \tag{8.38}$$

The effect of the smoothing steps is given by the next lemma.

Lemma 5. *The following smoothing property holds*:

$$\|\|R_j^m v\|\|_{s,j} \le C h_j^{\gamma(t-s)} m^{(t-s)/2} \|\|v\|\|_{t,j} \qquad \forall v \in V_j \quad and \quad 0 \le t \le s \le 2, \tag{8.39}$$

where $\gamma = 2$ for the standard smoother defined in Sect. 8.2.1 and $\gamma = 1$ for the nonstandard smoother defined in Sect. 8.2.2.

Proof. Let r_j be the spectral radius of $S_j^{-1} A_j$. Then $\lambda_j r_j \le 1$ by (8.7) and it follows from (8.9), (8.11) and the spectral theorem that

$$\|\|R_j^m v\|\|_{s,j}^2 = \langle S_j (S_j^{-1} A_j)^s (Id_j - \lambda_j S_j^{-1} A_j)^m v, (Id_j - \lambda_j S_j^{-1} A_j)^m v \rangle$$

$$\le \lambda_j^{t-s} \langle S_j (S_j^{-1} A_j)^t (\lambda_j S_j^{-1} A_j)^{(s-t)} (Id_j - \lambda_j S_j^{-1} A_j)^m v,$$

$$(Id_j - \lambda_j S_j^{-1} A_j)^m v \rangle$$

$$\leq Ch_j^{2\gamma(t-s)}\Big[\max_{0\leq x\leq 1} x^{(s-t)}(1-x)^{2m}\Big]\langle S_j(S_j^{-1}A_j)^t v, v\rangle$$

$$\leq Ch_j^{2\gamma(t-s)} m^{(t-s)} \|v\|_{t,j}^2. \qquad \square$$

In the case where $s = t$, the constant C in (8.39) can be taken to be 1, i.e., we have

$$\|R_j v\|_{t,j} \leq \|v\|_{t,j} \qquad \forall v \in V_j \quad \text{and} \quad t \in \mathbb{R}. \tag{8.40}$$

8.4.3 Relations Between Mesh-Dependent Norms and Sobolev Norms

Let $\tilde{E}_j : V_j \longrightarrow H_0^2(\Omega)$ be the enriching operator introduced in Sect. 4.2. From Lemma 2 and standard inverse estimates we have

$$\|\tilde{E}_j v\|_{L_2(\Omega)}^2 \leq C\Big(\sum_{T\in\mathcal{T}_h}\|\tilde{E}_j v - v\|_{L_2(T)}^2 + \|v\|_{L_2(T)}^2\Big) \leq C\|v\|_{L_2(\Omega)}^2 \quad \forall v \in V_j,$$

$$\tag{8.41}$$

and similarly

$$\|\tilde{E}_j v\|_{H^1(\Omega)} \leq C |v|_{H^1(\Omega)} \qquad \forall v \in V_j. \tag{8.42}$$

It follows from (3.25), (4.20), (8.35), (8.36), (8.41), and interpolation between Hilbert scales [17, 65, 88, 90] that

$$\|\tilde{E}_j v\|_{H^{2s}(\Omega)} \leq C \|v\|_{s,j} \qquad \forall v \in V_j, \ 0 \leq s \leq 1 \tag{8.43}$$

for the mesh-dependent norms associated with the standard smoother in Sect. 8.2.1. Similarly, it follows from (3.25), (4.20), (8.35), (8.37) and (8.42) that

$$\|\tilde{E}_j v\|_{H^{1+s}(\Omega)} \leq C \|v\|_{s,j} \qquad \forall v \in V_j, \ 0 \leq s \leq 1 \tag{8.44}$$

for the mesh-dependent norms associated with the nonstandard smoother in Sect. 8.2.2.

We can connect the scale of Sobolev spaces to the finite element space V_j by an operator $J_j : L_2(\Omega) \longrightarrow V_j$ defined by

$$J_j v = \Pi_j(Q_j v) \qquad \forall v \in L_2(\Omega), \tag{8.45}$$

where $\Pi_j : C(\bar{\Omega}) \longrightarrow V_j$ is the nodal interpolation operator for the Lagrange finite element space V_j, and Q_j is the orthogonal projection from $L_2(\Omega)$ onto $\tilde{W}_j$, the C^1 finite element space that appeared in the construction of $\tilde{E}_j$ (cf. Sect. 4.2).

Since the dofs of V_h are also dofs of $\tilde{W}_h$, we can obtain the following estimates by direct element by element calculations:

$$\|\Pi_j \tilde{w}\|_{L_2(\Omega)} \leq C \|\tilde{w}\|_{L_2(\Omega)} \qquad \forall \tilde{w} \in \tilde{W}_h, \tag{8.46}$$

$$|\Pi_j \tilde{w}|_{H^1(\Omega)} \leq C |\tilde{w}|_{H^1(\Omega)} \qquad \forall \tilde{w} \in \tilde{W}_h. \tag{8.47}$$

Lemma 6. *We have*

$$J_j(\tilde{E}_j v) = v \qquad\qquad \forall\, v \in V_j, \tag{8.48}$$

$$\|J_j \zeta\|_h \le C |\zeta|_{H^2(\Omega)} \qquad \forall\, \zeta \in H_0^2(\Omega), \tag{8.49}$$

$$|J_j \zeta|_{H^1(\Omega)} \le C |\zeta|_{H^1(\Omega)} \qquad \forall\, \zeta \in H_0^1(\Omega), \tag{8.50}$$

$$\|J_j \zeta\|_{L_2(\Omega)} \le C \|\zeta\|_{L_2(\Omega)} \qquad \forall\, \zeta \in L_2(\Omega). \tag{8.51}$$

Proof. The relation (8.48) follows immediately from (4.22) and (8.45), and the bound (8.51) is a direct consequence of (8.45) and (8.46).

The estimates (8.49) and (8.50) follow from (4.21), (8.45), (8.47) and two well-known bounds [19] for Q_j :

$$\|Q_j \zeta\|_{H^2(\Omega)} \le C \|\zeta\|_{H^2(\Omega)} \qquad \forall\, \zeta \in H_0^2(\Omega),$$

$$\|Q_j \zeta\|_{H^1(\Omega)} \le C \|\zeta\|_{H^1(\Omega)} \qquad \forall\, \zeta \in H_0^1(\Omega). \qquad \square$$

The following lemma establishes the links between mesh-dependent norms and Sobolev norms.

Lemma 7. *We have*

$$C_1 \||v\||_{s,j} \le \|\tilde{E}_j v\|_{H^{2s}(\Omega)} \le C_2 \||v\||_{s,j} \qquad \forall\, v \in V_j,\ s \in [0,1] \setminus \left\{ \frac{1}{4}, \frac{3}{4} \right\} \tag{8.52}$$

for the mesh-dependent norms associated with the standard smoother in Sect. 8.2.1, and

$$C_1 \||v\||_{s,j} \le \|\tilde{E}_j v\|_{H^{1+s}(\Omega)} \le C_2 \||v\||_{s,j} \qquad \forall\, v \in V_j,\ s \in [0,1] \setminus \left\{ \frac{1}{2} \right\} \tag{8.53}$$

for the mesh-dependent norms associated with the nonstandard smoother in Sect. 8.2.2.

Proof. For the mesh-dependent norms associated with the standard smoother, it follows from (8.49), (8.51) and interpolation theory for Sobolev spaces [88, 90] that

$$\||J_j \zeta\||_{s,j} \le C \|\zeta\|_{H^{2s}(\Omega)} \qquad \forall\, \zeta \in H_0^{2s}(\Omega) \text{ and } s \in [0,1] \setminus \left\{ \frac{1}{4}, \frac{3}{4} \right\},$$

and hence, in view of (8.48),

$$\||v\||_{s,j} = \||J_j \tilde{E}_j v\||_{s,j} \le C \|\tilde{E}_j v\|_{H^{2s}(\Omega)}. \tag{8.54}$$

The estimate (8.52) follows from (8.43) and (8.54).
Similarly, we can prove (8.53) using (8.44) and (8.48)–(8.50). $\square$

Remark 12. The results of Lemma 7 are also valid for the exceptional values provided the corresponding Sobolev space $H^t(\Omega)$ is replaced by the space $\tilde{H}^t(\Omega)$ (or $H_{00}^t(\Omega)$) (cf. [88, 90]).

8.4.4 Approximation Properties

The effects of the coarse grid correction is measured by appropriate norms of the operator $Id_j - I_{j-1}^j P_j^{j-1}$. We begin with a technical lemma which is an analog of Theorem 5 and Theorem 6.

Lemma 8. *Let $\phi \in H^{-2+\alpha}(\Omega)$, $\zeta \in H_0^2(\Omega)$ satisfy*

$$a(\zeta, v) = \langle \phi, v \rangle \qquad \forall v \in H_0^2(\Omega),$$

and $\zeta_{j-1}^\dagger \in V_{j-1}$ satisfy

$$a_{j-1}(\zeta_{j-1}^\dagger, v) = \langle \phi, \tilde{E}_j v \rangle \qquad \forall v \in V_{j-1} \subset V_j.$$

Then we have

$$\|\zeta - \tilde{E}_j \zeta_{j-1}^\dagger\|_{H^{2-\alpha}(\Omega)} \le C h^{2\alpha} \|\phi\|_{H^{-2+\alpha}(\Omega)}. \tag{8.55}$$

Proof. We will focus on the case where $\alpha \in (\frac{1}{2}, 1)$. We have an analog of (6.8):

$$\|\zeta - \zeta_{j-1}^\dagger\|_{h_{j-1}} \le \|\zeta - \Pi_{j-1}^\sharp \zeta\|_{h_{j-1}} + C \max_{w \in V_{j-1} \setminus \{0\}} \frac{a_{j-1}(\Pi_{j-1}^\sharp \zeta - \zeta_{j-1}^\dagger, w)}{\|w\|_{h_{j-1}}} \tag{8.56}$$

where $\Pi_{j-1}^\sharp$ is the nodal interpolation operator for the *quadratic* Lagrange finite element space associated with $\mathcal{T}_{j-1}$, and also an analog of (6.9):

$$a_{j-1}(\Pi_{j-1}^\sharp \zeta - \zeta_{j-1}^\dagger, w) = \sum_{T \in \mathcal{T}_{j-1}} \int_T \nabla^2(\Pi_{j-1}^\sharp \zeta - \zeta) : \nabla^2(\tilde{E}_j w) \, dx$$

$$+ \sum_{e \in \mathcal{E}_{j-1}} \int_e \left\{\!\!\left\{ \frac{\partial^2 w}{\partial n_e^2} \right\}\!\!\right\} \left[\!\!\left[\frac{\partial \Pi_{j-1}^\sharp \zeta}{\partial n_e} \right]\!\!\right] ds$$

$$+ \sigma \sum_{e \in \mathcal{E}_{j-1}} \frac{1}{|e|} \int_e \left[\!\!\left[\frac{\partial \Pi_{j-1}^\sharp \zeta}{\partial n_e} \right]\!\!\right] \left[\!\!\left[\frac{\partial(w - \tilde{E}_j w)}{\partial n_e} \right]\!\!\right] ds$$

$$\tag{8.57}$$

$$- \sum_{e \in \mathcal{E}_{j-1}^i} \int_e \left[\!\!\left[\frac{\partial^2 \Pi_{j-1}^\sharp \zeta}{\partial n_e \partial t_e} \right]\!\!\right] \frac{\partial(w - \tilde{E}_j w)}{\partial t_e} \, ds$$

$$- \sum_{e \in \mathscr{E}^i_{j-1}} \int_e \left[\!\!\left[\frac{\partial^2 \Pi^\sharp_{j-1} \zeta}{\partial n_e^2} \right]\!\!\right] \left\{\!\!\left\{ \frac{\partial (w - \tilde{E}_j w)}{\partial n_e} \right\}\!\!\right\} ds.$$

It follows from (3.30), (4.17), (6.6), (8.56) and (8.57) that we have the following analog of (6.7):

$$\|\zeta - \zeta^\dagger_{j-1}\|_{h_{j-1}} \le C h^\alpha_{j-1} \|\phi\|_{H^{-2+\alpha}(\Omega)}. \tag{8.58}$$

Now we turn to the error estimate in the lower order Sobolev norm. By duality, we can write

$$\|\zeta - \tilde{E}_j \zeta^\dagger_{j-1}\|_{H^{2-\alpha}(\Omega)} = \sup_{\psi \in H^{-2+\alpha}(\Omega)\setminus\{0\}} \frac{\langle \psi, \zeta - \tilde{E}_j \zeta^\dagger_{j-1}\rangle}{\|\psi\|_{H^{-2+\alpha}(\Omega)}}. \tag{8.59}$$

Let $\xi \in H^2_0(\Omega)$ satisfy

$$a(\xi, v) = \langle \psi, v\rangle \qquad \forall\, v \in H^2_0(\Omega),$$

and $\xi^*_j \in V_j$ satisfy

$$a_j(\xi^*_j, v) = \langle \psi, \tilde{E}_j v\rangle \qquad \forall\, v \in V_j.$$

Then we have, by Theorem 4,

$$\|\xi - \xi^*_j\|_{h_j} \le C h^\alpha \|\psi\|_{H^{-2+\alpha}(\Omega)}. \tag{8.60}$$

We also have an analog of (6.18):

$$\langle \psi, \zeta - \tilde{E}_j \zeta^\dagger_{j-1}\rangle$$

$$= \sum_{T \in \mathscr{T}_j} \int_T \nabla^2(\xi - \Pi^\sharp_j \xi) : \nabla^2(\zeta - \zeta^\dagger_{j-1})\, dx$$

$$+ \sum_{T \in \mathscr{T}_j} \int_T \nabla^2(\xi - \xi^*_j) : \nabla^2(\zeta^\dagger_{j-1} - \Pi^\sharp_j \zeta)\, dx$$

$$+ \sum_{e \in \mathscr{E}_j} \int_e \left\{\!\!\left\{ \frac{\partial^2 (\Pi^\sharp_j \xi - \xi^*_j)}{\partial n_e^2} \right\}\!\!\right\} \left[\!\!\left[\frac{\partial \zeta^\dagger_{j-1}}{\partial n_e} \right]\!\!\right] ds$$

$$- \sum_{e \in \mathscr{E}^i_j} \int_e \left[\!\!\left[\frac{\partial^2 \Pi^\sharp_j \xi}{\partial n_e^2} \right]\!\!\right] \left\{\!\!\left\{ \frac{\partial(\zeta - \zeta^\dagger_{j-1})}{\partial n_e} \right\}\!\!\right\} ds - \sum_{e \in \mathscr{E}_j} \int_e \left[\!\!\left[\frac{\partial^2 \Pi^\sharp_j \xi}{\partial n_e \partial t_e} \right]\!\!\right] \frac{\partial(\zeta - \zeta^\dagger_{j-1})}{\partial t_e}\, ds$$

$$+\sum_{e\in\mathscr{E}_j}\int_e\left\{\!\!\left\{\frac{\partial^2(\Pi_j^\sharp\zeta-\zeta_{j-1}^\dagger)}{\partial n_e^2}\right\}\!\!\right\}\left[\!\!\left[\frac{\partial\xi_j^*}{\partial n_e}\right]\!\!\right]ds-\sum_{e\in\mathscr{E}_j^i}\int_e\left[\!\!\left[\frac{\partial^2\Pi_j^\sharp\zeta}{\partial n_e^2}\right]\!\!\right]\left\{\!\!\left\{\frac{\partial(\xi-\xi_j^*)}{\partial n_e}\right\}\!\!\right\}ds$$

$$-\sum_{e\in\mathscr{E}_j^i}\int_e\left[\!\!\left[\frac{\partial^2\Pi_j^\sharp\zeta}{\partial n_e\partial t_e}\right]\!\!\right]\frac{\partial(\xi-\xi_j^*)}{\partial t_e}ds-\sigma\sum_{e\in\mathscr{E}_j}\frac{1}{|e|}\int_e\left[\!\!\left[\frac{\partial\xi_j^*}{\partial n_e}\right]\!\!\right]\left[\!\!\left[\frac{\partial\zeta_{j-1}^\dagger}{\partial n_e}\right]\!\!\right]ds,$$

where $\Pi_j^\sharp$ is the nodal interpolation operator for the quadratic Lagrange finite element space associated with $\mathscr{T}_j$. It then follows from (3.30), (6.10), (8.58) and (8.60) that

$$\langle\psi,\zeta-\tilde{E}_j\zeta_{j-1}^\dagger\rangle\le Ch^{2\alpha}\|\phi\|_{H^{-2+\alpha}(\Omega)}\|\psi\|_{H^{-2+\alpha}(\Omega)},$$

which together with (8.59) implies (8.55).

The case where $\alpha\in[1,2]$ can be handled in a similar fashion by using the nodal interpolation operator for the *cubic* Lagrange finite element. $\qquad\square$

We can now establish the approximation properties.

Lemma 9. *We have*

$$\|(Id_j-I_{j-1}^jP_j^{j-1})v\|_{1-\alpha,j}\le Ch_j^{2\alpha}\|v\|_{1+\alpha,j}\qquad\forall v\in V_j\qquad(8.61)$$

for the mesh-dependent norm associated with the nonstandard smoother in Sect. 8.2.2.

Proof. Let $v_j\in V_j$ be arbitrary. We will prove (8.61) by a duality argument. From (8.53) and duality, we have

$$\|(Id_j-I_{j-1}^jP_j^{j-1})v\|_{1-\alpha,j}\le C\|\tilde{E}_j(Id_j-I_{j-1}^jP_j^{j-1})v\|_{H^{2-\alpha}(\Omega)}$$

$$=C\sup_{\phi\in H^{-2+\alpha}(\Omega)\backslash\{0\}}\frac{\langle\phi,\tilde{E}_j(Id_j-I_{j-1}^jP_j^{j-1})v\rangle}{\|\phi\|_{H^{-2+\alpha}(\Omega)}}.\qquad(8.62)$$

Let $\phi\in H^{-2+\alpha}(\Omega)$ be arbitrary, $\zeta\in H_0^2(\Omega)$ satisfy

$$a(\zeta,v)=\langle\phi,v\rangle\qquad\forall v\in H_0^2(\Omega),\qquad(8.63)$$

and $\zeta_j^*\in V_j$ satisfy

$$a_j(\zeta_j^*,v)=\langle\phi,\tilde{E}_jv\rangle\qquad\forall v\in V_j.\qquad(8.64)$$

It follows from Theorem 6 and (8.19) that

$$\|\zeta - \tilde{E}_j \zeta_j^*\|_{H^{2-\alpha}(\Omega)} \leq Ch^{2\alpha}\|\phi\|_{H^{-2+\alpha}(\Omega)}. \tag{8.65}$$

Using (8.5) and (8.64) we can write

$$
\begin{aligned}
\langle \phi, \tilde{E}_j (Id_j - I_{j-1}^j P_j^{j-1})v \rangle &= a_j \big(\zeta_j^*, (Id_j - I_{j-1}^j P_j^{j-1})v \big) \\
&= a_j(\zeta_j^*, v) - a_{j-1}(P_j^{j-1}\zeta_j^*, P_j^{j-1}v) \\
&= a_j(\zeta_j^* - I_{j-1}^j \zeta_{j-1}^\dagger, v),
\end{aligned}
\tag{8.66}
$$

where $\zeta_{j-1}^\dagger = P_j^{j-1}\zeta_j^*$ satisfies

$$a_{j-1}(\zeta_{j-1}^\dagger, w) = a_j(\zeta_j^*, I_{j-1}^j w) = \langle \phi, \tilde{E}_j w \rangle \qquad \forall\, w \in V_{j-1}. \tag{8.67}$$

By Lemma 8, we have

$$\|\zeta - \tilde{E}_j \zeta_{j-1}^\dagger\|_{H^{2-\alpha}(\Omega)} \leq Ch^{2\alpha}\|\phi\|_{H^{-2+\alpha}(\Omega)}. \tag{8.68}$$

It then follows from (8.38), (8.53), (8.65) and (8.68) that

$$
\begin{aligned}
|a_j(\zeta_j^* - I_{j-1}^j \zeta_{j-1}^\dagger, v)| &\leq \|\zeta_j^* - I_{j-1}^j \zeta_{j-1}^\dagger\|_{1-\alpha,j}\|v\|_{1+\alpha,j} \\
&\leq C\|\tilde{E}_j(\zeta_j^* - I_{j-1}^j \zeta_{j-1}^\dagger)\|_{H^{2-\alpha}(\Omega)}\|v\|_{1+\alpha,j} \\
&\leq C\big(\|\tilde{E}_j \zeta_j^* - \zeta\|_{H^{2-\alpha}(\Omega)} + \|\zeta - \tilde{E}_j I_{j-1}^j \zeta_{j-1}^\dagger\|_{H^{2-\alpha}(\Omega)}\big)\|v\|_{1+\alpha,j} \\
&\leq Ch^{2\alpha}\|\phi\|_{H^{-2+\alpha}(\Omega)}\|v\|_{1+\alpha,j},
\end{aligned}
$$

which together with (8.62) and (8.66) implies (8.61).   □

The following result for the standard smoother can be established by similar arguments.

Lemma 10. *We have*

$$\|(Id_j - I_{j-1}^j P_j^{j-1})v\|_{1-\frac{\alpha}{2},j} \leq Ch_j^{2\alpha}\|v\|_{1+\frac{\alpha}{2},j} \qquad \forall\, v \in V_j \tag{8.69}$$

for the mesh-dependent norm associated with the standard smoother in Sect. 8.2.1.

8.4.5 Analysis of the Two-Grid Algorithm

In the two-grid algorithm the coarse grid residual equation (8.16) is solved exactly. Therefore (cf. (8.29)) the error propagation operator $\mathbb{E}_j^{\mathrm{TG}}$ of the two-grid algorithm is given by

$$\mathbb{E}_j^{\mathrm{TG}} = R_j^m (Id_j - I_{j-1}^j P_j^{j-1}) R_j^m. \tag{8.70}$$

Lemma 11. *We have*

$$\||\mathbb{E}_j^{\mathrm{TG}} v\||_{1,j} \leq C_{\mathrm{TG}} m^{-\alpha} \||v\||_{1,j} \qquad \forall\, v \in V_j \qquad\qquad (8.71)$$

for the mesh-dependent norm associated with the nonstandard smoother, and

$$\||\mathbb{E}_j^{\mathrm{TG}} v\||_{1,j} \leq C_{\mathrm{TG}} m^{-(\alpha/2)} \||v\||_{1,j} \qquad \forall\, v \in V_j \qquad\qquad (8.72)$$

for the mesh-dependent norm associated with the standard smoother.

Proof. Let $v \in V_j$ be arbitrary. It follows from (8.39) (with $\gamma = 1$), (8.61) and (8.70) that

$$\begin{aligned}
\||\mathbb{E}_j^{\mathrm{TG}} v\||_{1,j} &= \||R_j^m (Id_j - I_{j-1}^j P_j^{j-1}) R_j^m v\||_{j,1} \\
&\leq C h_j^{-\alpha} m^{-(\alpha/2)} \||(Id_j - I_{j-1}^j P_j^{j-1}) R_j^m v\||_{1-\alpha,1} \\
&\leq C h_j^\alpha m^{-(\alpha/2)} \||R_j^m v\||_{1+\alpha,j} \\
&\leq C m^{-\alpha} \||v\||_{1,j},
\end{aligned}$$

which yields (8.71).

The estimate (8.72) can be established by similar arguments using (8.39) (with $\gamma = 2$) and (8.69). $\square$

Therefore the two-grid algorithm is a contraction for m sufficiently large (but independent of j). Moreover, the algorithm based on the nonstandard smoother that takes advantage of a multigrid Poisson solve in the smoothing steps is more effective than the standard smoother. This is another important advantage of C^0 interior penalty methods. Note that all other existing multigrid algorithms for fourth order problems [20, 22, 80, 85, 98, 100, 101] share the less effective estimate (8.72).

8.4.6 Analysis of the W-Cycle Algorithm

We can establish the convergence property of the W-cycle algorithm using Lemma 11 and a perturbation argument. First we note that, by (3.22), (3.25), (8.35) and a direct calculation,

$$\||I_{j-1}^j v\||_{1,j} \leq C_{\mathrm{CF}} \||v\||_{1,j-1} \qquad \forall\, v \in V_{j-1}, \qquad\qquad (8.73)$$

and hence, by (8.22),

$$\||P_j^{j-1} v\||_{1,j-1} \leq C_{\mathrm{CF}} \||v\||_{1,j} \qquad \forall\, v \in V_j. \qquad\qquad (8.74)$$

Theorem 9. *Given any* $C_* > C_{\text{TG}}$, *there exists a positive integer* m_* *independent of* j *such that*

$$\|\mathbb{E}_j^{\text{W}} v\|_{1,j} \leq \frac{C_*}{m^{\alpha/\gamma}} \|v\|_{1,j} \qquad \forall v \in V_j \tag{8.75}$$

provided $m \geq m_*$, *where* $\gamma = 1$ *for the nonstandard smoother and* $\gamma = 2$ *for the standard smoother.*

Proof. We will prove (8.75) by mathematical induction. The case $j = 0$ holds for any m_* since $\mathbb{E}_0^{\text{W}} = 0$. Assume that $j \geq 1$ and (8.75) is valid for $j - 1$. Let $v \in V_j$ be arbitrary. From (8.30), (8.40), (8.70), Lemma 11, (8.73), (8.74) and the induction hypothesis, we have

$$\|\mathbb{E}_j^{\text{W}} v\|_{1,j} \leq \left(C_{\text{TG}} m^{-(\alpha/\gamma)} + C_*^2 C_{\text{CF}}^2 m^{-2(\alpha/\gamma)} \right) \|v\|_{1,j}.$$

If we choose $m_* > 0$ so that

$$m_*^{-(\alpha/\gamma)} \leq \frac{C_* - C_{\text{TG}}}{C_{\text{CF}}^2 C_*^2},$$

then for $m \geq m_*$ we have

$$C_{\text{TG}} m^{-(\alpha/\gamma)} + C_*^2 C_{\text{CF}}^2 m^{-2(\alpha/\gamma)} \leq \left(C_{\text{TG}} + C_*^2 C_{\text{CF}}^2 m_*^{-(\alpha/\gamma)} \right) m^{-(\alpha/\gamma)} \leq \frac{C_*}{m^{\alpha/\gamma}}$$

and hence (8.75) also holds for j. $\qquad\qquad\qquad\qquad\qquad\qquad\qquad\qquad\square$

Remark 13. It follows from (8.75) that the W-cycle algorithm is a contraction with a contraction number independent of grid levels, provided that m is sufficiently large (but independent of the grid levels). Thus the W-cycle algorithm is *uniformly convergent.*

8.4.7   Results for the V-Cycle and F-Cycle Algorithms

The convergence property in Theorem 9 can also be established for the V-cycle and F-cycle algorithms using the additive multigrid theory developed in [26,27]. Details can be found in [35].

Note that the F-cycle algorithm is very attractive for C^0 interior penalty methods (and other discontinuous Galerkin methods). It is more robust than the V-cycle algorithm in the sense that it takes fewer smoothing steps for the F-cycle algorithm to become uniformly convergent. It is less expensive than the W-cycle algorithm. But when both F-cycle and W-cycle algorithms are convergent, they have almost identical performance.

Remark 14. Multigrid convergence results for discontinuous Galerkin methods for second order problems can be found in [40].

9 Domain Decomposition Methods

Domain decomposition methods [32, 70, 86, 89] provide another efficient approach
to solving the discrete problems resulting from C^0 interior penalty methods. We
will focus on additive Schwarz domain decomposition preconditioners which have
a high level of built-in parallelism and can be used in a preconditioned conjugate
gradient algorithm that solves the discrete problem efficiently.

   Below we will discuss an abstract theory for the additive Schwarz preconditioners and consider an overlapping domain decomposition preconditioner [37] for
C^0 interior penalty methods. A Bramble-Pasciak-Schatz nonoverlapping domain
decomposition preconditioner for C^0 interior penalty methods is studied in [38].
Other domain decomposition algorithms for discontinuous Galerkin methods can
be found in [4, 5, 11, 48, 53, 54, 66].

9.1 An Abstract Theory of Additive Schwarz Preconditioners

Let V be a finite dimensional vector space and the linear operator $A : V \longrightarrow V'$ be
symmetric positive definite (SPD), i.e.,

$$\langle Av, w \rangle = \langle Aw, v \rangle \qquad \forall\, v, w \in V, \tag{9.1}$$

$$\langle Av, v \rangle > 0 \qquad \forall\, v \in V. \tag{9.2}$$

An abstract additive Schwarz preconditioner $B : V' \longrightarrow V$ for A is defined by
the formula

$$B = \sum_{j=0}^{J} I_j A_j^{-1} I_j^t, \tag{9.3}$$

where the linear operators $A_j : V_j \longrightarrow V_j'$ are SPD for $0 \le j \le J$, and the
auxiliary vector spaces $V_0, V_1, \ldots, V_J$ are connected to V by the linear operators
$I_j : V_j \longrightarrow V$ for $0 \le j \le J$.

   The algebraic theory of additive Schwarz preconditioners [14, 32, 49, 56, 70, 78,
86, 89, 96, 99] is given in the following theorem.

Theorem 10. *Under the condition that*

$$V = \sum_{j=0}^{J} I_j V_j, \tag{9.4}$$

*the operator B is SPD and the eigenvalues of BA are positive. Moreover, we have
the following characterizations of the maximum and minimum eigenvalues of BA :*

$$\lambda_{\max}(BA) = \max_{v \in V \setminus \{0\}} \frac{\langle Av, v \rangle}{\displaystyle \min_{\substack{v = \sum_{j=0}^{J} I_j v_j \\ v_j \in V_j}} \sum_{j=0}^{J} \langle A_j v_j, v_j \rangle}, \tag{9.5}$$

$$\lambda_{\min}(BA) = \min_{v \in V \setminus \{0\}} \frac{\langle Av, v \rangle}{\displaystyle \min_{\substack{v = \sum_{j=0}^{J} I_j v_j \\ v_j \in V_j}} \sum_{j=0}^{J} \langle A_j v_j, v_j \rangle}. \tag{9.6}$$

Proof. It is clear that B is symmetric and $\langle \gamma, B\gamma \rangle \geq 0 \quad \forall \gamma \in V'$. Suppose $\langle \gamma, B\gamma \rangle = 0$ for some $\gamma \in V'$. We have

$$0 = \left\langle \gamma, \sum_{j=0}^{J} I_j A_j^{-1} I_j^t \gamma \right\rangle = \sum_{j=0}^{J} \langle I_j^t \gamma, A_j^{-1} I_j^t \gamma \rangle,$$

which implies $I_j^t \gamma = 0$ for $0 \leq j \leq J$ because $A_j : V_j \longrightarrow V_j'$ is SPD. It then follows from (9.4) that, given any $v \in V$, we have

$$\langle \gamma, v \rangle = \left\langle \gamma, \sum_{j=0}^{J} I_j v_j \right\rangle = \sum_{j=0}^{J} \langle I_j^t \gamma, v_j \rangle = 0,$$

which implies $\gamma = 0$ and hence B is SPD.

Note that the operator $BA : V \longrightarrow V$ is SPD with respect to the inner product $\langle B^{-1} \cdot, \cdot \rangle$, and hence all the eigenvalues of BA are positive. Moreover we can characterize the maximum and minimum eigenvalues of BA by the following Raleigh quotient formulas (cf. [55]):

$$\lambda_{\max}(BA) = \max_{v \in V \setminus \{0\}} \frac{\langle Av, v \rangle}{\langle B^{-1} v, v \rangle} \quad \text{and} \quad \lambda_{\min}(BA) = \min_{v \in V \setminus \{0\}} \frac{\langle Av, v \rangle}{\langle B^{-1} v, v \rangle}.$$

We can therefore establish (9.5) and (9.6) by showing that

$$\langle B^{-1} v, v \rangle = \min_{\substack{v = \sum_{j=0}^{J} I_j v_j \\ v_j \in V_j}} \sum_{j=0}^{J} \langle A_j v_j, v_j \rangle \quad \forall v \in V. \tag{9.7}$$

Indeed we have, by the Cauchy-Schwarz inequality for the Euclidean inner product and for the inner product $\langle A \cdot, \cdot \rangle$,

$$\langle B^{-1}v, v \rangle = \langle B^{-1}v, \sum_{j=0}^{J} I_j v_j \rangle$$

$$= \sum_{j=0}^{J} \langle A_j v_j, A_j^{-1} I_j^t B^{-1}v \rangle$$

$$\leq \sum_{j=0}^{J} \langle A_j v_j, v_j \rangle^{\frac{1}{2}} \langle I_j^t B^{-1}v, A_j^{-1} I_j^t B^{-1}v \rangle^{\frac{1}{2}}$$

$$= \left(\sum_{j=0}^{J} \langle A_j v_j, v_j \rangle \right)^{\frac{1}{2}} \left(\sum_{j=0}^{J} \langle I_j^t B^{-1}v, A_j^{-1} I_j^t B^{-1}v \rangle \right)^{\frac{1}{2}}$$

$$= \left(\sum_{j=0}^{J} \langle A_j v_j, v_j \rangle \right)^{\frac{1}{2}} \left(\langle B^{-1}v, \sum_{j=0}^{J} I_j A_j^{-1} I_j^t B^{-1}v \rangle \right)^{\frac{1}{2}}$$

$$= \left(\sum_{j=0}^{J} \langle A_j v_j, v_j \rangle \right)^{\frac{1}{2}} \langle B^{-1}v, v \rangle^{\frac{1}{2}},$$

which implies

$$\langle B^{-1}v, v \rangle \leq \sum_{j=0}^{J} \langle A_j v_j, v_j \rangle \tag{9.8}$$

whenever $v = \sum_{j=0}^{J} I_j v_j$ and $v_j \in V_j$ for $0 \leq j \leq J$.
On the other hand, for the special decomposition

$$v = B B^{-1}v = \sum_{j=0}^{J} I_j (A_j^{-1} I_j^t B^{-1}v) = \sum_{j=0}^{J} I_j v_j,$$

where $v_j = A_j^{-1} I_j^t B^{-1}v \in V_j$, we have

$$\sum_{j=0}^{J} \langle A_j v_j, v_j \rangle = \sum_{j=0}^{J} \langle I_j^t B^{-1}v, A_j^{-1} I_j^t B_j^{-1}v \rangle$$

$$= \left\langle B^{-1}v, \sum_{j=0}^{J} I_j A_j^{-1} I_j^t B_j^{-1}v \right\rangle = \langle B^{-1}v, v \rangle. \tag{9.9}$$

The relation (9.7) follows from (9.8) and (9.9). $\qquad \square$

By combining the elegant formulas (9.5) and (9.6) with analytical tools that provide two-sided estimates for (9.7) in terms of $\langle Av, v \rangle$, we can obtain concrete condition number estimates for a specific additive Schwarz preconditioner.

9.2 A Two-Level Additive Schwarz Domain Decomposition Preconditioner

In this section we consider a two-level additive Schwarz preconditioner for the discrete problem resulting from a C^0 interior penalty method for the model problem (2.1). This overlapping domain decomposition preconditioner was first introduced in [49] for conforming finite element methods for second order problems.

9.2.1 Set-Up

Let Ω be divided into J nonoverlapping polygonal subdomains that are aligned with the triangulation $\mathcal{T}_h$. By extending these subdomains we obtain J overlapping subdomains $\Omega_1, \ldots, \Omega_J$ which are also aligned with $\mathcal{T}_h$. The overlap among the subdomains is measured by δ. Furthermore we assume there is a coarse triangulation $\mathcal{T}_H$ of Ω aligned with $\mathcal{T}_h$. (A typical example is given by Fig. 3, where $\mathcal{T}_h$ is represented by the middle figure, $\mathcal{T}_H$ is represented by the left figure, and one of the overlapping subdomain is depicted in the right figure.)

As a consequence of the geometric assumptions, we can construct a partition of unity $\theta_j \in C^\infty(\bar{\Omega})$ for $1 \leq j \leq J$ (cf. [82]) such that

$$0 \leq \theta_j \leq 1, \tag{9.10}$$

$$\theta_j = 0 \quad \text{on } \Omega \setminus \Omega_j, \tag{9.11}$$

$$\sum_{j=1}^{J} \theta_j = 1 \quad \text{on } \bar{\Omega}, \tag{9.12}$$

$$\|\nabla \theta_j\|_{L_\infty(\Omega)} \leq \frac{C}{\delta} \quad \text{and} \quad \|\nabla^2 \theta_j\|_{L_\infty(\Omega)} \leq \frac{C}{\delta^2}. \tag{9.13}$$

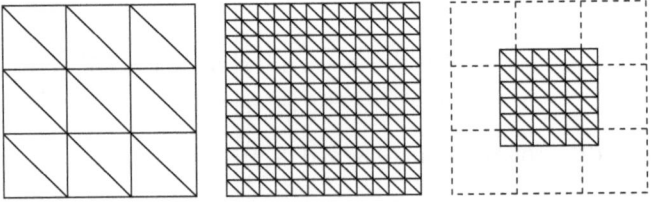

Fig. 3 An overlapping domain decomposition

We also assume that there exists a positive integer N_c such that

the closure of each subdomain can have nonempty intersection with

the closures of at most N_c many other subdomains. (9.14)

Recall V_h is the P_k ($k \geq 2$) Lagrange finite element space associated with $\mathscr{T}_h$. The operator $A_h : V_h \longrightarrow V_h'$ represents the bilinear form (3.17), i.e.,

$$\langle A_h w, v \rangle = a_h(w, v) \qquad \forall v, w \in V_h. \tag{9.15}$$

It is the operator to be preconditioned.

We will denote by $V_0 = V_H$ the P_2 Lagrange finite element space associated with $\mathscr{T}_H$. The operator $A_0 = A_H : V_H \longrightarrow V_H'$ represents the bilinear form $a_H(\cdot, \cdot)$ associated with the coarse triangulation $\mathscr{T}_H$.

The subdomain space V_j ($1 \leq j \leq J$) is defined by

$$V_j = \{v \in V_h : v = 0 \text{ on } \Omega \setminus \Omega_j\}. \tag{9.16}$$

The bilinear form $a_j(\cdot, \cdot)$ on V_j is the analog of $a_h(\cdot, \cdot)$ for the subdomain Ω_j, i.e.,

$$a_j(w, v) = \sum_{T \in \mathscr{T}_{h,j}} \int_T (\nabla^2 w : \nabla^2 v)\, dx + \sum_{e \in \mathscr{E}_{h,j}} \int_e \left\{\!\!\left\{ \frac{\partial^2 w}{\partial n_e^2} \right\}\!\!\right\} \left[\!\!\left[\frac{\partial v}{\partial n_e} \right]\!\!\right] ds$$

$$+ \sum_{e \in \mathscr{E}_{h,j}} \int_e \left\{\!\!\left\{ \frac{\partial^2 v}{\partial n_e^2} \right\}\!\!\right\} \left[\!\!\left[\frac{\partial w}{\partial n_e} \right]\!\!\right] ds + \sigma \sum_{e \in \mathscr{E}_{h,j}} \frac{1}{|e|} \int_e \left[\!\!\left[\frac{\partial w}{\partial n_e} \right]\!\!\right] \left[\!\!\left[\frac{\partial v}{\partial n_e} \right]\!\!\right] ds,$$

$$\tag{9.17}$$

where $\mathscr{T}_{h,j}$ is the set of the triangles in $\mathscr{T}_h$ that are subsets of Ω_j, $\mathscr{E}_{h,j}$ is the set of the edges of the triangles in $\mathscr{T}_{h,j}$, the jumps $[\![\cdot]\!]$ and averages are defined by (3.5) and (3.11) if $e \subset \Omega_j$, and by (3.7) and (3.12) if $e \subset \partial\Omega_j$.

The operator $A_j : V_j \longrightarrow V_j'$ ($1 \leq j \leq J$) is defined by

$$\langle A_j w, v \rangle = a_j(w, v) \qquad \forall v, w \in V_j. \tag{9.18}$$

The auxiliary spaces $V_1, \ldots, V_J$ are connected to V_h by the natural injections I_j for $1 \leq j \leq J$. For the coarse space V_H, we define $I_0 : V_H \longrightarrow V_h$ by

$$I_0 = \Pi_h \circ \tilde{E}_H, \tag{9.19}$$

where $\tilde{E}_H : V_H \longrightarrow H_0^2(\Omega)$ is the enriching operator introduced in Sect. 4.2 and Π_h is the nodal interpolation operator for V_h.

The two-level additive Schwarz preconditioner $B_{\mathrm{TL}} : V_h' \longrightarrow V_h$ is then defined by

$$B_{\mathrm{TL}} = \sum_{j=0}^{J} I_j A_j^{-1} I_j^t. \tag{9.20}$$

Remark 15. If we use the natural injection from V_H into V_h as I_0, then the different scalings that appear in the penalty terms of the C^0 interior penalty methods on $\mathscr{T}_h$ and $\mathscr{T}_H$ will adversely affect the performance of the preconditioner B_{TL}. Numerical evidence to this effect can be found in [37]. This problem is eliminated by the operator I_0 defined by (9.19). (See the estimate (9.23) below.)

Given any $v \in V_h$, we have, by (9.12),

$$v = \Pi_h v = \Pi_h \sum_{j=1}^{J} \theta_j v = \sum_{j=1}^{J} \Pi_h(\theta_j v).$$

Therefore the condition (9.4) is satisfied because $v_j = \Pi_h(\theta_j v) \in V_j$ by (9.11) and (9.16).

9.3   Estimate for $\lambda_{\max}(B_{\mathrm{TL}} A_h)$

Let $v \in V_h$ be arbitrary and $v = \sum_{j=0}^{J} I_j v_j$, where $v_j \in V_j$. It follows from the Cauchy-Schwarz inequality and the condition (9.14) that

$$\langle A_h v, v \rangle \leq 2\langle A_h I_0 v_0, I_0 v_0 \rangle + 2\left\langle A_h \sum_{j=1}^{J} v_j, \sum_{j=1}^{J} v_j \right\rangle$$

$$\leq 2\langle A_h I_0 v_0, I_0 v_0 \rangle + C \sum_{j=1}^{J} \langle A_h v_j, v_j \rangle. \tag{9.21}$$

In view of (3.17), (3.22), (3.25) (and the analog for $a_j(\cdot, \cdot)$), (9.15), (9.17) and (9.18), we have

$$\langle A_h v_j, v_j \rangle \leq C \langle A_j v_j, v_j \rangle \qquad \forall \, v_j \in V_j \quad \text{and} \quad 1 \leq j \leq J. \tag{9.22}$$

The following lemma provides the key properties of I_0.

Lemma 12. *We have*

$$\|v_H - I_0 v_H\|_{L_2(\Omega)} + H|v_H - I_0 v_H|_{H^1(\Omega)} + H^2\|I_0 v_H\|_h \leq CH^2\|v_H\|_H$$

for all $v_H \in V_H$, where $\|\cdot\|_H$ is the analog of $\|\cdot\|_h$ (cf. (3.22)) for $\mathscr{T}_H$.

Proof. It follows from (3.30), Lemma 2 and Corollary 2 that

$$\|v_H - I_0 v_H\|_{L_2(\Omega)} \leq \|v_H - \tilde{E}_H v_H\|_{L_2(\Omega)} + \|\Pi_h \tilde{E}_H v_H - \tilde{E}_H v_H\|_{L_2(\Omega)}$$

$$\leq C\big(H^2\|v_H\|_H + h^2|\tilde{E}_H v_H|_{H^2(\Omega)}\big)$$

$$\leq CH^2\|v_H\|_H,$$

and the estimate for $|v_H - I_0 v_H|_{H^1(\Omega)}$ can be established similarly. From (3.30) and Corollary 2 we have

$$\sum_{T \in \mathcal{T}_h} |I_0 v_H|^2_{H^2(T)} = \sum_{T \in \mathcal{T}_h} |\Pi_h \tilde{E}_H v_H|^2_{H^2(T)} \leq C|\tilde{E}_H v_H|^2_{H^2(\Omega)} \leq C\|v_H\|^2_H.$$

Finally, we obtain from (3.30), the trace theorem with scaling and Corollary 2

$$\sum_{e \in \mathcal{E}_h} \frac{1}{|e|} \| [\![\partial(I_0 v_H)/\partial n_e]\!] \|^2_{L_2(e)} = \sum_{e \in \mathcal{E}_h} \frac{1}{|e|} \| [\![\partial(\Pi_h \tilde{E}_H v_H - \tilde{E}_H v_H)/\partial n_e]\!] \|^2_{L_2(e)}$$

$$\leq C|\tilde{E}_H v_H|^2_{H^2(\Omega)} \leq C\|v_H\|^2_H \qquad (9.23)$$

and the estimate for $\|I_0 v_H\|_h$ follows. $\qquad\qquad\qquad\qquad\qquad\qquad\qquad$ □

Remark 16. The estimate (9.23) does *not* hold if I_0 is just the natural injection from V_H into V_h.

We can now use (3.25) and Lemma 12 to conclude that

$$\langle A_h I_0 v_0, I_0 v_0 \rangle \leq C\|I_0 v_0\|^2_h \leq C\|v_0\|^2_H \leq C\langle A_0 v_0, v_0 \rangle. \qquad (9.24)$$

Combining (9.21), (9.22) and (9.24), we find

$$\langle A_h v, v \rangle \leq C \sum_{j=0}^{J} \langle A_j v_j, v_j \rangle,$$

which implies

$$\langle A_h v, v \rangle \leq C \min_{\substack{v = \sum_{j=0}^{J} I_j v_j \\ v_j \in V_j}} \sum_{j=0}^{J} \langle A_j v_j, v_j \rangle \qquad \forall v \in V_h. \qquad (9.25)$$

The estimates (9.5) and (9.25) yield the following result.

Lemma 13. *There exists a positive constant C independent of h, H, δ and J such that*

$$\lambda_{\max}(B_{\mathrm{TL}} A_h) \leq C.$$

9.4   Estimate for $\lambda_{\min}(B_{\mathrm{TL}} A_h)$

First we introduce an operator $J_h^H : V_h \longrightarrow V_H$ defined by

$$J_h^H = \Pi_H \circ \tilde{E}_h. \tag{9.26}$$

The following analog of Lemma 12 can be established by similar arguments.

Lemma 14. *We have*

$$\|v - J_h^H v\|_{L_2(\Omega)} + H|v - J_h^H v|_{H^1(\Omega)} + H^2 \|J_h^H v\|_H$$
$$\leq C H^2 \|v\|_h \qquad \forall\, v \in V_h.$$

The following lemma provides a lower bound for the eigenvalues of $B_{\mathrm{TL}} A_h$.

Lemma 15. *There exists a positive constant C independent of h, H, δ and J such that*

$$\lambda_{\min}(B_{\mathrm{TL}} A_h) \geq C\left(1 + \frac{H^4}{\delta^4}\right)^{-1}.$$

Proof. Let $v \in V_h$ be arbitrary, $v_0 = J_h^H v$ and $v_j = \Pi_h\big(\theta_j(v - I_0 v_0)\big)$. It follows from (9.11) and (9.12) that $v_j \in V_j$ for $0 \leq j \leq J$ and $v = \sum_{j=0}^J I_j v_j$.

In order to apply (9.6), we need to estimate the energy of v_j by the energy of v. We begin with v_0. From (3.25) and Lemma 14, we have

$$\langle A_0 v_0, v_0 \rangle \leq C \|v_0\|_H^2 = C\|J_h^H v\|_H^2 \leq C\|v\|_h^2 \leq C\langle A_h v, v\rangle. \tag{9.27}$$

Next we consider v_j for $1 \leq j \leq J$. Let $w = v - I_0 v_0$ and $w_j = \theta_j w$, so that $v_j = \Pi_h w_j$. It follows from Lemma 12 and Lemma 14 that

$$\begin{aligned}
\|w\|_{L_2(\Omega)} &= \|v - I_0 J_h^H v\|_{L_2(\Omega)} \\
&\leq \|v - J_h^H v\|_{L_2(\Omega)} + \|I_0 J_h^H v - J_h^H v\|_{L_2(\Omega)} \\
&\leq C\big(H^2 \|v\|_h + H^2 \|J_h^H v\|_H\big) \\
&\leq C H^2 \|v\|_h.
\end{aligned} \tag{9.28}$$

Similarly we have

$$|w|_{H^1(\Omega)} \leq CH\|v\|_h \quad \text{and} \quad \|w\|_h \leq C\|v\|_h. \tag{9.29}$$

Let $T \in \mathscr{T}_{h,j}$ be arbitrary and $\tilde{\theta}_{j,T}$ be the P_1 interpolant of θ_j on T that agrees with θ_j at the vertices of T. The following interpolation error estimates are standard [32, 44]:

$$\|\tilde{\theta}_{j,T}\|_{L_\infty(T)} \le \|\theta_j\|_{L_\infty(T)}, \tag{9.30}$$

$$\|\nabla\tilde{\theta}_{j,T}\|_{L_\infty(T)} \le C\|\nabla\theta_j\|_{L_\infty(T)}, \tag{9.31}$$

$$\|\tilde{\theta}_{j,T} - \theta_j\|_{L_\infty(T)} \le C h_T^2 \|\nabla^2\theta_j\|_{L_\infty(T)}. \tag{9.32}$$

It follows from (3.30), (9.10), (9.13), (9.30)–(9.32), a standard inverse estimate and the product rule that

$$
\begin{aligned}
|v_j|^2_{H^2(T)} &\le 2|\Pi_h(\tilde{\theta}_{j,T}w)|^2_{H^2(T)} + 2|\Pi_h((\theta_j - \tilde{\theta}_{j,T})w)|^2_{H^2(T)} \\
&\le C\big[|\tilde{\theta}_{j,T}w|^2_{H^2(T)} + h_T^{-4}\|\Pi_h((\theta_j - \tilde{\theta}_{j,T})w)\|^2_{L_2(T)}\big] \\
&\le C\big[\|\tilde{\theta}_{j,T}\|^2_{L_\infty(T)}|w|^2_{H^2(T)} + \|\nabla\tilde{\theta}_{j,T}\|^2_{L_\infty(T)}|w|^2_{H^1(T)} \\
&\quad + h_T^{-4}\|(\theta_j - \tilde{\theta}_{j,T})\|^2_{L_\infty(T)}\|w\|^2_{L_2(T)}\big] \\
&\le C\big[|w|^2_{H^2(T)} + \frac{1}{\delta^2}|w|^2_{H^1(T)} + \frac{1}{\delta^4}\|w\|^2_{L_2(T)}\big].
\end{aligned}
\tag{9.33}
$$

For any $e \in \mathscr{E}_{h,j}$, we have

$$
\frac{1}{|e|}\|\,[\![\partial v_j/\partial n_e]\!]\,\|^2_{L_2(e)}
$$

$$
\le \frac{2}{|e|}\|\,[\![\partial w_j/\partial n_e]\!]\,\|^2_{L_2(e)} + \frac{2}{|e|}\|\,[\![\partial(\Pi_h w_j - w_j)/\partial n_e]\!]\,\|^2_{L_2(e)}. \tag{9.34}
$$

The first term on the right-hand side of (9.34) can be estimated as follows:

$$
\begin{aligned}
\frac{1}{|e|}\|\,[\![\partial w_j/\partial n_e]\!]\,\|^2_{L_2(e)} &= \frac{1}{|e|}\|\,[\![\partial(\theta_j w)/\partial n_e]\!]\,\|^2_{L_2(e)} \\
&= \frac{1}{|e|}\|\theta_j\,[\![\partial w/\partial n_e]\!]\,\|^2_{L_2(e)} \\
&\le \frac{1}{|e|}\|\,[\![\partial w/\partial n_e]\!]\,\|^2_{L_2(e)} \\
&\le \frac{2}{|e|}\|\,[\![\partial v/\partial n_e]\!]\,\|^2_{L_2(e)} + \frac{2}{|e|}\|\,[\![\partial(I_0 v_0)/\partial n_e]\!]\,\|^2_{L_2(e)}.
\end{aligned}
\tag{9.35}
$$

For the second term on the right-hand side of (9.34), we can use (3.30), the trace theorem with scaling, (9.10) and (9.13) to obtain

$$
\begin{aligned}
\frac{1}{|e|}\|\,[\![\partial(\Pi_h w_j - w_j)/\partial n_e]\!]\,\|^2_{L_2(e)} &\le C \sum_{T \in \mathscr{T}_e} |w_j|^2_{H^2(T)} \\
&\le C \sum_{T \in \mathscr{T}_e} \big[\|\theta_j\|^2_{L_\infty(T)}|w|^2_{H^2(T)} + \|\nabla\theta_j\|^2_{L_\infty(T)}|w|^2_{H^1(T)}\big]
\end{aligned}
\tag{9.36}
$$

$$\leq C \sum_{T \in \mathscr{T}_e} \left[|w|^2_{H^2(T)} + \frac{1}{\delta^2} |w|^2_{H^1(T)} \right].$$

It follows from (3.22), (3.25), (9.17), (9.18), Lemma 12, (9.28)–(9.29), (9.33)–(9.36) that

$$\sum_{j=1}^{J} \langle A_j v_j, v_j \rangle \leq C \sum_{j=1}^{J} \left(\sum_{T \in \mathscr{T}_{h,j}} |v_j|^2_{H^2(T)} + \sum_{e \in \mathscr{E}_{h,j}} \frac{1}{|e|} \| [\![\partial v_j / \partial n_e]\!] \|^2_{L_2(e)} \right)$$

$$\leq C \left(\|v\|^2_h + \|I_0 v_0\|^2_h + \sum_{T \in \mathscr{T}_h} \left[|w|^2_{H^2(T)} + \frac{1}{\delta^2} |w|^2_{H^1(T)} + \frac{1}{\delta^4} \|w\|^2_{L_2(T)} \right] \right)$$

$$\leq C \left(1 + \frac{H^4}{\delta^4} \right) \|v\|^2_h = C \left(1 + \frac{H^4}{\delta^4} \right) \langle A_h v, v \rangle. \tag{9.37}$$

Combining (9.27) and (9.37) we find

$$\sum_{j=0}^{J} \langle A_j v_j, v_j \rangle \leq C \left(1 + \frac{H^4}{\delta^4} \right) \langle A_h v, v \rangle,$$

which implies

$$\min_{\substack{v=\sum_{j=0}^{J} I_j v_j \\ v_j \in V_j}} \sum_{j=0}^{J} \langle A_j v_j, v_j \rangle \leq C \left(1 + \frac{H^4}{\delta^4} \right) \langle A_h v, v \rangle \qquad \forall \, v \in V_h. \tag{9.38}$$

The estimate for $\lambda_{\min}(B_{\mathrm{TL}} A_h)$ follows from (9.6) and (9.38).

9.5 Condition Number Estimates

In view of Lemma 13 and Lemma 15, the following result is immediate.

Theorem 11. *There exists a positive constant C independent of h, H, δ and J such that*

$$\kappa(B_{\mathrm{TL}} A_h) = \frac{\lambda_{\max}(B_{\mathrm{TL}} A_h)}{\lambda_{\min}(B_{\mathrm{TL}} A_h)} \leq C \left(1 + \frac{H^4}{\delta^4} \right).$$

In particular, in the case of generous overlap where δ is comparable to H, the two-level additive Schwarz preconditioner is an optimal preconditioner. Note that this is made possible by the correct definition of I_0 (cf. Remark 16).

In the case of small overlap (say $\delta = h$), the magnitude of H/δ becomes significant and one can improve the condition number estimate under additional shape regularity assumptions on the subdomains $\Omega_1, \ldots, \Omega_J$. This was first discussed in [50] for conforming finite element methods for second order problems. For the C^0 interior penalty method the condition number estimate for $B_{TL}A_h$ can be improved to

$$\kappa(B_{TL}A_h) = \frac{\lambda_{\max}(B_{TL}A_h)}{\lambda_{\min}(B_{TL}A_h)} \le C\left(1 + \frac{H^3}{\delta^3}\right). \tag{9.39}$$

Details can be found in [37].

Remark 17. The condition number estimate (9.39) is sharp [33,37].

10 Concluding Remarks

For simplicity we have restricted our discussion of C^0 interior penalty methods to polygonal domains. But these methods can be naturally combined with the isoparametric technique [13, 32, 44, 67] to handle domains with smooth boundaries. This is again due to the fact that the underlying finite element spaces are standard finite element spaces for second order problems where the isoparametric methodology has proven successes. Details can be found in [31].

Finally we mention that C^0 interior penalty methods have also been applied to many other problems [28, 30, 36, 52, 72, 94, 95]. The techniques developed in this article are also relevant for these applications and also other discontinuous Galerkin methods for fourth order problems [10, 75, 76, 87].

Acknowledgements The author would like to thank the National Science Foundation for supporting her research on C^0 interior penalty methods through the grants DMS-03-11790, DMS-07-13835 and DMS-10-16332. She would also like to thank Shiyuan Gu, Thirupathi Gudi, Li-yeng Sung and Kening Wang for helpful comments. This article was completed during the author's visit at the Institute for Mathematics and its Applications, which was supported by funds provided by the National Science Foundation.

References

1. R.A. Adams and J.J.F. Fournier. *Sobolev Spaces (Second Edition)*. Academic Press, Amsterdam, 2003.
2. S. Agmon, A. Douglis, and L. Nirenberg. Estimates near the boundary for solutions of elliptic partial differential equations satisfying general boundary conditions. I. *Comm. Pure Appl. Math.*, 12:623–727, 1959.
3. M. Ainsworth and J. T. Oden. *A Posteriori Error Estimation in Finite Element Analysis*. Wiley-Interscience, New York, 2000.
4. P.F. Antonietti and B. Ayuso. Schwarz domain decomposition preconditioners for discontinuous Galerkin approximations of elliptic problems: non-overlapping case. *M2AN Math. Model. Numer. Anal.*, 41:21–54, 2007.
5. P.F. Antonietti and B. Ayuso. Two-level Schwarz preconditioners for super penalty discontinuous Galerkin methods. *Commun. Comput. Phys.*, 5:398–412, 2009.

6. J.H. Argyris, I. Fried, and D.W. Scharpf. The TUBA family of plate elements for the matrix displacement method. *Aero. J. Roy. Aero. Soc.*, 72:701–709, 1968.

7. D.N. Arnold, F. Brezzi, B. Cockburn, and L.D. Marini. Unified analysis of discontinuous Galerkin methods for elliptic problems. *SIAM J. Numer. Anal.*, 39:1749–1779, 2001/02.

8. I. Babuška. The finite element method with Lagrange multipliers. *Numer. Math.*, 20:179–192, 1973.

9. C. Bacuta, J.H. Bramble, and J.E. Pasciak. Shift theorems for the biharmonic Dirichlet problem. In *Recent Progress in Computational and Applied PDEs*, pages 1–26. Kluwer/Plenum, New York, 2002.

10. G. Baker. Finite element methods for elliptic equations using nonconforming elements. *Math. Comp.*, 31:45–89, 1977.

11. A. Barker, S.C. Brenner, E.-H. Park, and L.-Y. Sung. Two-level additive Schwarz preconditioners for a weakly over-penalized symmetric interior penalty method. *J. Sci. Comput.*, 47:27–49, 2011.

12. G.P. Bazeley, Y.K. Cheung, B.M. Irons, and O.C. Zienkiewicz. Triangular elements in bending - conforming and nonconforming solutions. In *Proceedings of the Conference on Matrix Methods in Structural Mechanics*. Wright Patterson A.F.B., Ohio, 1965.

13. C. Bernardi. Optimal finite-element interpolation on curved domains. *SIAM J. Numer. Anal.*, 26:1212–1240, 1989.

14. P. Bjørstad and J. Mandel. On the spectra of sums of orthogonal projections with applications to parallel computing. *BIT*, 31:76–88, 1991.

15. H. Blum and R. Rannacher. On the boundary value problem of the biharmonic operator on domains with angular corners. *Math. Methods Appl. Sci.*, 2:556–581, 1980.

16. A. Bonito and R.H. Nochetto. Quasi-optimal convergence rate of an adaptive discontinuous Galerkin method. *SIAM J. Numer. Anal.*, 48:734–771, 2010.

17. J.H. Bramble. *Multigrid Methods*. Longman Scientific & Technical, Essex, 1993.

18. J.H. Bramble and S.R. Hilbert. Estimation of linear functionals on Sobolev spaces with applications to Fourier transforms and spline interpolation. *SIAM J. Numer. Anal.*, 7:113–124, 1970.

19. J.H. Bramble and J. Xu. Some estimates for a weighted L^2 projection. *Math. Comp.*, 56:463–476, 1991.

20. J.H. Bramble and X. Zhang. Multigrid methods for the biharmonic problem discretized by conforming C^1 finite elements on nonnested meshes. *SIAM J. Numer. Anal.*, 33:555–570, 1996.

21. J.H. Bramble and X. Zhang. The Analysis of Multigrid Methods. In P.G. Ciarlet and J.L. Lions, editors, *Handbook of Numerical Analysis, VII*, pages 173–415. North-Holland, Amsterdam, 2000.

22. S.C. Brenner. An optimal-order nonconforming multigrid method for the biharmonic equation. *SIAM J. Numer. Anal.*, 26:1124–1138, 1989.

23. S.C. Brenner. Two-level additive Schwarz preconditioners for nonconforming finite elements. In D.E. Keyes and J. Xu, editors, *Domain Decomposition Methods in Scientific and Engineering Computing*, pages 9–14. Amer. Math. Soc., Providence, 1994. Contemporary Mathematics 180.

24. S.C. Brenner. Two-level additive Schwarz preconditioners for nonconforming finite element methods. *Math. Comp.*, 65:897–921, 1996.

25. S.C. Brenner. Convergence of nonconforming multigrid methods without full elliptic regularity. *Math. Comp.*, 68:25–53, 1999.

26. S.C. Brenner. Convergence of the multigrid V-cycle algorithm for second order boundary value problems without full elliptic regularity. *Math. Comp.*, 71:507–525, 2002.

27. S.C. Brenner. Convergence of nonconforming V-cycle and F-cycle multigrid algorithms for second order elliptic boundary value problems. *Math. Comp.*, 73:1041–1066 (electronic), 2004.

28. S.C. Brenner, S. Gu, T. Gudi, and L.-Y. Sung. A C^0 interior penalty method for a biharmonic problem with essential and natural boundary conditions of Cahn-Hilliard type. *preprint*, 2010.

29. S.C. Brenner, T. Gudi, and L.-Y. Sung. An a posteriori error estimator for a quadratic C^0 interior penalty method for the biharmonic problem. *IMA J. Numer. Anal.*, 30:777–798, 2010.
30. S.C. Brenner and M. Neilan. A C^0 interior penalty method for a fourth order elliptic singular perturbation problem. *SIAM J. Numer. Anal.*, 49:869–892, 2011.
31. S.C. Brenner, M. Neilan, and L.-Y. Sung. Isoparametric C^0 interior penalty methods for plate bending problems on smooth domains. *preprint*, 2011.
32. S.C. Brenner and L.R. Scott. *The Mathematical Theory of Finite Element Methods (Third Edition)*. Springer-Verlag, New York, 2008.
33. S.C. Brenner and L.-Y. Sung. Lower bounds for two-level additive Schwarz preconditioners for nonconforming finite elements. In Z. Chen et al., editor, *Advances in Computational Mathematics*, Lecture Notes in Pure and Applied Mathematics 202, pages 585–604. Marcel Dekker, New York, 1998.
34. S.C. Brenner and L.-Y. Sung. C^0 interior penalty methods for fourth order elliptic boundary value problems on polygonal domains. *J. Sci. Comput.*, 22/23:83–118, 2005.
35. S.C. Brenner and L.-Y. Sung. Multigrid algorithms for C^0 interior penalty methods. *SIAM J. Numer. Anal.*, 44:199–223, 2006.
36. S.C. Brenner, L.-Y. Sung, and Y. Zhang. Finite element methods for the displacement obstacle problem of clamped plates. *Math. Comp.*, (to appear).
37. S.C. Brenner and K. Wang. Two-level additive Schwarz preconditioners for C^0 interior penalty methods. *Numer. Math.*, 102:231–255, 2005.
38. S.C. Brenner and K. Wang. An iterative substructuring algorithm for a C^0 interior penalty method. *preprint*, 2011.
39. S.C. Brenner, K. Wang, and J. Zhao. Poincaré-Friedrichs inequalities for piecewise H^2 functions. *Numer. Funct. Anal. Optim.*, 25:463–478, 2004.
40. S.C. Brenner and J. Zhao. Convergence of multigrid algorithms for interior penalty methods. *Appl. Numer. Anal. Comput. Math.*, 2:3–18, 2005.
41. F. Brezzi. On the existence, uniqueness and approximation of saddle-point problems arising from Lagrangian multipliers. *RAIRO Anal. Numér.*, 8:129–151, 1974.
42. F. Brezzi and M. Fortin. *Mixed and Hybrid Finite Element Methods*. Springer-Verlag, New York-Berlin-Heidelberg, 1991.
43. P.G. Ciarlet. Sur l'élément de Clough et Tocher. *RAIRO Anal. Numér.*, 8:19–27, 1974.
44. P.G. Ciarlet. *The Finite Element Method for Elliptic Problems*. North-Holland, Amsterdam, 1978.
45. P.G. Ciarlet and P.-A. Raviart. A mixed finite element method for the biharmonic equation. In *Mathematical aspects of finite elements in partial differential equations (Proc. Sympos., Math. Res. Center, Univ. Wisconsin, Madison, Wis., 1974)*, pages 125–145. Publication No. 33. Math. Res. Center, Univ. of Wisconsin-Madison, Academic Press, New York, 1974.
46. M. Dauge. *Elliptic Boundary Value Problems on Corner Domains*, Lecture Notes in Mathematics 1341. Springer-Verlag, Berlin-Heidelberg, 1988.
47. W. Dörfler. A convergent adaptive algorithm for Poisson's equation. *SIAM J. Numer. Anal.*, 33:1106–1124, 1996.
48. M. Dryja, J. Galvis, and M. Sarkis. BDDC methods for discontinuous Galerkin discretization of elliptic problems. *J. Complexity*, 23:715–739, 2007.
49. M. Dryja and O.B. Widlund. An additive variant of the Schwarz alternating method in the case of many subregions. Technical Report 339, Department of Computer Science, Courant Institute, 1987.
50. M. Dryja and O.B. Widlund. Domain decomposition algorithms with small overlap. *SIAM J. Sci. Comput.*, 15:604–620, 1994.
51. T. Dupont and R. Scott. Polynomial approximation of functions in Sobolev spaces. *Math. Comp.*, 34:441–463, 1980.
52. G. Engel, K. Garikipati, T.J.R. Hughes, M.G. Larson, L. Mazzei, and R.L. Taylor. Continuous/discontinuous finite element approximations of fourth order elliptic problems in structural and continuum mechanics with applications to thin beams and plates, and strain gradient elasticity. *Comput. Methods Appl. Mech. Engrg.*, 191:3669–3750, 2002.

53. X. Feng and O.A. Karakashian. Two-level additive Schwarz methods for a discontinuous Galerkin approximation of second order elliptic problems. *SIAM J. Numer. Anal.*, 39:1343–1365, 2001.

54. X. Feng and O.A. Karakashian. Two-level non-overlapping Schwarz preconditioners for a discontinuous Galerkin approximation of the biharmonic equation. *J. Sci. Comput.*, 22/23:289–314, 2005.

55. G.H. Golub and C.F. Van Loan. *Matrix Computations (third edition)*. The Johns Hopkins University Press, Baltimore, 1996.

56. M. Griebel and P. Oswald. On the abstract theory of additive and multiplicative Schwarz algorithms. *Numer. Math.*, 70:163–180, 1995.

57. P. Grisvard. *Elliptic Problems in Non Smooth Domains*. Pitman, Boston, 1985.

58. T. Gudi. A new error analysis for discontinuous finite element methods for linear elliptic problems. *Math. Comp.*, 79:2169–2189, 2010.

59. W. Hackbusch. *Multi-grid Methods and Applications*. Springer-Verlag, Berlin-Heidelberg-New York-Tokyo, 1985.

60. R.H.W. Hoppe, G. Kanschat, and T. Warburton. Convergence analysis of an adaptive interior penalty discontinuous Galerkin method. *SIAM J. Numer. Anal.*, 47:534–550, 2008/09.

61. C. Johnson. On the convergence of a mixed finite-element method for plate bending problems. *Numer. Math.*, 21:43–62, 1973.

62. O.A. Karakashian and F. Pascal. Convergence of adaptive discontinuous Galerkin approximations of second-order elliptic problems. *SIAM J. Numer. Anal.*, 45:641–665, 2007.

63. V. Kondratiev. Boundary value problems for elliptic equations in domains with conical or angular points. *Trans. Moscow Math. Soc.*, pages 227–313, 1967.

64. V.A. Kozlov, V.G. Maz'ya, and J. Rossmann. *Elliptic Boundary Value Problems in Domains with Point Singularities*. AMS, Providence, 1997.

65. S.G. Kreĭn, Ju.I. Petunin, and E.M. Semenov. Interpolation of Linear Operators. In *Translations of Mathematical Monographs 54*. American Mathematical Society, Providence, 1982.

66. C. Lasser and A. Toselli. An overlapping domain decomposition preconditioner for a class of discontinuous Galerkin approximations of advection-diffusion problems. *Math. Comp.*, 72:1215–1238, 2003.

67. M. Lenoir. Optimal isoparametric finite elements and error estimates for domains involving curved boundaries. *SIAM J. Numer. Anal.*, 23:562–580, 1986.

68. S. Li and K. Wang. Condition number estimates for C^0 interior penalty methods. In *Domain decomposition methods in science and engineering XVI*, volume 55 of *Lect. Notes Comput. Sci. Eng.*, pages 675–682. Springer, Berlin, 2007.

69. J. Mandel, S. McCormick, and R. Bank. Variational Multigrid Theory. In S. McCormick, editor, *Multigrid Methods, Frontiers In Applied Mathematics 3*, pages 131–177. SIAM, Philadelphia, 1987.

70. T.P.A. Mathew. *Domain Decomposition Methods for the Numerical Solution of Partial Differential Equations*. Springer-Verlag, Berlin, 2008.

71. T. Miyoshi. A finite element method for the solutions of fourth order partial differential equations. *Kumamoto J. Sci. (Math.)*, 9:87–116, 1972/73.

72. L. Molari, G. N. Wells, K. Garikipati, and F. Ubertini. A discontinuous Galerkin method for strain gradient-dependent damage: study of interpolations and convergence. *Comput. Methods Appl. Mech. Engrg.*, 195:1480–1498, 2006.

73. P. Morin, R.H. Nochetto, and K.G. Siebert. Data oscillation and convergence of adaptive FEM. *SIAM J. Numer. Anal.*, 38:466–488, 2000.

74. L.S.D. Morley. The triangular equilibrium problem in the solution of plate bending problems. *Aero. Quart.*, 19:149–169, 1968.

75. I. Mozolevski and E. Süli. A priori error analysis for the hp-version of the discontinuous Galerkin finite element method for the biharmonic equation. *Comput. Methods Appl. Math.*, 3:596–607 (electronic), 2003.

76. I. Mozolevski, E. Süli, and P.R. Bösing. hp-version a priori error analysis of interior penalty discontinuous Galerkin finite element approximations to the biharmonic equation. *J. Sci. Comput.*, 30:465–491, 2007.

77. S.A. Nazarov and B.A. Plamenevsky. *Elliptic Problems in Domains with Piecewise Smooth Boundaries*. de Gruyter, Berlin-New York, 1994.

78. S. Nepomnyaschikh. On the application of the bordering method to the mixed boundary value problem for elliptic equations and on mesh norms in $W_2^{1/2}(S)$. *Sov. J. Numer. Anal. Math. Modelling*, 4:493–506, 1989.

79. J. Nečas. *Les Méthodes Directes en Théorie des Équations Elliptiques*. Masson, Paris, 1967.

80. P. Peisker. A multilevel algorithm for the biharmonic problem. *Numer. Math.*, 46:623–634, 1985.

81. B. Rivière. *Discontinuous Galerkin methods for solving elliptic and parabolic equations*. Society for Industrial and Applied Mathematics (SIAM), Philadelphia, PA, 2008.

82. W. Rudin. *Functional Analysis (Second Edition)*. McGraw-Hill, New York, 1991.

83. Z. Shi. On the convergence of the incomplete biquadratic nonconforming plate element. *Math. Numer. Sinica*, 8:53–62, 1986.

84. Z. Shi. Error estimates of Morley element. *Chinese J. Numer. Math. & Appl.*, 12:9–15, 1990.

85. Z. Shi and Z. Xie. Multigrid methods for Morley element on nonnested meshes. *J. Comput. Math.*, 16:385–394, 1998.

86. B. Smith, P. Bjørstad, and W. Gropp. *Domain Decomposition*. Cambridge University Press, Cambridge, 1996.

87. E. Süli and I. Mozolevski. hp-version interior penalty DGFEMs for the biharmonic equation. *Comput. Methods Appl. Mech. Engrg.*, 196:1851–1863, 2007.

88. L. Tartar. *An Introduction to Sobolev Spaces and Interpolation Spaces*. Springer, Berlin, 2007.

89. A. Toselli and O.B. Widlund. *Domain Decomposition Methods - Algorithms and Theory*. Springer, New York, 2005.

90. H. Triebel. *Interpolation Theory, Function Spaces, Differential Operators*. North-Holland, Amsterdam, 1978.

91. U. Trottenberg, C. Oosterlee, and A. Schüller. *Multigrid*. Academic Press, San Diego, 2001.

92. R. Verfürth. A posteriori error estimation and adaptive mesh-refinement techniques. In *Proceedings of the Fifth International Congress on Computational and Applied Mathematics (Leuven, 1992)*, volume 50, pages 67–83, 1994.

93. R. Verfürth. *A Review of A Posteriori Error Estimation and Adaptive Mesh-Refinement Techniques*. Wiley-Teubner, Chichester, 1995.

94. G.N. Wells, K. Garikipati, and L. Molari. A discontinuous Galerkin formulation for a strain gradient-dependent damage model. *Comput. Methods Appl. Mech. Engrg.*, 193:3633–3645, 2004.

95. G.N. Wells, E. Kuhl, and K. Garikipati. A discontinuous Galerkin method for the Cahn-Hilliard equation. *J. Comput. Phys.*, 218:860–877, 2006.

96. J. Xu. Iterative methods by space decomposition and subspace correction. *SIAM Review*, 34:581–613, 1992.

97. A. Ženíšek. Interpolation polynomials on the triangle. *Numer. Math.*, 15:283–296, 1970.

98. S. Zhang. An optimal order multigrid method for biharmonic, C^1 finite element equations. *Numer. Math.*, 56:613–624, 1989.

99. X. Zhang. *Studies in Domain Decomposition: Multilevel Methods and the Biharmonic Dirichlet Problem*. PhD thesis, Courant Institute, 1991.

100. J. Zhao. Convergence of nonconforming V-cycle and F-cycle methods for the biharmonic problem using the Morley element. *Electron. Trans. Numer. Anal.*, 17:112–132, 2004.

101. J. Zhao. Convergence of V- and F-cycle multigrid methods for the biharmonic problem using the Hsieh-Clough-Tocher element. *Numer. Methods Partial Differential Equations*, 21:451–471, 2005.

Introduction to Applications of Numerical Analysis in Time Domain Computational Electromagnetism

Qiang Chen and Peter Monk

Abstract We discuss two techniques for the solution of the time domain Maxwell system. The first is a partial differential equation based approach using conforming finite elements and implicit time stepping that is suitable when stiff problems are encountered, and where the medium is inhomogeneous. In particular we analyze the use of edge elements and certain A-stable schemes using the Fourier-Laplace transform. For a homogeneous medium, an integral equation approach can be used and we describe and analyze the convolution quadrature method applied to the electric field integral equation. In either case we emphasize that the convergence analysis depends on energy estimates for the continuous problem.

1 Introduction

Electromagnetic wave propagation underlies many technologies that are central to modern life including, for example, cell phones, wi-fi and radar. The need to design better antennas, assess their safety, and ensure that electromagnetic emissions do not interfere with other devices all require the capability of simulating electromagnetic wave propagation. Computational Electromagnetics (CEM) seeks to provide the tools to do this and so to further the current explosion of applications.

Indeed there are so many potential applications, each with its own special peculiarities, that it is impossible to cover all of CEM in one introductory article. For example, depending on the time-scale of interest, one might solve the full Maxwell system, a reduced eddy current problem, or even a static problem. Here we shall only consider the full Maxwell system, limiting us to relatively short time simulations. Special problems also arise in applications to wave-guides [117],

Q. Chen · P. Monk (✉)
Department of Mathematical Sciences, University of Delaware, Newark, DE 19716, USA
e-mail: qchen@math.udel.edu; monk@math.udel.edu

J. Blowey and M. Jensen (eds.), *Frontiers in Numerical Analysis – Durham 2010*,
Lecture Notes in Computational Science and Engineering 85,
DOI 10.1007/978-3-642-23914-4_3, © Springer-Verlag Berlin Heidelberg 2012

Table 1 Fundamental variables in Maxwell's equations. All are functions of position $x \in \mathbb{R}^3$ and time t

E:	Electric field (vector)	H:	Magnetic field (vector)
B:	Magnetic induction (vector)	D:	Electric displacement (vector)
J:	Current density (vector)	ρ:	Charge density (scalar)
t:	Time	x:	Position

diffraction gratings and periodic structures [83], but we will also avoid these very interesting and important applications here.

Non-linear problems arise in many areas, for example at high field intensity [57], when simulating super-conduction [39], magnetic materials [26] or magneto hydrodynamics (MHD) [27] but we will also ignore them here. Fortunately, in most other areas, electromagnetic wave propagation is often described well by a linear system of differential equations which will be our focus.

We shall be interested in computing solutions of Maxwell's equations. These relate five vector functions (the electric field $E = E(x, t) \in \mathbb{R}^3$ etc) and one scalar function of position $x \in \mathbb{R}^3$ and time t given in Table 1. The basic Maxwell system is then:

$$\dot{D} - \nabla \times H = -J, \tag{1}$$

$$\dot{B} + \nabla \times E = 0, \tag{2}$$

$$\nabla \cdot D = \rho, \tag{3}$$

$$\nabla \cdot B = 0, \tag{4}$$

where these equations hold in the relevant domain of the field. Here $\dot{D} = \partial D / \partial t$ and similarly for $\dot{B}$. To close the system we make several assumptions on constitutive equations relating these quantities. At this stage we simply assume that there are bounded piecewise smooth functions $\epsilon = \epsilon(x)$ (called electric permittivity), $\mu = \mu(x)$ (called magnetic permeability), and $\sigma = \sigma(x)$ (called conductivity), such that

$$0 < \mu_{min} \leq \mu(x) \leq \mu_{max} < \infty$$

$$0 < \epsilon_{min} \leq \epsilon(x) \leq \epsilon_{max} < \infty$$

$$0 \leq \sigma(x) \leq \sigma_{max} < \infty$$

for all x. Then for a simple linear medium we assume the constitutive equations

$$D = \epsilon E, \quad B = \mu H \quad \text{and} \quad J = \sigma E + J^i \tag{5}$$

where J^i is the imposed current density assumed known (for example it might result from modeling a source of radiation). Later we shall see that these coefficients may be frequency dependent (and for metals it is possible that ϵ is negative).

Unless there are strong magnetic effects μ can be considered constant (magnets are one more place where our assumption of linearity breaks down). There are also interesting cases where both ϵ and μ can be effectively negative (so called meta-materials [93])).

A typical choice is to use the constitutive equations to eliminate D and H to arrive at the system

$$\epsilon \dot{E} + \sigma E - \nabla \times \mu^{-1} B = -J^i, \tag{6}$$

$$\dot{B} + \nabla \times E = 0, \tag{7}$$

$$\nabla \cdot \epsilon E = \rho \quad \text{and} \quad \nabla \cdot B = 0. \tag{8}$$

Note that, taking the divergence of (6) and using the fact that $\nabla \cdot (\nabla \times \mu^{-1} B) = 0$ (see for example [104]), we have

$$\dot{\rho} + \nabla \cdot (\sigma E + J^i) = 0, \tag{9}$$

which is called "change conservation".

It is quite possible, and even desirable, to base a numerical scheme on (6)–(8) but we have one more step to our final system. By taking the time derivative of (6) and using (7) we obtain the second order in time Maxwell system

$$\epsilon \ddot{E} + \sigma \dot{E} + \nabla \times \mu^{-1} \nabla \times E = -\dot{J}^i, \tag{10}$$

where $\ddot{E} = \partial^2 E / \partial t^2$. Equation (10) will be at the core of our time domain finite element solver (TDFE) that will be described in detail in Sect. 3. The magnetic induction can be reintroduced later if needed.

Another approach very often used in practice, is to formally take the Fourier transform in time of (10). We denote by $\hat{E}$, and $\hat{J}^i$ the Fourier transform of E, and J^i respectively so that

$$\hat{E}(x, \omega) = \int_{-\infty}^{\infty} E(x, t) \exp(i\omega t) \, dt$$

where ω is the transform parameter. Now recall that $\hat{\dot{E}} = -i\omega \hat{E}$ so that (10) becomes the time-harmonic Maxwell system

$$\omega^2 \left(\epsilon + \frac{i\sigma}{\omega} \right) \hat{E} - \nabla \times \mu^{-1} \nabla \times \hat{E} = -i\omega \hat{J}^i. \tag{11}$$

With suitable boundary conditions, we can solve (11), using, for example, finite element methods [104]. However, when $\sigma = 0$ (i.e. in a non-conducting medium), we need to solve a matrix problem in which the matrix may be indefinite (and is increasingly indefinite as ω increases). As a result it is not known how to solve the resulting linear system in an optimal way [63], and indeed designing a solver for this problem is still the subject of research.

We can avoid this problem by working in the time domain. In addition, if broadband results are needed (i.e. the solution for a wide range of ω), it may be more efficient to solve Maxwell's equations once in the time domain rather than solve multiple time-harmonic problems at different frequencies.

We now need to complete our discussion of Maxwell's equations by discussing boundary conditions and formulating the boundary value problems to be the focus of this paper. At least at microwave and radar frequencies, metals are often modeled as perfect conductors so that

$$E \times n = 0 \tag{12}$$

on the surface of the metal, where n is the unit normal pointing into the metal. Another useful boundary condition that models an imperfect conductor is the impedance boundary condition

$$H \times n - Z E_T = 0 \tag{13}$$

where $Z > 0$ is the surface impedance and $E_T = (n \times E) \times n$ and the unit normal n again points out of the domain of computation.

We will first consider an approximation to a scattering problem. Suppose D is a bounded Lipschitz polyhedron with connected complement and boundary Γ and D_1 is a second bounded Lipschitz polyhedron with boundary Σ containing Γ in its interior. We solve Maxwell's equations in the domain $\Omega = D_1 \setminus \overline{D}$. It will be convenient to keep the topology simple (although this is not a requirement for the method, but makes some theoretical aspects easier to discuss) so we assume that Σ and Γ (disjoint by construction) are both connected, and that Ω is simply connected. The first problem we shall consider is to seek $E = E(x, t)$ that satisfies

$$\epsilon \ddot{E} + \sigma \dot{E} + \nabla \times \mu^{-1} \nabla \times E = -\dot{J}^i \tag{14}$$

in Ω for $0 < t \leq T$, where T is some fixed final time, subject to the boundary conditions

$$(\mu^{-1} \nabla \times E) \times n + Z \dot{E}_T = 0 \text{ on } \Sigma, \tag{15}$$

$$E \times n = 0 \text{ on } \Gamma, \tag{16}$$

for $0 < t \leq T$ where (15) is obtained from the time derivative of (13) using (7) and the constitutive relation between H and B. The initial data is

$$E(x, 0) = E_0, \ \dot{E}(x, 0) = E_1 \text{ in } \Omega \tag{17}$$

where E_0 and E_1 are given functions with $\nabla \cdot \epsilon E_j = 0$, $j = 0, 1$. Usually we choose $E_0 = E_1 = 0$ and in this case we shall show that (14)–(17) has a solution provided J^i is smooth enough.

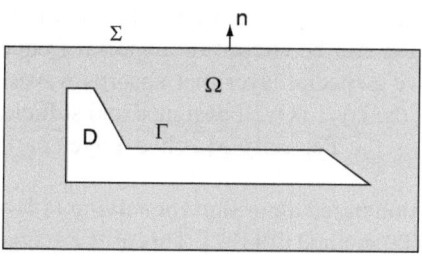
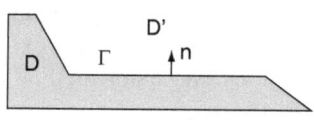

Fig. 1 We shall use several domains for Maxwell's equations in this paper. The first is the region Ω outside Γ (boundary of D) and inside Σ shown in the left panel. The unit normal on the boundary points out of Ω. Later we will simplify the problem to computing in D (right figure), and in the last two sections we will solve the problem simultaneously D and $D' = \mathbb{R}^3 \setminus \overline{D}$

We shall see that the impedance boundary condition and conductivity pose no complications to time domain solvers, so in later sections, for simplicity we shall take $\Omega = D$ and solve (14) in Ω subject to the boundary condition (16) and initial conditions (17). See Fig. 1, a cartoon of these problems.

Particularly in radar or antenna applications, it is necessary to consider wave propagation in an unbounded domain. Suppose we have a bounded object D (for example an aircraft) illuminated by a field due to the current source $\boldsymbol{J}^i$. In the absence of the aircraft (for this application the background parameters are $\sigma = 0$, $\epsilon = \epsilon_0$, $\mu = \mu_0$ where μ_0 and ϵ_0, the electromagnetic parameters for free space), the source $\boldsymbol{J}^i$ would give rise to a field $(\boldsymbol{E}^i, \boldsymbol{H}^i)$ which we assume vanishes for $t < 0$ in the neighborhood of D. Then we want to find scattered field $\boldsymbol{E}^s$ such that

$$\epsilon_0 \ddot{\boldsymbol{E}}^s + \nabla \times \mu_0^{-1} \nabla \times \boldsymbol{E}^s = 0 \text{ on } D' = \mathbb{R}^3 \setminus \bar{D}, \tag{18}$$

for $0 < t \leq T$ such that

$$\boldsymbol{E}^s \times \boldsymbol{n} = -\boldsymbol{E}^i \times n \text{ on } \Gamma = \partial D \tag{19}$$

for $0 < t \leq T$. The initial conditions are then

$$\boldsymbol{E}^s(\boldsymbol{x}, 0) = 0 \quad \dot{\boldsymbol{E}}^s(\boldsymbol{x}, 0) = 0 \text{ in } D'. \tag{20}$$

We shall show that (18)–(20) can be solved using time domain boundary integral equations (TDBIE) that only involve calculations on Γ.

Scattering problems can also be approximated using the solution $\boldsymbol{E}$ of (14)–(17) by selecting Σ to be sufficiently far from Γ and by selecting $Z = \sqrt{\mu_0/\epsilon_0}$, the impedance of free space. In this case (15) is a crude absorbing boundary condition, and the approximation improves as Σ moves further from Γ. Better absorbing boundary conditions can be used to replace (13), and there has been a great deal of work in this area. However, this is beyond the scope of this paper. We direct the reader to the classical paper of Engquist and Majda [59] and to the recent work of Hagstrom, Warbuton and Givoli [76] for more details. Alternatively, the electromagnetic parameters σ, ϵ and μ can be chosen in a special frequency dependent and anisotropic manner to give a special layer that absorbs waves and again allows for a good approximation if the layer is well designed and sufficiently far from D (the main method for this is the Perfectly Matched Layer or PML [25, 41, 104]).

The principal partial differential equation based algorithm for solving (14)–(17) is the finite difference time domain (FDTD) method of [132]. This uses a staggered finite difference grid in space and leap-frog time-stepping. It has been enormously elaborated and tested over the years and is very popular in electrical engineering (see for example the book of Taflove [121]). There are numerous commercial codes that use this method as the underlying solver [130]. It is a difficult method to beat, but it suffers from two general deficiencies: 1) it is most accurate when used with a rectilinear grid, but this complicates fitting curved boundaries, 2) it is usually second order accurate in the time and space step.

At first sight, the finite element method offers a simple way around the two drawbacks of FDTD mentioned above. The use of tetrahedral or mapped hexahedral elements allows for much greater flexibility in geometry modeling, and higher order finite element methods are easy to construct. But as we shall see, the Finite Element Time Domain (FETD) approach has a significant drawback particularly when applied to wave propagation in simple media: a matrix equation must be solved at each step. This is sometimes seen as undesirable for two reasons: 1) the time taken to solve a linear system at every timestep will increase solution time compared to a simple explicit method, 2) the solution of a hyperbolic problem should only depend on data within the light cone, but solving a matrix at each timestep couples all the unknowns at each timestep and hence could allow superluminal signal propagation.

The first objection is certainly a strong incentive to modify the method or seek a new method. We shall give examples of two approaches to solving this problem in the paper, but in some cases it maybe that an explicit timestepping scheme imposes such a small timestep that it becomes too expensive to make progress. In this case a fully implicit method may be a reasonable choice, and it is the one we shall emphasize here. Furthermore, even with an implicit scheme we cannot take arbitrarily large timesteps otherwise dispersion error will build up [2].

The second objection to FETD (concerning the light cone) is also very reasonable on physical grounds, but if we can obtain a robust and accurate method, the computed solution may still be sufficiently accurate to be of use even if the light cone property is not exactly respected.

The use of FETD methods based on conforming finite elements has a long history. Early work used standard continuous piecewise linear elements, but it soon became apparent that special care is needed with these elements in order to produce a robust method because of the possibility of exciting spurious modes in the solution [13, 14, 44]. An alternative is to use the edge elements of Nédélec [106, 107] that can control spurious modes as well as result in a discretization that respects charge conservation. These elements are used by several groups in electrical engineering

[90,91,112,113] and are also advocated in the recent review [42] that complements this paper.

The plan of this paper is as follows. In the Sect. 2 we relate the Maxwell system to the wave equation to point out that all the difficulties encountered in approximating the wave equation are also encountered when solving Maxwell's equations. Furthermore lessons learned about the wave equation have immediate consequences for Maxwell's equations. In Sect. 3 we describe how conforming finite elements can be used to discretize Maxwell's equations in space and derive semi-discrete error estimates via an energy argument. We also derive fully discrete error estimates using the Fourier-Laplace transform. We briefly describe mass-lumping, discontinuous Galerkin (DG) methods and frequency dependent media (where implicit methods are often used). In Sects. 4 and 5 we show a different approach to approximating Maxwell's equations in a homogeneous medium using time domain boundary integral equations. We start with some rather standard background material in Sect. 4 and then, in Sect. 5 and again using the Fourier-Laplace transform, we sketch how to derive error estimates for a non-standard approach to timestepping the boundary integral equations. We try to draw some conclusions and make some comments in Sect. 6.

Concerning notation, we use $\| \cdot \|$ to denote the Euclidean norm of a vector, and $\| \cdot \|_Y$ to denote the Y-norm of a function. We do not use boldface to distinguish between scalar and vector quantities - this should be clear from the context. We use a generic constant C that may be everywhere different.

Although the FETD and TDBIE schemes in this paper are somewhat outside mainstream research, many of the techniques used are quite standard and so we hope this paper will be useful to all students interested in computational electromagnetism. Out of a desire for simplicity of exposition, we have also not been thorough in stating the function spaces needed for some of the results. We hope that a mathematical audience will not find this too upsetting. Throughout we have tried to give pointers to the main papers and techniques in the area.

Finally, no introductory paper can do justice to all of the many methods and analytical studies of computational wave propagation. We have tried to reference some of this vast literature (mainly from the mathematical point of view), but undoubtedly have left out a large number of excellent and relevant contributions. For this we apologize.

2   The Wave Equation in One Space Dimension

We can learn a good deal about solving the time domain Maxwell system by studying the wave equation. To see why the wave equation is relevant, let $\epsilon = \epsilon_0$, $\mu = \mu_0$, $\sigma = 0$ in (14)–(17) and suppose $J^i = 0$ (this choice of coefficients describes electromagnetic wave propagation in a vacuum or air). We also assume that the initial data satisfies $\nabla \cdot \epsilon_0 E_j = 0$, $j = 0, 1$, then taking the divergence of (14) we see that $\nabla \cdot \epsilon \ddot{E} = 0$ so $\nabla \cdot \epsilon E = 0$ for all t. Since $\epsilon = \epsilon_0$ is constant, $\nabla \cdot E = 0$ for all t. Then using the identity

$$\nabla \times \nabla \times E = -\Delta E + \nabla \nabla \cdot E \qquad (21)$$

we see that (14) may be written as

$$\epsilon_0 \mu_0 \ddot{E} = \Delta E .$$

Thus, in free space each component of the electric field satisfies the scalar wave equation.

Now, to further simplify the problem, suppose that $E = (u(x_1, t), 0, 0)^T$ where x_1 is the first component of x. Dropping the subscript 1, we then obtain the second order wave equation

$$\frac{1}{c^2} \ddot{u} = u_{xx} \qquad (22)$$

where $c = (\epsilon_0 \mu_0)^{-1/2}$. The Cauchy problem of solving (22) for $x \in \mathbb{R}$ and $0 < t \leq T$ when

$$u(x, 0) = f(x), \quad \dot{u}(x, 0) = 0 \quad \text{for all} \quad x \in \mathbb{R}$$

for some given $f \in C_0^\infty(\mathbb{R})$ has the well-known solution

$$u(x, t) = \frac{1}{2}(f(x + ct) + f(x - ct)) \qquad (23)$$

as can be checked by direct substitution. It is clear from (23) that u consists of left and right going solutions that move with wave speed c (in this case c denotes the speed of light), and so there is finite speed of propagation (also true for Maxwell's equations with frequency independent coefficients, see e.g. [87]).

Boundary conditions complicate the picture a little since waves can be reflected and absorbed there. Let us consider the problem of approximating the boundary value problem of finding $u(x, t)$ such that

$$\frac{1}{c^2} \ddot{u} = u_{xx} \quad \text{for} \quad x \in (0, 1), \ 0 < t \leq T \qquad (24)$$

subject to the boundary condition (corresponding to (12))

$$u(0, t) = u(1, t) = 0, \quad 0 \leq t < T, \qquad (25)$$

and initial conditions

$$u(x, 0) = f(x), \quad \dot{u}(x, 0) = 0, \quad x \in (0, 1). \qquad (26)$$

The analysis of numerical methods for linear hyperbolic problems usually rests, more or less explicitly, on a discrete energy inequality motivated by an energy inequality for the wave equation. The standard way to derive such an inequality for the second order wave equation is to multiply (24) by $\dot{u}$ and integrate over $x \in (0, 1)$ to obtain

$$\int_0^1 \left(\frac{1}{c^2} \ddot{u}\dot{u} - u_{xx}\dot{u} \right) dx = 0.$$

Integrating the spatial derivative term by parts and using the boundary conditions we get

$$\int_0^1 \left(\frac{1}{c^2} \ddot{u}\dot{u} + u_x\dot{u}_x \right) dx = 0.$$

or

$$\frac{1}{2}\frac{d}{dt} \int_0^1 \left(\frac{1}{c^2}|\dot{u}|^2 + |u_x|^2 \right) dx = 0.$$

This gives the standard conservation of energy result for the wave equation:

$$\int_0^1 \left(\frac{1}{c^2}|\dot{u}(x,t)|^2 + |u_x(x,t)|^2 \right) dx = \int_0^1 \left(\frac{1}{c^2}|\dot{u}(x,0)|^2 + |u_x(x,0)|^2 \right) dx.$$

We shall shortly see how a discrete version of this equality (allowing also for a non-homogeneous source term) can be used to derive stability and error estimates for a particular discretization of the wave equation. In the discrete case the integral will be replaced by a weighted sum and the time derivative by a suitable finite difference in time.

2.1 The Finite Difference Method

We should perhaps immediately introduce a finite element method in space to discretize the wave equation, but for the purposes of this introductory material it is easier to consider a finite difference method. Here a standard finite difference approach to this problem (this is also a "mass-lumped" finite element method [47]) is to use a uniform mesh $\{x_j\}_{j=0}^J$, $x_j = jh$, $0 < j \le J$, where the step size $h = 1/J$. We start by discretizing only in space by seeking $u_j(t)$ that we hope will approximate $u(x_j, t)$, for each j. Obviously $u_0(t) = u_J(t) = 0$ from the boundary conditions (25). Using centered differences in space we obtain the semi-discrete problem of finding $\{u_j(t)\}_{j=0}^J$, such that

$$\frac{1}{c^2}\ddot{u}_j = \frac{u_{j+1} - 2u_j + u_{j-1}}{h^2}, \quad 1 \le j \le J - 1, \quad 0 < t \le T, \qquad (27)$$

with

$$u_0(t) = u_J(t) = 0, \quad 0 \le t \le T, \qquad (28)$$

and

$$u_j(0) = f(x_j), \quad 1 \le j \le J - 1; \quad \dot{u}_j(0) = 0, \quad 1 \le j \le J - 1. \tag{29}$$

In principle one could now try any time-stepping scheme for systems of ordinary differential equations on (27)–(29), but the standard choice for this problem is the leap-frog scheme. We use a uniform grid in time and select a time-step $\Delta t > 0$. Then we seek $u_j^n \simeq u(x_j, t_n)$ where $t_n = n\Delta t$, for $1 \le j \le J$ and $0 \le n \le T/\Delta t$ such that

$$\frac{1}{c^2} \left(\frac{u_j^{n+1} - 2u_j^n + u_j^{n-1}}{\Delta t^2} \right) = \frac{u_{j+1}^n - 2u_j^n + u_{j-1}^n}{h^2} \tag{30}$$

for $1 \le j \le J - 1$, $n \ge 0$ and where $u_j^{-1} = u_j^1$ (implementing the condition $\dot{u}(x, 0) = 0$).

These equations have some nice features: they are centered in space and time so the local truncation error is $O(\Delta t^2 + h^2)$ (to see this, simply plug the exact solution into (30) and use Taylor's series about $x = x_j$ and $t = t_n$). In addition, if $\{u_j^n\}_{j=0}^J$ and $\{u_j^{n-1}\}_{j=0}^J$ are known, we can solve (30) explicitly for each j to find $\{u_j^{n+1}\}_{j=0}^J$:

$$u_j^{n+1} = 2u_j^n - u_j^{n-1} + \frac{\Delta t^2 c^2}{h^2} \left(u_{j+1}^n - 2u_j^n + u_{j-1}^n \right).$$

No equations have to be solved so the method is said to be explicit.

We now proceed to investigate the convergence of this scheme. Let $w_j^n = u_j^n - u(x_j, t_n)$. The errors $\left\{ w_j^n \right\}_{j=0}^J$ satisfy a system like (30) but with inhomogeneous right hand side arising from the local truncation error. In particular, if we define $\mathbf{w}^{n+1} = (w_1^{n+1}, w_2^{n+1}, \ldots, w_{J-1}^{n+1})^T \in \mathbb{R}^{J-1}$ and similarly for other discrete quantities we can readily see that $\mathbf{w}^{n+1}$, $\mathbf{w}^n$ and $\mathbf{w}^{n-1}$ are related by

$$\frac{1}{c^2} \frac{(\mathbf{w}^{n+1} - 2\mathbf{w}^n + \mathbf{w}^{n-1})}{\Delta t^2} = -\frac{1}{h^2} A\mathbf{w}^n + \mathbf{r}^n, \quad n \ge 0 \tag{31}$$

where $\mathbf{w}^0 = 0$ (we can assume we interpolate exactly the initial data) and $|r_j^n| = O(\Delta t^2 + h^2)$ for each j and n. The matrix A is the $(J-1) \times (J-1)$ symmetric tridiagonal array corresponding to centered finite differences

$$A = \begin{pmatrix} 2 & -1 & 0 & \cdots & 0 \\ -1 & 2 & -1 & \cdots & 0 \\ 0 & & \ddots & & -1 \\ \vdots & & & \ddots & \vdots \\ 0 & & & -1 & 2 \end{pmatrix}.$$

To obtain an energy estimate we need to multiply (31) by the discrete analogue of $\dot{w}$ and integrate over space. We do this by multiplying (31) by $(\mathbf{w}^{n+1} - \mathbf{w}^{n-1})^T$ to obtain

$$\frac{1}{c^2}\left[\frac{\|\mathbf{w}^{n+1} - \mathbf{w}^n\|^2}{\Delta t^2} - \frac{\|\mathbf{w}^n - \mathbf{w}^{n-1}\|^2}{\Delta t^2}\right] = -\frac{1}{h^2}\mathbf{w}^{n+1^T} A\mathbf{w}^n + \frac{1}{h^2}\mathbf{w}^{n^T} A\mathbf{w}^{n-1}$$

$$+ (\mathbf{w}^{n+1} - \mathbf{w}^{n-1})^T \mathbf{r}^n, \tag{32}$$

where $\|\mathbf{y}\| = \sqrt{\mathbf{y}^T\mathbf{y}}$ and where we have used the fact that A is symmetric. Adding this equality over n and noting the telescoping sum we have

$$\frac{1}{c^2}\left[\frac{\|\mathbf{w}^{n+1} - \mathbf{w}^n\|^2}{\Delta t^2} - \frac{\|\mathbf{w}^1 - \mathbf{w}^0\|^2}{\Delta t^2}\right] = -\frac{1}{h^2}\mathbf{w}^{n+1^T} A\mathbf{w}^n + \frac{1}{h^2}\mathbf{w}^{1^T} A\mathbf{w}^0 \tag{33}$$

$$+ \sum_{m=1}^{n}\left[(\mathbf{w}^{m+1}-\mathbf{w}^m) + (\mathbf{w}^m-\mathbf{w}^{m-1})\right]^T \mathbf{r}^m.$$

This is the discrete analogue of conservation of energy.

Each term on the right hand side of (33) must now be estimated starting with $\mathbf{w}^{n+1^T} A\mathbf{w}^n$ which we write as $\mathbf{w}^{n+1^T} A\mathbf{w}^n = (\mathbf{w}^{n+1} - \mathbf{w}^n)^T A\mathbf{w}^n + \mathbf{w}^{n^T} A\mathbf{w}^n$ to obtain:

$$h\frac{\|\mathbf{w}^{n+1} - \mathbf{w}^n\|^2}{c^2\Delta t^2} + \frac{\mathbf{w}^{n^T} A\mathbf{w}^n}{h} + \frac{(\mathbf{w}^{n+1} - \mathbf{w}^n)^T A\mathbf{w}^n}{h} \tag{34}$$

$$= h\frac{\|\mathbf{w}^1 - \mathbf{w}^0\|^2}{c^2\Delta t^2} + \frac{\mathbf{w}^{1^T} A\mathbf{w}^0}{h}$$

$$+ h\sum_{m=1}^{n}\left[(\mathbf{w}^{m+1} - \mathbf{w}^m) + (\mathbf{w}^m - \mathbf{w}^{m-1})\right]^T \mathbf{r}^m.$$

where the factor h has been multiplied through since $h^{1/2}\|\mathbf{w}^{n+1} - \mathbf{w}^n\|$ is the discrete analogue of the $L^2(0, 1)$ norm. Using the Cauchy-Schwarz inequality $|\mathbf{x} \cdot \mathbf{y}| \le \|\mathbf{x}\|\,\|\mathbf{y}\|$, and arithmetic geometric mean inequality

$$\|\mathbf{x}\|\,\|\mathbf{y}\| \le \frac{1}{2\alpha}\|\mathbf{x}\|^2 + \frac{\alpha}{2}\|\mathbf{y}\|^2$$

for any $\alpha > 0$ we obtain

$$|(\mathbf{w}^{n+1} - \mathbf{w}^n)^T A\mathbf{w}^n| \le \|\mathbf{w}^{n+1} - \mathbf{w}^n\|\,\|A\mathbf{w}^n\|$$

$$\le \frac{h^2}{2\alpha}\frac{\|\mathbf{w}^{n+1} - \mathbf{w}^n\|^2}{c^2\Delta t^2} + \frac{\alpha}{2}\frac{c^2\Delta t^2}{h^2}\|A\mathbf{w}^n\|^2.$$

Noting that A is positive semi-definite and has spectral radius less than 4 (by Gerschgorin's Theorem), we have $\|A\mathbf{w}^n\|^2 \leq 4\mathbf{w}^{n^T} A\mathbf{w}^n$. Thus, (34) becomes

$$\left(1 - \frac{1}{2\alpha}\right) h \frac{\|\mathbf{w}^{n+1} - \mathbf{w}^n\|^2}{c^2 \Delta t^2} + \left(1 - 2\alpha \frac{\Delta t^2 c^2}{h^2}\right) \frac{\mathbf{w}^{n^T} A\mathbf{w}^n}{h}$$

$$= h \frac{\|\mathbf{w}^1 - \mathbf{w}^0\|^2}{c^2 \Delta t^2} + \frac{\mathbf{w}^{1^T} A\mathbf{w}^0}{h} + h \sum_{m=1}^{n} [(\mathbf{w}^{m+1} - \mathbf{w}^m) + (\mathbf{w}^m - \mathbf{w}^{m-1})]^T \mathbf{r}^m.$$

We can make the coefficients on the left hand side positive if we choose α such that

$$1 - \frac{1}{2\alpha} > 0, \quad 1 - 2\alpha \frac{\Delta t^2}{h^2} c^2 > 0$$

so

$$c^2 \frac{\Delta t^2}{h^2} < \frac{1}{2\alpha} < 1.$$

In particular, for stability (at least by this argument) we need to choose Δt and h to satisfy a "Courant-Friedrichs-Lewy" or CFL condition

$$\Delta t < h/c.$$

When using explicit time stepping, we expect some condition restricting the time step in terms of the space step size. The only way to avoid a CFL condition is to use an implicit method in time. We shall see that a different analysis suggests that the best choice is actually $\Delta t = h/c$ in one dimension. If $\Delta t > h/c$ we will generally see instability.

Continuing with our analysis, we can assume that $\mathbf{w}^0 = 0$ (i.e. we interpolate the initial data exactly) and let us assume that $\alpha, h, \Delta t$ are chosen so that there is a constant $\beta > 0$ with

$$1 - \frac{1}{2\alpha} > \beta \quad 1 - 2\alpha \frac{\Delta t^2}{h^2} c^2 > \beta$$

then our energy estimate becomes

$$\beta \left(\frac{h\|\mathbf{w}^{n+1} - \mathbf{w}^n\|^2}{c^2 \Delta t^2} + \frac{\mathbf{w}^{n^T} A\mathbf{w}^n}{h} \right)$$

$$= \frac{h}{c^2} \frac{\|\mathbf{w}^1\|^2}{\Delta t^2} + h \sum_{m=1}^{n} [(\mathbf{w}^{m+1} - \mathbf{w}^m) + (\mathbf{w}^m - \mathbf{w}^{m-1})]^T \cdot \mathbf{r}^m \qquad (35)$$

Now we need to perform a discrete Gronwall type argument to obtain our first estimate. This argument is adapted to hyperbolic problems in order to obtain polynomial growth of the constants appearing in the estimates as functions of T.

Suppose $T > 0$ is the final time of the calculation and let $N = T/\Delta t$. Denote by n^* the timestep at which

$$\mathscr{E}^* := h^{1/2} \frac{\|\mathbf{w}^{n^*+1} - \mathbf{w}^{n^*}\|}{c\Delta t} + \sqrt{\frac{\mathbf{w}^{n^{*T}} A \mathbf{w}^{n^*}}{h}}$$

$$= \max_{0 \le n \le N} \left[\frac{h^{1/2} \|\mathbf{w}^{n+1} - \mathbf{w}^n\|}{c\Delta t} + \sqrt{\frac{\mathbf{w}^{n^T} A \mathbf{w}^n}{h}} \right].$$

Then (35) may be written, choosing $n = n^*$, as

$$\frac{1}{2}\beta\mathscr{E}^{*2} \le \frac{h\|\mathbf{w}^1\|^2}{c^2\Delta t^2} + h\sum_{m=1}^{n^*}(\|\mathbf{w}^{m+1} - \mathbf{w}^m\| + \|\mathbf{w}^m - \mathbf{w}^{m-1}\|)\|\mathbf{r}^m\|$$

$$\le \frac{h\|\mathbf{w}^1\|^2}{c^2\Delta t^2} + 2\mathscr{E}^* c\Delta t \sum_{m=1}^{n^*} h^{1/2}\|\mathbf{r}^m\|.$$

Using the arithmetic geometric mean inequality there is a constant $C > 0$ such that

$$\mathscr{E}^{*2} \le C \left(\frac{h\|\mathbf{w}^1\|^2}{c^2\Delta t^2} + \left(c\Delta t \sum_{m=1}^{n^*} h^{1/2}\|\mathbf{r}^m\| \right)^2 \right).$$

However, $\mathbf{r}^m$ is the vector of local truncation errors and so $|r_j^m| = O(\Delta t^2 + h^2)$ for each j and m. Since there are $O(1/h)$ spatial points, $h^{1/2}\|\mathbf{r}^m\| = O(\Delta t^2 + h^2)$ and assuming $\frac{\sqrt{h}\|\mathbf{w}^1\|^2}{c\Delta t} = O(h^2 + \Delta t^2)$ (i.e. the first step is sufficiently accurate) we have proved that

$$\max_{0 \le n \le N} h^{1/2} \frac{\|\mathbf{w}^{n+1} - \mathbf{w}^n\|}{c\Delta t} + \sqrt{\frac{(\mathbf{w}^n)^T A \mathbf{w}^n}{h}} \le CT(\Delta t^2 + h^2). \qquad (36)$$

To obtain an estimate for $\mathbf{w}^n$ itself we can either use the term $(\mathbf{w}^n)^T A \mathbf{w}^n / h$ and a discrete Poincaré inequality or we can obtain a slightly less attractive bound (because of the presence of an extra factor T) in the following way. Note that $\mathbf{w}^n = \sum_{m=0}^{n-1}(\mathbf{w}^{m+1} - \mathbf{w}^m) + \mathbf{w}^0$ so using the triangle inequality and (36)

$$h^{1/2}\|\mathbf{w}^n\| \le c\Delta t \sum_{m=0}^{n-1} h^{1/2} \frac{\|\mathbf{w}^{n+1} - \mathbf{w}^m\|}{c\Delta t}$$

$$= c\Delta t(n-1)CT(\Delta t^2 + h^2)$$

$$= c\, t_n T(\Delta t^2 + h^2).$$

We have proved the following theorem:

Theorem 1. *If the CFL condition $\Delta t < h/c$ is satisfied and if $\mathbf{u}^1$ is computed sufficiently accurately, the error $w_j^n = u_j^n - u(x_j, t_n)$ satisfies*

$$\max_{0 \le n \le N} h^{1/2} \|\mathbf{w}^n\| = O(\Delta t^2 + h^2)$$

and this error bound increases quadratically with the final time T where $T = N\Delta t$.

Remark 1. Using the fact that A corresponds to second order spatial differences and $A = LL^T$ where L is a matrix corresponding to first order differences, we could use (36) to show that

$$h^{1/2} \frac{\|L^T \mathbf{w}^n\|}{h} = O(\Delta t^2 + h^2)$$

with a constant growing linearity with T. This corresponds to an H^1 error estimate in finite element theory. A discrete Poincaré inequality then shows that $\|\mathbf{w}^n\| \le C \|L^T \mathbf{w}^n\|/h$ and so we can estimate $\|\mathbf{w}^n\|$ with the same order as in our theorem but a better T dependence.

This rather labored and trivial example illustrates a few points:

1. Error analysis follows from a local truncation or consistency analysis followed by an energy argument which, for a linear problem, implies stability.
2. Explicit methods require a CFL condition that limits the time step in terms of the spatial stepsize and wave speed.
3. The error grows polynomially with the final time T.

2.2 Dispersion Analysis

The analysis of the previous section shows we get optimal order of convergence but that the error increases with the duration T of the simulation. So we need to refine the spatial and temporal mesh if we increase T to ensure a fixed global error. We can see this more clearly by considering the pure Cauchy problem (22). Using a Fourier decomposition of the initial data, the solution can be decomposed into plane wave components each having the following form

$$u = \exp(i(kx - \omega t)) \tag{37}$$

where k and ω are related by the "dispersion relation"

$$\omega = \pm ck$$

relating the spatial wave number k and temporal wave number ω (also the Fourier transform variable mentioned in the introduction). The group velocity is then

$$c_g = \frac{|\omega|}{|k|} = c,$$

and is independent of frequency (for a plane wave the temporal frequency is $f = \omega/2\pi$, the spatial wave length is $2\pi/k$).

Now let us consider the discrete problem (30) where now $-\infty < j < \infty$. We can seek discrete plane wave solutions

$$u_j^n = \exp\left(i\left(kx_j - \omega_{h,\Delta t}t_n\right)\right).$$

Substituting into (30) and dividing by u_j^n we obtain

$$\frac{1}{c^2\Delta t^2}\left(\exp(-i\omega_{h,\Delta t}\Delta t) - 2 + \exp(i\omega_{h,\Delta t}\Delta t)\right)$$

$$= \frac{1}{h^2}\left(\exp(ik\,h) - 2 + \exp(-ik\,h)\right)$$

or

$$\sin^2\left(\frac{w_{h,\Delta t}\Delta t}{2}\right) = \frac{c^2\Delta t^2}{h^2}\sin^2\left(\frac{k\,h}{2}\right).$$

To avoid exponentially growing solutions we need $c^2\Delta t^2/h^2 \leq 1$ so the CFL condition arises naturally (and now equality is allowed). Assuming that the CFL condition is satisfied

$$\omega_{h,\Delta t} = \pm\frac{2}{\Delta t}\sin^{-1}\left(\frac{c\Delta t}{h}\sin\left(\frac{kh}{2}\right)\right)$$

gives the "discrete dispersion relation", and the discrete phase velocity

$$c_{h,\Delta t} = \left|\frac{\omega_{h,\Delta t}}{k}\right|$$

is no longer independent of k. Different Fourier components travel at different speeds. For h and Δt small, Taylor series show that

$$\omega_{h,\Delta t} = \pm\left(ck - \frac{ck^3}{24}(h^2 - c^2\Delta t^2) + \dots\right).$$

This reveals the unexpected fact that the first term in the error series vanishes if $\Delta t = h/c$. So the error is least (actually zero in this 1D example!) at the maximum timestep. This is an artifact of the 1D wave equation, and in higher dimension we cannot obtain perfect solutions. But even in 2 or 3 dimensions, for finite difference methods with a well chosen time-stepping strategy, the maximum allowable time-step is often best (in particular this is true for the FDTD scheme for Maxwell's equations).

If $c\Delta t/h < 1$ then generally $w_{h,\Delta t} \neq \pm ck$. For small h and Δt, the numerical phase velocity is less than the true phase velocity and higher frequencies (large k) have a phase velocity that is more in error. This causes numerical dispersion of the wave. Since there is a velocity error, the solution error will grow with time as we have seen from the error estimates.

For a fixed order method, numerical dispersion cannot be avoided in higher dimensions [16]. However, the phase errors can be decreased by using higher order methods. Ultimately (at large enough T) phase error will accumulate sufficiently to ruin the numerical solution. Given a desired final time and frequency content of the solution, we have to choose the order of the method, space and time steps to give acceptable error up to the final time T.

To illustrate these observations, let us now look at some numerical results. We solve (30) with initial data $f(x) = \exp(-80(x - 1/2)^2)$ and wave speed $c = 1$. To illustrate dispersion error we need to avoid a perfect solution so we choose $\Delta t = 1/30$ and $h = 1/26$. For small times the solution splits into left and right going pulses as in (23). These then reflect from the ends of the interval. After one time unit the reflected solution should coalesce to agree with the negative of the original initial data, while after two time units it should agree precisely with the initial data. Comparing initial data and the solution at $t = 0.2, 2, 4$ and 6 gives a way to visualize phase error effects on the solution (Fig. 2).

In Fig. 3 we show the phase velocity for two choices of the CFL parameter. To allow for a clearer interpretation we recall that for wave number k the wavelength is $2\pi/k$ and hence the number of grid points per wavelength is $G = 2\pi/kh$. For a fixed CFL parameter α we set $\Delta t = \alpha h$ and then plot the error in the phase velocity $1 - \omega_{h,\Delta t}/k$. For a small number of grid points per wavelength (around 3) the error is very large, decreasing as G increases.

There remains the problem of how to choose the spatial discretization for a wave calculation. There is no hard and fast rule, but it is necessary to consider the data for the problem. By using a Fourier transform it may be possible to find the range of important frequencies in the data, and then decide on the largest frequency that needs to be accurately modeled. This translates into a wavelength in each part of the domain (assuming piecewise constant coefficients) and in each part of the domain we need to have "sufficiently many grid points per wavelength". For low order schemes a "rule of thumb" is that we need roughly 10 grid points per wavelength to represent the wave (for conforming finite element methods, Ainsworth [5] shows that even if very high order methods are used, at least π grid points per wavelength are needed). But as we have seen, phase error accumulates as the simulation proceeds, so the actual mesh density is dependent on the accuracy required at the final time. Higher order methods have greatly improved phase error (see for example [4, 5, 7, 47]) and for that reason should be considered. There is a well developed theory of variable order methods for the finite element approximation of the time-harmonic problem suggesting how the mesh parameter h and order of the finite element approximation need to be chosen in that case [101] but this needs to be extended to the time domain. Isogeometric elements have also been considered [38] and may well prove to be a very attractive method due to their superior spectral properties.

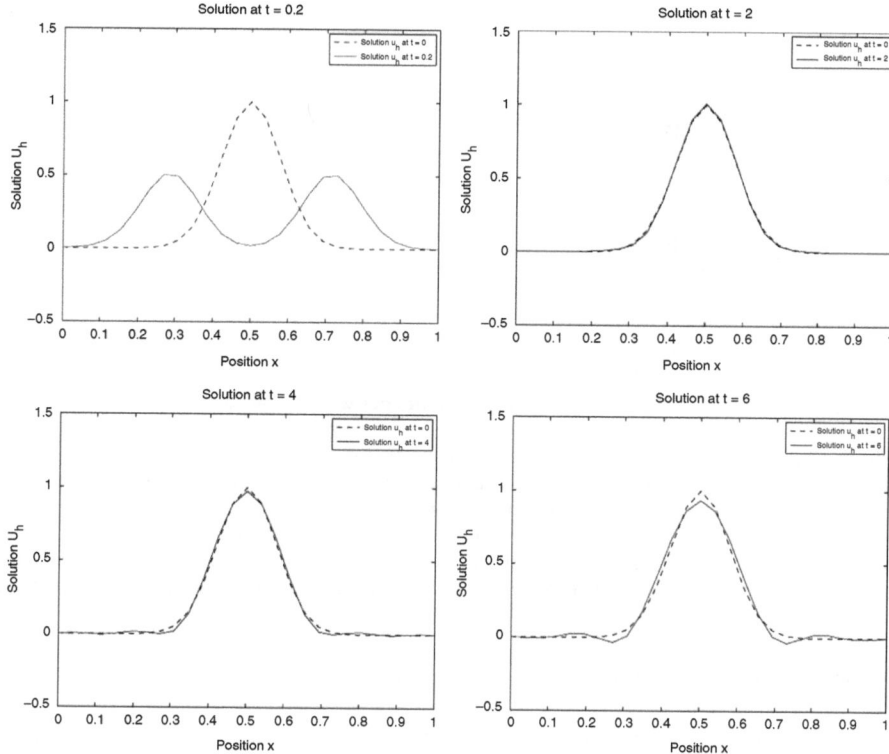

Fig. 2 In each panel we show the solution at a particular time (solid line) and the initial data (dashed line) for the discrete problem described in the text. In the top left panel $t = .2$ and the single bell curve at $t = 0$ has split into a left going and right going part with half the amplitude. In the top right panel $t = 2$ and the initial solution and final solution show close agreement as expected. By $t = 4$, shown in the bottom left panel, the effect of dispersion is building and ripples show in the solution. This error grows with time as can be seen in the bottom right panel at $t = 6$

3 Time Dependent Finite Element Method

We are now going to develop and analyze two partial differential equation based approaches to solving the problem of finding the electric field $\boldsymbol{E} = \boldsymbol{E}(\boldsymbol{x}, t) \in \mathbb{R}^3$ that satisfies the second order Maxwell system discussed in the introduction:

$$\epsilon \ddot{\boldsymbol{E}} + \sigma \dot{\boldsymbol{E}} + \nabla \times \mu^{-1} \nabla \times \boldsymbol{E} = \boldsymbol{F} \quad \text{in } \Omega \text{ for } 0 < t \leq T, \tag{38}$$

for some final time T where $\boldsymbol{F}$ is a given source function ($\boldsymbol{F} = -\dot{\boldsymbol{J}}^i$), subject to the boundary conditions

$$\boldsymbol{E} \times \boldsymbol{n} = 0 \text{ on } \Gamma, \ 0 < t \leq T, \tag{39}$$

$$(\mu^{-1} \nabla \times \boldsymbol{E}) \times \boldsymbol{n} + Z \dot{\boldsymbol{E}}_T = 0 \text{ on } \Sigma, \ 0 < t \leq T. \tag{40}$$

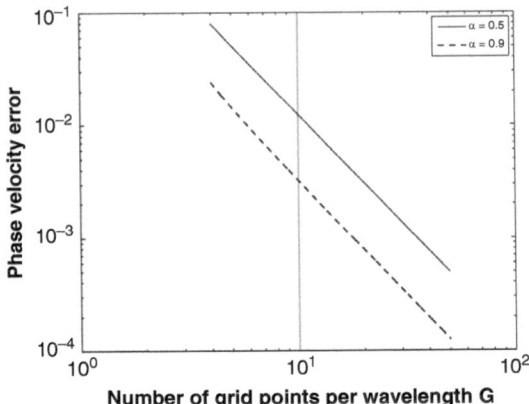

Fig. 3 Phase error $1 - \omega_{h,\Delta t}/k$ against number of grid points per wavelength G for two values of the CFL parameter α. This one dimensional example is unusual in that $\alpha = 1$ corresponds to a perfect method, but when phase error is present and depending on the duration of the simulation the number of grid points per wavelength needs to be chosen to obtain acceptable accuracy for the highest wave number needed in a simulation. The vertical line marks 10 grid points per wavelength, usually considered the minimum needed for a low order method like the one we have described in this section

We shall assume the homogeneous initial conditions

$$E(x,0) = \dot{E}(x,0) = 0 \quad \text{for } x \in \Omega. \tag{41}$$

Proceeding formally we can obtain a weak form suitable for discretization (and analysis!) by multiplying (38) by a smooth test vector $\boldsymbol{\xi} \in \left(C^\infty(\mathbb{R}^3)\right)^3$ and integrating over Ω, then using the identity that for $\boldsymbol{u}, \boldsymbol{\xi} \in H(\text{curl}; \Omega)$ (see for example [104])

$$\int_\Omega \nabla \times \boldsymbol{u} \cdot \boldsymbol{\xi} \, dV = \int_{\partial\Omega} \boldsymbol{n} \times \boldsymbol{u} \cdot \boldsymbol{\xi}_T \, dA + \int_\Omega \boldsymbol{u} \cdot \nabla \times \boldsymbol{\xi} \, dV$$

where $\boldsymbol{n}$ is the unit outward normal to Ω and $\boldsymbol{\xi}_T = (\boldsymbol{n} \times \boldsymbol{\xi}) \times \boldsymbol{n}$. We obtain

$$(\epsilon \ddot{\boldsymbol{E}}, \boldsymbol{\xi}) + (\sigma \dot{\boldsymbol{E}}, \boldsymbol{\xi}) + (\mu^{-1} \nabla \times \boldsymbol{E}, \nabla \times \boldsymbol{\xi})$$
$$- \int_{\partial\Omega} (\mu^{-1} \nabla \times \boldsymbol{E}) \times \boldsymbol{n} \cdot \boldsymbol{\xi}_T \, dA = (\boldsymbol{F}, \boldsymbol{\xi}) \tag{42}$$

where $(\boldsymbol{u}, \boldsymbol{v}) = \int_\Omega \boldsymbol{u} \cdot \boldsymbol{v} \, dV$. On Σ we can use (40) but on Γ we need to enforce the essential boundary condition (39). So we need to assume $\boldsymbol{\xi} \in C_0^\infty(\mathbb{R}^3 \backslash \bar{D})$ implying that $\boldsymbol{\xi} \times \boldsymbol{n} = 0$ on Γ. Then (42) becomes

$$(\epsilon \ddot{\boldsymbol{E}}, \boldsymbol{\xi}) + (\sigma \dot{\boldsymbol{E}}, \boldsymbol{\xi}) + (\mu^{-1} \nabla \times \boldsymbol{E}, \nabla \times \boldsymbol{\xi}) + Z \langle \dot{\boldsymbol{E}}_T, \boldsymbol{\xi}_T \rangle = (\boldsymbol{F}, \boldsymbol{\xi})$$

for all $\boldsymbol{\xi} \in C_0^\infty(\mathbb{R}^3 \backslash \bar{D})$, where $\langle \boldsymbol{u}, \boldsymbol{\xi} \rangle = \int_\Sigma \boldsymbol{u} \cdot \boldsymbol{\xi} \, dA$.

We are thus lead to consider the energy space

$$H(\text{curl}\,;\Omega) = \left\{ u \in L^2(\Omega) \mid \nabla \times u \in L^2(\Omega) \right\}.$$

However, to take care of boundary conditions we actually need a nonstandard subspace of this space:

$$X = \left\{ u \in H(\text{curl};\Omega) \mid n \times u = 0 \text{ on } \Gamma,\ u_T \in L^2(\Sigma) \right\}.$$

We thus seek a suitably smooth in time solution $E(t) \in X$ such that

$$(\epsilon \ddot{E}, \xi) + (\sigma \dot{E}, \xi) + (\mu^{-1} \nabla \times E, \nabla \times \xi) + \langle Z \dot{E}_T, \xi_T \rangle = (F, \xi) \qquad (43)$$

for all $\xi \in X$ and $0 < t \leq T$, subject to the initial conditions (41).

3.1 Energy Estimates and Existence Theory

Assuming a suitable solution exists, let us look at some energy equalities for the variational problem (43). As we saw in the last section, a good choice of test functions is $\xi = \dot{E}$ but here we show how to derive more estimates by taking the more general choice $\xi = \dot{E} \exp(-\lambda t)$ for some $\lambda \geq 0$ (we are indebted to Noel Walkington for pointing out this choice of test function to us). Using this test function and noticing that

$$\frac{1}{2} \frac{d}{dt} (\epsilon \dot{E}, e^{-\lambda t} \dot{E}) = (\epsilon \ddot{E}, e^{-\lambda t} \dot{E}) - \frac{\lambda}{2} (\epsilon \dot{E}, e^{-\lambda t} \dot{E})$$

with a similar equality for the curl term, we obtain

$$\frac{1}{2} \frac{d}{dt} \left[(\epsilon \dot{E}, e^{-\lambda t} \dot{E}) + (\mu^{-1} \nabla \times E, e^{-\lambda t} \nabla \times E) \right] + \frac{\lambda}{2} (\epsilon \dot{E}, e^{-\lambda t} \dot{E})$$

$$+ \frac{\lambda}{2} (\nabla \times E, \nabla \times E\, e^{-\lambda t}) + (\sigma \dot{E}, e^{-\lambda t} \dot{E}) + Z \langle \dot{E}_T, e^{-\lambda t} \dot{E}_T \rangle = (F, e^{-\lambda t} \dot{E}).$$

Using the notation $\|u\|_{L^2_\epsilon(\Omega)}^2 = \int_\Omega \epsilon u \cdot u \, dV$, and similarly for $\|u\|_{L^2_\sigma(\Omega)}$ and other weighted norms, we get, by integration in time,

$$\frac{e^{-\lambda t}}{2} \left[\|\dot{E}(t)\|_{L^2_\epsilon(\Omega)}^2 + \|\nabla \times E(t)\|_{L^2_{\mu^{-1}}(\Omega)}^2 \right]$$

$$+ \int_0^t e^{-\lambda \tau} \left[\frac{\lambda}{2} \|\dot{E}(\tau)\|_{L^2_\epsilon(\Omega)}^2 + \|\dot{E}(\tau)\|_{L^2_\sigma(\Omega)}^2 + \frac{\lambda}{2} \|\nabla \times E(\tau)\|_{L^2_{\mu^{-1}}(\Omega)}^2 \right] d\tau$$

$$+ \int_0^t e^{-\lambda \tau} \|\dot{E}_T(\tau)\|_{L^2_Z(\Sigma)}^2 d\tau = \int_0^t (F, e^{-\lambda \tau} \dot{E}) d\tau, \qquad (44)$$

where we have used the initial data (41). This equality can now be used to provide bounds on $\dot{E}$ and $\nabla \times E$ in various norms. For example choosing $\lambda = 0$ and using a Gronwall type argument as in the previous section we get

$$\max_{0 \le t \le T} \left(\|\dot{E}(t)\|_{L^2_\epsilon(\Omega)} + \|\nabla \times E(t)\|_{L^2_{\mu^{-1}}(\Omega)} \right) \le C \int_0^T \|F(\tau)\|_{L^2_{\epsilon^{-1}}(\Omega)} d\tau,$$

where C is independent of T, E and F. Alternatively, from (44) and using the arithmetic geometric mean inequality

$$\frac{\lambda}{2} \int_0^t e^{-\lambda \tau} \left[\|\dot{E}(\tau)\|^2_{L^2_\epsilon(\Omega)} + \|\nabla \times E(\tau)\|^2_{L^2_{\mu^{-1}}(\Omega)} \right] d\tau \le \int_0^t (F, e^{-\lambda \tau} \dot{E}) d\tau,$$

$$\le \int_0^t e^{-\lambda \tau} \left(\frac{1}{\lambda} \|F\|^2_{L_{\epsilon^{-1}}(\Omega)} + \frac{\lambda}{4} \|\dot{E}\|^2_{L^2_\epsilon(\Omega)} \right) d\tau.$$

Choosing $\lambda = 1/T$ gives,

$$\int_0^T \left(\|\dot{E}(\tau)\|^2_{L^2_\epsilon(\Omega)} + \|\nabla \times E(\tau)\|^2_{L^2_{\mu^{-1}}(\Omega)} \right) d\tau \le C T^2 \int_0^T \|F(\tau)\|^2_{L^2_{\epsilon^{-1}}(\Omega)} d\tau.$$

Motivated by the exponentially weighted energy estimates, and to provide error estimates for a general class of time stepping methods, we shall present an analysis based on the Fourier-Laplace transform used by [48] for error estimates for the semi-discrete wave equation (see also [19]). In doing so, we shall make critical use of some time stepping estimates of Lubich [96].

Let $\hat{E}$ denote the Fourier-Laplace transform of E defined by

$$\hat{E}(x, s) = \int_0^\infty E(x, t) e^{-st} dt$$

where $s = \varphi + i\omega$, $\varphi, \omega \in \mathbb{R}$, and $\varphi \ge \varphi_0$ for some fixed $\varphi_0 > 0$. Recalling that E satisfies (43) and that $\hat{\dot{E}} = s\hat{E}$, we see that $\hat{E} \in X$, the Fourier-Laplace transform of E, satisfies

$$s^2(\epsilon \hat{E}, \xi) + s(\sigma \hat{E}, \xi) + (\mu^{-1} \nabla \times \hat{E}, \nabla \times \xi) + s\langle Z \hat{E}_T, \xi_T \rangle = (\hat{F}, \xi) \quad (45)$$

for all $\xi \in X$. This is the Fourier-Laplace analogue of equation (11). It will prove convenient to use the s weighted norm on X given by

$$\|u\|^2_{X,s} = \|su\|^2_{L^2_\epsilon(\Omega)} + \|\nabla \times u\|^2_{L^2_{\mu^{-1}}(\Omega)} + \|su\|^2_{L^2_\sigma(\Omega)} + \|su_T\|^2_{L^2_Z(\Sigma)}$$

which has an obvious counterpart in our energy estimates, see (44).

It is now easy to prove that (45) has a solution. Let us define the sesquilinear form $a : X \times X \to \mathbb{C}$ by

$$a(\hat{E}, \xi) = s^2(\epsilon\hat{E}, \overline{\xi}) + s(\sigma\hat{E}, \overline{\xi}) + (\mu^{-1}\nabla\times\hat{E}, \overline{\nabla\times\xi}) + \langle sZ\hat{E}_T, \overline{\xi}_T \rangle \quad (46)$$

for any $\hat{E} \in X$ and $\xi \in X$, so that the solution $\hat{E} \in X$ of (45) satisfies

$$a(\hat{E}, \xi) = (\hat{F}, \overline{\xi}) \quad \forall \xi \in X.$$

The sesquilinear form is coercive, since choosing $\xi = s\hat{E}$ (corresponding to the exponentially weighted time derivative test function in the energy estimates) we have:

$$a(\hat{E}, s\hat{E}) = |s|^2 s(\epsilon\hat{E}, \overline{\hat{E}}) + |s|^2(\sigma\hat{E}, \overline{\hat{E}}) + \overline{s}(\mu^{-1}\nabla\times\hat{E}, \overline{\nabla\times\hat{E}})$$
$$+ |s|^2\langle Z\hat{E}_T, \overline{\hat{E}}_T \rangle. \quad (47)$$

Recalling that $s = \varphi + i\omega$, we have

$$\Re(a(\hat{E}, s\hat{E})) = \varphi\left[\|s\hat{E}\|^2_{L^2_\epsilon(\Omega)} + \|\nabla\times\hat{E}\|^2_{L^2_{\mu^{-1}}(\Omega)}\right] + \|s\hat{E}\|^2_{L^2_\sigma(\Omega)} + \|s\hat{E}_T\|^2_{L^2_Z(\Sigma)}$$
$$\geq \min(\varphi, 1)\|\hat{E}\|^2_{X,s} \geq \min(\varphi_0, 1)\|\hat{E}\|^2_{X,s}. \quad (48)$$

In the application in this section we can assume $0 < \varphi_0 \leq 1$. In some places, a useful choice is $\varphi_0 = O(1/T)$ which can be made small for large T.

The sesquilinear form $a(\cdot, \cdot)$ is also continuous since

$$|a(\hat{E}, \xi)| \leq \|s\hat{E}\|_{L^2_\epsilon(\Omega)}\|s\xi\|_{L^2_\epsilon(\Omega)} + \frac{1}{|s|}\|s\hat{E}\|_{L^2_\sigma(\Omega)}\|s\xi\|_{L^2_\sigma(\Omega)}$$

$$+ \|\nabla\times\hat{E}\|_{L^2_{\mu^{-1}}(\Omega)}\|\nabla\times\xi\|_{L^2_{\mu^{-1}}(\Omega)} + \frac{1}{|s|}\|s\hat{E}_T\|_{L^2_Z(\Sigma)}\|s\xi_T\|_{L^2_Z(\Sigma)}$$

$$\leq \max(1, 1/\varphi_0)\|\hat{E}\|_{X,s}\|\xi\|_{X,s}.$$

Now using the Lax-Milgram Lemma (see for example [104]) we see that (45) has a solution and

$$\varphi_0\|\hat{E}\|^2_{X,s} \leq |a(\hat{E}, s\hat{E})| = |(\epsilon^{-1/2}\hat{F}, s\epsilon^{1/2}\hat{E})| \leq \|\hat{F}\|_{L^2_{\epsilon^{-1}}(\Omega)}\|\hat{E}\|_{X,s},$$

and so we have, again assuming $\varphi_0 \leq 1$,

Theorem 2. *For any $\hat{F} \in L^2(\Omega)$, (45) has a unique solution $\hat{E} \in X$ such that*

$$\|\hat{E}\|_{X,s} \leq \frac{1}{\varphi_0}\|\hat{F}\|_{L^2_{\epsilon^{-1}}(\Omega)}.$$

Remark 2. Note that $\varphi \|\hat{\boldsymbol{E}}\|_{L^2_\epsilon(\Omega)} \leq \|\hat{\boldsymbol{E}}\|_{X,s}$. So the estimate of this theorem can be used to bound the L^2_ϵ norm of $\hat{\boldsymbol{E}}$.

Now we can transform back to the time domain to obtain a solution of the time Maxwell system. To do this we use Parseval's Theorem. This is not valid for the Laplace transform, but, following [19] and using the fact that $s = \varphi + i\omega$ for fixed φ and noting that $\hat{g}(s)$ is the standard Fourier transform of $\exp(-\varphi t)g(t) \in L^2(-\infty, \infty)$ where $g(t) = 0$ for $t < 0$, then

$$\int_{\varphi+i\mathbb{R}} |\hat{g}(s)|^2 ds = \int_0^\infty e^{-2\varphi t} |g(t)|^2 dt. \tag{49}$$

For a Hilbert space Y let

$$L^2(\mathbb{R}, Y) = \left\{ f : \mathbb{R} \to Y \mid \int_{-\infty}^\infty \|f(t)\|_Y^2 \, dt < \infty \right\},$$

then define (see e.g. [19] for more details),

$$H_\varphi^p(\mathbb{R}_+, Y) = \{ f : \mathbb{R} \to Y \mid \exp(-\varphi t)\partial^\ell f / \partial t^\ell \in L^2(\mathbb{R}, Y), \ 0 \leq \ell \leq p,$$
$$f = 0 \text{ for } t < 0 \}.$$

Using the estimate in Theorem 2 and (49) gives a weak solution $\boldsymbol{E}$ of the time domain problem. In particular if $\boldsymbol{F} \in H_\varphi^q(\mathbb{R}_+, L^2_{\epsilon-1}(\Omega))$ then $\boldsymbol{E}$ is such that

$$\boldsymbol{E} \in H_{\varphi_0}^{q+1}(\mathbb{R}_+, L^2_\epsilon(\Omega)), \ \nabla \times \boldsymbol{E} \in H_{\varphi_0}^q(\mathbb{R}_+, L^2_{\mu-1}(\Omega)), \ \boldsymbol{E}_T \in H_{\varphi_0}^{q+1}(\mathbb{R}_+, L^2_Z(\Sigma)).$$

For our variational problem to make sense we need that $\ddot{\boldsymbol{E}}(t)$ and $\boldsymbol{F}(t)$ to be well defined for each t. It suffices that $q = 2$ so that $\boldsymbol{E} \in H_{\varphi_0}^3(\mathbb{R}_+, L^2_\epsilon(\Omega))$ so that $\ddot{\boldsymbol{E}} \in C^0(\mathbb{R}_+, L^2_\epsilon(\Omega))$. For more details on the Fourier-Laplace approach to the wave equation and Maxwell's equations see [17, 19, 73, 96, 122]. An alternative approach to existence theory is to use semigroup theory which is beyond the scope of this introductory text [23, 135].

We have not implemented an explicit approximation to the divergence (this can be done via Lagrange multipliers, see [23]). But it turns out that (43) automatically gives control of the divergence. Recall our assumption that Ω is simply connected and Σ and Γ are disjoint closed surfaces and each is individually connected (more complex topology can be handled by more elaborate choices of the upcoming space S [11]). Then define the electrostatic potentials by

$$S = \{ p \in H^1(\Omega) \mid p = 0 \text{ on } \Gamma, \ p = c, \text{ constant on } \Sigma \}, \tag{50}$$

so that $\nabla S \subset X$ and moreover if $\nabla \times \boldsymbol{u} = 0$ then $\boldsymbol{u} = \nabla p$ for some $p \in S$ (see [9]) so that $\ker(\text{curl}) = \nabla S$. We thus have a small part of the de Rham diagram

$$S \xrightarrow{\nabla} X. \tag{51}$$

Choosing $\boldsymbol{\xi} = \nabla \eta$, $\eta \in S$, in (43) shows that

$$(\epsilon \ddot{\boldsymbol{E}}, \nabla \eta) + (\sigma \dot{\boldsymbol{E}}, \nabla \eta) = (\boldsymbol{F}, \nabla \eta) \quad \text{for all } \eta \in S. \tag{52}$$

Integrating by parts (or interpreting this formula using distributional derivatives) we have, recalling that $\boldsymbol{F} = -\dot{\boldsymbol{J}}^i$,

$$\nabla \cdot (\epsilon \ddot{\boldsymbol{E}}) + \nabla \cdot (\sigma \dot{\boldsymbol{E}} + \dot{\boldsymbol{J}}^i) = 0$$

which is the time derivative of the standard expression for charge conservation (see (9)). Thus, our variational formula enforces energy and charge conservation.

3.2   Semi-Discrete Problem

We obtain a semi-discrete approximation scheme by discretizing in space using a family of finite dimensional subspaces $X_h \subset X$ parameterized by $h > 0$. We will give concrete examples of X_h shortly, but we can proceed by seeking $\boldsymbol{E}_h = \boldsymbol{E}_h(t) \in X_h$ that satisfies

$$(\epsilon \ddot{\boldsymbol{E}}_h, \boldsymbol{\xi}_h) + (\sigma \dot{\boldsymbol{E}}_h, \boldsymbol{\xi}) + (\mu^{-1} \nabla \times \boldsymbol{E}_h, \nabla \times \boldsymbol{\xi}_h)$$
$$+ \langle Z \dot{\boldsymbol{E}}_{h,T}, \boldsymbol{\xi}_{h,T} \rangle = (\boldsymbol{F}, \boldsymbol{\xi}_h) \quad \forall \boldsymbol{\xi}_h \in X_h, \tag{53}$$

subject to the initial conditions (41). This is a system of second order linear ordinary differential equations and so has a solution. Since the method is conforming and semi-discrete, the energy analysis still holds. But discrete charge conservation can only be realized if we have a sufficiently large subspace $S_h \subset S$ such that $\nabla S_h \subset X_h$ and this is one reason for our upcoming choice of edge finite elements.

   We can now use an energy analysis to provide error estimates for the semi-discrete scheme. For any $\boldsymbol{\eta}_h = \boldsymbol{\eta}_h(t) \in X_h$ sufficiently smooth in time, let $\boldsymbol{e}_h = \boldsymbol{\eta}_h - \boldsymbol{E}_h$ then

$$(\epsilon \ddot{\boldsymbol{e}}_h, \dot{\boldsymbol{e}}_h) + (\sigma \dot{\boldsymbol{e}}_h, \dot{\boldsymbol{e}}_h) + (\mu^{-1} \nabla \times \boldsymbol{e}_h, \nabla \times \dot{\boldsymbol{e}}_h) + \langle Z \dot{\boldsymbol{e}}_{h,T}, \dot{\boldsymbol{e}}_{h,T} \rangle =$$
$$(\epsilon (\ddot{\boldsymbol{\eta}}_h - \ddot{\boldsymbol{E}}), \dot{\boldsymbol{e}}_h) + (\sigma (\dot{\boldsymbol{\eta}}_h - \dot{\boldsymbol{E}}), \dot{\boldsymbol{e}}_h) + (\mu^{-1} \nabla \times (\boldsymbol{\eta}_h - \boldsymbol{E}), \nabla \times \dot{\boldsymbol{e}}_h)$$
$$+ \langle Z (\dot{\boldsymbol{\eta}}_h - \dot{\boldsymbol{E}})_T, \dot{\boldsymbol{e}}_{h,T} \rangle$$

where we have used the fact that $\boldsymbol{E}_h$ and $\boldsymbol{E}$ both satisfy (53) when we choose the test function $\boldsymbol{\xi} = \boldsymbol{\xi}_h = \dot{\boldsymbol{e}}_h \in X_h$. This equality may then be rewritten as

$$\frac{1}{2}\frac{d}{dt}\left(\|\dot{e}_h\|_{L^2_\epsilon(\Omega)}^2 + \|\nabla\times e_h\|_{L^2_{\mu^{-1}}(\Omega)}^2\right) + \|\dot{e}_h\|_{L^2_\sigma(\Omega)}^2 + \|\dot{e}_{h,T}\|_{L^2_Z(\Sigma)}^2$$

$$= (\epsilon(\ddot{\eta}_h - \ddot{E}), \dot{e}_h) + (\sigma(\dot{\eta}_h - \dot{E}), \dot{e}_h) + (\mu^{-1}\nabla\times(\eta_h - E), \nabla\times\dot{e}_h)$$

$$+ \langle Z(\dot{\eta}_h - \dot{E})_T, \dot{e}_{h,T}\rangle.$$

Integrating in time and using the Cauchy-Schwarz inequality, together with the fact that we can choose $\eta_h(0) = E_h(0) = 0$ and $\dot{\eta}_h(0) = \dot{E}_h(0) = 0$ we have (also integrating also the curl term by parts on the right hand side)

$$\frac{1}{2}\left(\|\dot{e}_h(t)\|_{L^2_\epsilon(\Omega)}^2 + \|\nabla\times e_h(t)\|_{L^2_{\mu^{-1}}(\Omega)}^2\right) + \int_0^t\left(\|\dot{e}_h\|_{L^2_\sigma(\Omega)}^2 + \|\dot{e}_{h,T}\|_{L^2_Z(\Sigma)}^2\right)d\tau$$

$$\le \int_0^t\left\{\|\ddot{\eta}_h - \ddot{E}\|_{L^2_\epsilon(\Omega)}\|\dot{e}_h\|_{L^2_\epsilon(\Omega)} + \|\nabla\times(\dot{\eta}_h - \dot{E})\|_{L^2_{\mu^{-1}}(\Omega)}\|\nabla\times e_h\|_{L^2_{\mu^{-1}}(\Omega)}\right.$$

$$\left. + \|\dot{\eta}_h - \dot{E}\|_{L^2_\sigma(\Omega)}\|\dot{e}_h\|_{L^2_\sigma(\Omega)} + \|(\dot{\eta}_h - \dot{E})_T\|_{L^2_Z(\Sigma)}\|\dot{e}_{h,T}\|_{L^2_Z(\Sigma)}\right\}d\tau$$

$$+ \|\nabla\times(\eta_h - E)(t)\|_{L^2_{\mu^{-1}}(\Omega)}\|\nabla\times e_h(t)\|_{L^2_{\mu^{-1}}(\Omega)}.$$

Note that if we could choose the function η_h such that $(\mu^{-1}\nabla\times(\eta_h - E), \nabla\times\xi_h) = 0$ for all $\xi_h \in X_h$, then we could avoid the integration by parts step, avoid the curl terms on the right hand side of this inequality, and hence improve the final error estimate. This is possible if we use the edge elements that we shall describe shortly.

Let

$$(\mathcal{E}(e_h))^2 = \frac{1}{2}\left(\|\dot{e}_h(t)\|_{L^2_\epsilon(\Omega)}^2 + \|\nabla\times e_h(t)\|_{L^2_{\mu^{-1}}(\Omega)}^2\right)$$

$$+ \int_0^t\left(\|\dot{e}_h(\tau)\|_{L^2_\sigma(\Omega)}^2 + \|\dot{e}_{h,T}(\tau)\|_{L^2_Z(\Sigma)}^2\right)d\tau. \tag{54}$$

We now apply a Gronwall type argument as before to derive the error estimate

$$\max_{0\le t\le T}\mathcal{E}(\eta_h - E_h)(t) \le C\mathcal{R}(\eta_h - E)$$

for some constant C independent of T, h, η_h and E where

$$\mathcal{R}(\eta_h - E) = \int_0^T\|\ddot{\eta}_h - \ddot{E}\|_{L^2_\epsilon(\Omega)} + \|\nabla\times(\dot{\eta}_h - \dot{E})\|_{L^2_{\mu^{-1}}(\Omega)}d\tau$$

$$+ \left(\int_0^T\|\dot{\eta}_h - \dot{E}\|_{L^2_\sigma(\Omega)}^2 d\tau + \int_0^T\|(\dot{\eta}_h - \dot{E})_T\|_{L^2_Z(\Sigma)}^2 d\tau\right)^{1/2}$$

$$+ \max_{0\le t\le T}\|\nabla\times(\eta_h - E)(t)\|_{L^2_{\mu^{-1}}(\Omega)}. \tag{55}$$

Using the triangle inequality

$$\mathcal{E}(E - E_h)(t) \leq \mathcal{E}(E - \eta_h)(t) + \mathcal{E}(\eta_h - E_h)(t).$$

We have proved the following error estimate.

Lemma 1. *The semi-discrete solution satisfies*

$$\max_{0 \leq t \leq T} \mathcal{E}(E - E_h)(t) \leq \max_{0 \leq t \leq T} \mathcal{E}(E - \eta_h) + \mathcal{R}(E - \eta_h) \quad \forall \eta_h \in X_h.$$

Using the fact that

$$\frac{1}{2} \frac{d}{dt}(\epsilon(E - E_h), (E - E_h)) = (\epsilon(\dot{E} - \dot{E}_h), (E - E_h)),$$

and integrating both sides, we have

$$\frac{1}{2} \max_{0 \leq t \leq T} \|(E - E_h)(t)\|_{L^2_\epsilon(\Omega)} \leq \int_0^T \|\dot{E} - \dot{E}_h\|_{L^2_\epsilon(\Omega)} d\tau$$

$$\leq T \max_{0 \leq t \leq T} \|(\dot{E} - \dot{E}_h)(t)\|_{L^2_\epsilon(\Omega)} \leq T\mathcal{E}(E - E_h).$$

So we have the corollary

Corollary 1. *The semi-discrete solution satisfies*

$$\max_{0 \leq t \leq T} \|(E - E_h)(t)\|_{L^2_\epsilon(\Omega)} \leq T \left(\max_{0 \leq t \leq T} \mathcal{E}(E - \xi_h) + \mathcal{R}(\eta_h - E) \right) \quad (56)$$

for any $\eta_h \in X_h$.

To derive other estimates we can use the Fourier-Laplace transform. The spatially discrete Fourier-Laplace domain solution $\hat{E}_h \in X_h$ of (53) satisfies

$$a(\hat{E}_h, \xi_h) = (\hat{F}, \bar{\xi}_h) \quad \forall \xi_h \in X_h. \quad (57)$$

By Cea's Lemma (see for example, [104]) we have the existence of $\hat{E}_h$ and, assuming $\varphi_0 \leq 1$ the following theorem holds:

Theorem 3. *Problem (57) has a unique solution $\hat{E}_h \in X_h$ and this solution satisfies the error estimate*

$$\|\hat{E} - \hat{E}_h\|_{X,s} \leq C \frac{|s|}{\varphi_0^2} \|\hat{E} - \xi_h\|_{X,s}$$

for any $\xi_h \in X_h$ where C is independent of $\hat{E}$, $\hat{E}_h$, s, and h.

Remark 3.

- This result reduces the error estimate to an approximation problem. The right hand side is small if $\hat{E}$ can be well approximated by a function in X_h.
- The factor $|s|$ on the right hand side is typical for hyperbolic problems [52]. It results in an unbalanced estimate in the time domain since the time domain norm on the left hand side involves one less time derivative than the right hand side.

Now we can transform back to the time domain to obtain a semi-discrete error estimate (for other norms see [48]) using (49).

Theorem 4. *The semi-discrete solution $\hat{E}_h \in X_h$ of (53) satisfies the error estimate*

$$
\int_0^\infty e^{-2\varphi t} \left(\|\dot{E} - \dot{E}_h\|_{L^2_\epsilon(\Omega)}^2 + \|\nabla\times(E - E_h)\|_{L^2_{\mu^{-1}}(\Omega)}^2 + \|(\dot{E} - \dot{E}_h)_T\|_{L^2_Z(\Sigma)}^2 \right) dt
$$

$$
\leq \frac{1}{\varphi_0^2} \int_0^\infty e^{-2\varphi t} \left(\|\ddot{E} - \ddot{\xi}_h\|_{L^2_\epsilon(\Omega)}^2 + \|\nabla\times(\dot{E} - \dot{\xi}_h)\|_{L^2_{\mu^{-1}}(\Omega)}^2 + \|\dot{E} - \dot{\xi}_h\|_{L^2_\sigma(\Omega)}^2 \right.
$$

$$
\left. + \|(\ddot{E} - \ddot{\xi}_h)_T\|_{L^2_Z(\Sigma)}^2 \right) dt \tag{58}
$$

for all $\xi_h = \xi_h(t) \in X_h$.

Remark 4.

- An estimate in terms of powers of h can be obtained once we have given details of X_h.
- We could have established this estimate using the exponential weighted energy analysis culminating in (44) outlined previously by choosing $\lambda = \varphi$. We presented the Fourier-Laplace approach because of its relevance to the next section.

3.3 Discretization in Time

Whatever the choice of X_h, we can now proceed to a fully discrete problem by applying a time-discretization technique to the system of ordinary differential equations given by (53). Using the popular leap-frog discretization (see Sect. 2) we seek $E_h^n \in X_h$ for $n \geq 0$ that approximates $E(\cdot, t_n)$, $t_n = n\,\Delta t$, and satisfies

$$
\frac{1}{\Delta t^2}(\epsilon(E_h^{n+1} - 2E_h^n + E_h^{n-1}), \xi_h) + \frac{1}{2\Delta t}(\sigma(E_h^{n+1} - E_h^{n-1}), \xi_h) \tag{59}
$$

$$
+ (\mu^{-1}\nabla\times E_h^n, \nabla\times\xi_h) + \frac{1}{2\Delta t}\langle Z(E_h^{n+1} - E_h^{n-1})_T, \xi_{h,T}\rangle = (F^n, \xi_h)
$$

for all $\xi_h \in X_h$ together with the initial condition $E_h^n = 0$ for $n \leq 0$. Although a semi-discrete error analysis was given some time ago [102, 103], it is only relatively recently that this fully discrete scheme has been analyzed [42]. In the previous section we saw that once we have evaluated the local truncation error, we can obtain a fully discrete error estimate by mimicking the energy proof to obtain stability. This approach is given in detail in the excellent review [42] for edge elements and we direct the reader there for details.

Equation (59) reveals the major disadvantage of the conforming finite element approach described here. Suppose we have a basis $\{\boldsymbol{\xi}_j^h\}_{j=1}^{J_h}$ of X_h then, of course there are scalar coefficients E_j^n, $j = 1, \cdots, J_h$, such that

$$E_h^n = \sum_{j=1}^{J_h} E_j^n \boldsymbol{\xi}_j^h \tag{60}$$

and similarly for E_h^{n+1} and E_h^{n-1}. Substituting (60) into (59) and choosing the test function $\boldsymbol{\xi}_h = \boldsymbol{\xi}_i^h$ we obtain a matrix equation equivalent to (59). More precisely, let

$$\mathbf{E}^n = (E_1^n, E_2^n, \dots, E_{J_h}^n)^T$$

with similar expressions for $\mathbf{E}^{n+1}, \mathbf{E}^{n-1}$. Let S, B, M^ϵ and M^σ denote the $J_h \times J_h$ matrices with

$$M_{i,j}^\epsilon = (\epsilon \boldsymbol{\xi}_j^h, \boldsymbol{\xi}_i^h), \qquad\qquad M_{ij}^\sigma = (\sigma \boldsymbol{\xi}_j^h, \boldsymbol{\xi}_i^h),$$

$$S_{i,j} = (\mu^{-1} \nabla \times \boldsymbol{\xi}_j^h, \nabla \times \boldsymbol{\xi}_i^h), \quad B_{i,j} = Z \langle \boldsymbol{\xi}_{j,T}^h, \boldsymbol{\xi}_{i,T}^h \rangle,$$

and let $\mathbf{F}^n = (F_1^n, F_2^n, \dots, F_{J_h}^n)^T$ where $F_i^n = (\boldsymbol{F}, \boldsymbol{\xi}_i^h)$, $1 \le i \le J_h$. Then

$$\left(M^\epsilon + \frac{\Delta t}{2} M^\sigma + \frac{\Delta t}{2} B \right) \mathbf{E}^{n+1} = 2M^\epsilon \mathbf{E}^n - M^\epsilon \mathbf{E}^{n-1} - \Delta t^2 S \mathbf{E}^n$$

$$+ \frac{\Delta t}{2} M^\sigma \mathbf{E}^{n-1} + \frac{\Delta t}{2} B \mathbf{E}^{n-1} + \mathbf{F}^n. \tag{61}$$

Even though we applied an "explicit" time stepping scheme, we need to solve a linear system of equations at each time step having the matrix on the left hand side of (61). This has lead many authors to abandon edge elements for time dependent problems and instead develop discontinuous Galerkin methods to discretize (38), see for example [36, 37, 46, 64, 69–71]. This is a major modern trend. However, several practical codes use edge elements [93, 112, 113], and it is that approach we take first.

We have chosen to use $X_h \subset X$. Because we must solve a matrix problem at each time step, explicit methods no longer hold an overwhelming attraction. In addition, there are some cases when (53) can become very stiff and impose a crushing CFL condition. Causes can be small elements in the mesh, exotic materials present, or the use of high order finite elements. So both engineers [112–114] and mathematicians [89, 94, 98] have considered the use of implicit methods. The usual choice is backward Euler [45], a Newmark type method [113] or the trapezoidal rule [94].

Now suppose that we wish to use an alternative multistep method. To do this, we need to take a short detour into multistep methods! For a linear ordinary differential equation

$$y' = Ay + G, \quad y(0) = 0$$

where $y \in \mathbb{R}^n$ and $A \in \mathbb{R}^{n \times n}$, and G is a smooth vector function, a general k-step multistep method takes the form

$$\sum_{j=0}^{k} \alpha_j y_{n-j} = \Delta t \sum_{j=0}^{k} \beta_j (Ay_{n-j} + G_{n-j}) \tag{62}$$

where $y_n \approx y(t_n)$ and where the coefficients $\{\alpha_j\}$ and $\{\beta_j\}$ define the method. We will assume $\alpha_0/\beta_0 > 0$ (so the method is implicit). A method like this can be conveniently summarized using generating functions. Let $z \in \mathbb{C}, z \neq 0$, and define

$$Y^{\Delta t} = \sum_{n=0}^{N} z^{-n} y_n, \quad G^{\Delta t} = \sum_{n=0}^{N} z^{-n} G_n$$

where $|z|$ is large enough that these sum's converge. We assume $y_n = 0$ and $G_n = 0$ if $n \leq 0$.

Multiplying (62) by z^{-n} and adding we obtain

$$\sum_{n=0}^{\infty} \sum_{j=0}^{k} \alpha_j z^{-j} (y_{n-j} z^{-n+j}) = \Delta t \sum_{n=0}^{\infty} \sum_{j=0}^{k} \beta_j z^{-j} \left(A Y_{n-j} z^{-n+j} + G_{n-j} z^{-n+j} \right).$$

So if we define

$$\delta(z) = \frac{\sum_{j=0}^{k} \alpha_j z^{-j}}{\sum_{j=0}^{k} \beta_j z^{-j}} \tag{63}$$

we can summarize (62) by writing

$$\frac{\delta(z)}{\Delta t} Y^{\Delta t} = A Y^{\Delta t} + G^{\Delta t} \tag{64}$$

where this holds for all $z \in \mathbb{C}$ with $|z|$ large enough. Although this nicely summarizes the method, practically (62) is the way to compute the solution and (64) just a convenient way of writing the scheme.

In fact, only two choices of multistep method are most relevant:

1. *Backward Euler (BE).* Here

$$y_n - y_{n-1} = \Delta t (Ay_n + G_n)$$

so $\delta(z) = 1 - z^{-1}$.

2. *Backward Differentiation Formula (BDF2)*. Here

$$3y_n - 4y_{n-1} + y_{n-2} = 2\Delta t(Ay_n + G_n),$$

and $\delta(z) = (3 - 4z^{-1} + z^{-2})/2$.

We have used the notation of the z-transform ($z = \zeta^{-1}$ in the notation of Lubich [96]) to draw attention to the similarity between the approach here and analogue to digital signal processing [109].

The fully discrete Maxwell problem is to seek $E_h^n \in X_h$, $n = 0, 1, \ldots$ with $E_h^n = 0$ for $n \le 0$ such that if

$$E^{\Delta t} = \sum_{n=0}^{\infty} z^{-n} E_h^n, \quad F^{\Delta t} = \sum_{n=0}^{\infty} z^{-n} F^n$$

then

$$\frac{\delta^2(z)}{\Delta t^2}(\epsilon E^{\Delta t}, \xi_h) + \frac{\delta(z)}{\Delta t}(\sigma E^{\Delta t}, \xi_h) + (\mu^{-1}\nabla \times E^{\Delta t}, \nabla \times \xi_h)$$

$$+ \frac{\delta(z)}{\Delta t}\langle Z E_T^{\Delta t}, \xi_{h,T}\rangle = (F^{\Delta t}, \xi_h) \quad \forall \xi_h \in X_h. \tag{65}$$

Using, for example, BE time stepping gives

$$\frac{1}{\Delta t^2}(\epsilon(E_h^{n+1} - 2E_h^n + E_h^{n-1}), \xi_h) + \frac{1}{\Delta t}(\sigma(E_h^{n+1} - E_h^n), \xi_h)$$

$$+ (\mu^{-1}\nabla \times E_h^{n+1}, \nabla \times \xi_h) + \frac{Z}{\Delta t}\langle(E_h^{n+1} - E_h^n)_T, \xi_{h,T}\rangle = (F^{n+1}, \xi_h) \quad \forall \xi_h \in X_h.$$

so given E_h^n and E_h^{n-1} we compute $E_h^{n+1} \in X_h$ by solving

$$\frac{1}{\Delta t^2}(\epsilon E_h^{n+1}, \xi_h) + \frac{1}{\Delta t}(\sigma E_h^{n+1}, \xi_h) + (\mu^{-1}\nabla \times E_h^{n+1}, \nabla \times \xi_h)$$

$$+ \frac{Z}{\Delta t}\langle(E_h^{n+1})_T, \xi_{h,T}\rangle = \frac{1}{\Delta t^2}(\epsilon(2E_h^n - E_h^{n-1}), \xi_h) + \frac{1}{\Delta t}(\sigma E_h^n, \xi_h)$$

$$+ \frac{Z}{\Delta t}\langle(E_h^n)_T, \xi_{h,T}\rangle + (F^{n+1}, \xi_h)$$

for all $\xi_h \in X_h$. At each time step we must solve a matrix equation for E_h^{n+1}, but the matrix is symmetric positive definite (compared to the indefinite, non Hermitian matrix resulting from the time harmonic Maxwell problem). Note that error estimates for the BE scheme for Maxwell's equations were first proved in [45] using energy arguments.

We are now going to prove fully discrete error estimates using the convolution quadrature theory of Lubich [95, 96]. In particular we shall use the following theorem (in fact we quote a special case of Theorem 3.1 in [96]).

Theorem 5. *Let Y denote a Hilbert space, and $K(s) : Y \to Y'$, $\Re(s) \geq \varphi_0$, be analytic and satisfy the bound*

$$\|K(s)\| \leq C|s|^\mu, \quad Re(s) > \varphi_0,$$

for some $\mu > 0$. Let $\delta(z)$ correspond to an A-stable multistep method of order $p = 1, 2$ such that

$$\delta(z) \text{ has no poles on the unit circle} \tag{66}$$

then for $m \geq p + 2 + \mu$ and smooth data $\boldsymbol{F} : [0, T] \to Y'$ with $\boldsymbol{F}(0) = \boldsymbol{F}'(0) = F^{(m-1)}(0) = 0$ we have

$$\|K(\partial_t^{\Delta t})\boldsymbol{F}(t) - K(\partial_t)\boldsymbol{F}(t)\| \leq C \Delta t^p \int_0^t \|F^{(m)}(\tau)\| d\tau.$$

Remark 5. The theorem excludes the trapezoidal rule, although this is known to work perfectly well for the problem under consideration here [94]. In our case $\mu = 1$ and so $m \geq p + 3$. If $p = 2$ we need $m = 5$ and $F(0) = F'(0) = \cdots = F^{(4)}(0) = 0$! This is surely very smooth data in time.

To apply this theorem we define suitable solution operators. Using the sesquilinear form from (46), let

$$V(s) : X \to X',$$

where X' is the dual space of X, be such that for $\boldsymbol{u} \in X$, $V(s)u \in X'$ satisfies

$$(V(s)\boldsymbol{u}, \overline{\boldsymbol{\xi}}) = a(\boldsymbol{u}, \boldsymbol{\xi}) \quad \forall \boldsymbol{\xi} \in X.$$

Then from (48) we have the coercivity bound

$$|(V(s)\boldsymbol{u}, \overline{s\boldsymbol{u}})| = |a(\boldsymbol{u}, s\boldsymbol{u})| \geq \varphi_0 \|\boldsymbol{u}\|_{X,s}^2.$$

Theorem 2 shows that $V(s)^{-1}$ exists. Choosing $\boldsymbol{u} = V(s)^{-1}\boldsymbol{\xi}$, the above inequality shows that

$$\varphi_0 \|V^{-1}(s)\boldsymbol{\xi}\|_{X,s}^2 \leq |\overline{s}(\boldsymbol{\xi}, \overline{V^{-1}(s)\boldsymbol{\xi}})| \leq |s| \|\overline{\boldsymbol{\xi}}\|_{X',s} \|V^{-1}(s)\boldsymbol{\xi}\|_{X,s}.$$

Unfortunately, Lubich's Theorem 5 above requires us to use s-independent norms. So we define

$$\|\boldsymbol{u}\|_X^2 = \|\boldsymbol{u}\|_{L_\epsilon^2(\Omega)}^2 + \|\nabla \times \boldsymbol{u}\|_{L_{\mu-1}^2(\Omega)}^2 + \|\boldsymbol{u}\|_{L_\sigma^2(\Omega)}^2 + \|\boldsymbol{u}_T\|_{L_Z^2(\Sigma)}^2.$$

For each fixed s this is equivalent to the $\| \cdot \|_{X,s}$ norm since

$$\|u\|^2_{X,s} = \varphi_0^2 \frac{|s|^2}{\varphi_0^2} \|u\|^2_{L^2_\epsilon(\Omega)} + \| \nabla \times u\|^2_{L^2_{\mu^{-1}}(\Omega)} + \varphi_0^2 \frac{|s|^2}{\varphi_0^2} \|u\|^2_{L^2_\sigma(\Omega)} + \varphi_0^2 \frac{|s|^2}{\varphi_0^2} \|u_T\|^2_{L^2_Z(\Sigma)}$$

$$\geq \varphi_0^2 \|u\|^2_X.$$

under our assumption that $\varphi_0 \leq 1$. In the same way,

$$\|u\|^2_{X,s} \leq \frac{|s|^2}{\varphi_0^2} \|u\|^2_X.$$

Thus, we have the following:

Corollary 2. *Using the $\| \cdot \|_{X,s}$ norm, the operator $V : X \to X'$ defined by (62) is invertible, analytic for s with $\varphi \geq \varphi_0$ and,*

$$\|V^{-1}(s)\| \leq \frac{|s|}{\varphi_0}.$$

The same conclusion holds using the $\| \cdot \|_X$ norm except that the bound becomes

$$\|V^{-1}(s)\| \leq \frac{|s|}{\varphi_0^2}.$$

The semi-discrete problem can also be written as an operator equation. Define $V_h : X_h \to X_h$ by

$$(V_h(s)u_h, \overline{\xi}_h) = a(u_h, \xi_h) \quad \text{for all } \xi_h \in X_h$$

then since $X_h \subset X$ and

$$(V_h(s)u_h, \overline{\xi}_h) = (V(s)u_h, \overline{\xi}_h)$$

the coercivity of $V(s)$ implies the coercivity of $V_h(s)$ and, as a map from X to X' using the $\| \cdot \|_X$ norm,

$$\|V_h^{-1}(s)\| \leq \frac{|s|}{\varphi_0^2}. \tag{67}$$

We may write the solution $\hat{E} \in X$ of (45) as

$$\hat{E} = V^{-1}(s)\hat{F}$$

and the discrete Fourier-Laplace transform solution $\hat{E}_h \in X_h$ of (57)

$$\hat{E}_h = V_h^{-1}(s)P_h\hat{F}$$

where P_h denotes the $L^2(\Omega)$ projection onto X_h. Following Lubich [96], the time domain solution

$$E = V^{-1}(\partial_t)F$$

$V^{-1}(\partial_t)F$ is understood as a convolution so that

$$E = V^{-1}(\partial_t)F = \mathscr{L}^{-1}(V^{-1}(s)\hat{F}) \qquad (68)$$

where $\mathscr{L}^{-1}$ is the inverse Laplace transform. As we saw in Sect. 3.1, this provides a rather round about way to prove the existence of a solution to Maxwell's equations.
  The solution $E_h \in X_h$ of the semi-discrete problem is given in the same way

$$E_h = V_h^{-1}(\partial_t)P_h\hat{F}. \qquad (69)$$

The fully discrete approximation in time is then

$$E^{\Delta t}(t) = V^{-1}(\partial_t^{\Delta t})F(t) = \sum_{j \geq 0} w_j P_h F(t - j\,\Delta t)$$

and the weights are operators given by the identity [96]

$$\sum_{j=0}^{\infty} w_j z^{-j} = V^{-1}\left(\frac{\delta(z)}{\Delta t}\right)$$

for all $|z|$ large enough. An important point about convolution quadrature is that if two convolution operators $K_1(s), K_2(s)$ both analytic in s for $Re(s) \geq \varphi_0$ and satisfying the bound

$$\|K_j\| \leq C|s|^{\mu_j}, \quad \mu \geq 0, \ Re(s) \geq \varphi_0,$$

we have

$$K_1(\partial_t^{\Delta t})K_2(\partial_t^{\Delta t}) = (K_1 K_2)(\partial_t^h).$$

Hence

$$V_h(\partial_t^{\Delta t})E_h = P_h F,$$

and we see that (69) does indeed correspond to solving (53).
  We can expand

$$E_h^{\Delta t} - E = (E_h^{\Delta t} - E_h) + (E_h - E)$$

where we will evaluate these terms at the timestep. Then $(E_h - E)$ is estimated via the semi-discrete approximation in Corollary 1. The term

$$E_h^{\Delta t} - E_h = V_h^{-1}(\partial_t^{\Delta t})P_h F - V_h^{-1}(\partial_t)P_h F$$

and this can be estimated by Lubich's result (Theorem 5) with $K(s) = V_h^{-1}(s)$. We have thus sketched how to prove the following theorem:

Theorem 6. *Let the time discretization be a pth order A-stable multistep method satisfying (66). For smooth compatible data* $\boldsymbol{F}$ *(satisfying the conditions of Theorem 5 with* $m = p + 3$*) we have, for* $0 \leq t_n \leq T$,

$$\|\boldsymbol{E}_h^n - \boldsymbol{E}(\cdot, n \Delta t)\|_{L_{\varepsilon}^2(\Omega)} = O(\Delta t^p + \mathcal{R}(\boldsymbol{E} - \boldsymbol{\eta}_h) + \mathcal{E}(\boldsymbol{E} - \boldsymbol{\eta}_h))$$

for all $\boldsymbol{\eta}_h \in X_h$ *where* $\mathcal{R}$ *and* $\mathcal{E}$ *appear on the right hand side of (56).*

3.4 Finite Element Spaces

So far we have not been at all specific about how to construct the subspaces $X_h \subset X$. It appears that provided the terms on the right hand side of the estimate in Theorem 6 can be made small, we can make use of any choice of elements. We now give a very brief classical introduction to finite elements (see [10, 11, 29, 81, 82] for the more enlightening but also more elaborate finite element exterior calculus viewpoint).

In general, to construct a finite element space X_h we need a family of meshes indexed by $h > 0$. Suppose that Ω is covered by a collection of open sets or elements $\mathcal{T}_h = \{\Omega_k^h\}_{k=1}^{N_h}$. Let $h_k = \text{diam}(\Omega_k^h)$ for each k (the diameter of the smallest circumscribed sphere). Then $h = \max_k h_k$. As h decreases the largest element in the mesh becomes smaller and the mesh becomes finer and finer. These elements need to be a conforming finite element mesh [43] in the sense that:

1. $\overline{\Omega} = \cup_k \overline{\Omega}_k^h$, where overline denotes closure.
2. $\Omega_j^h \cap \Omega_k^h = \emptyset$ if $j \neq k$.
3. If $\overline{\Omega}_j^h \cap \overline{\Omega}_k^h \neq \emptyset$ then one of the following conditions hold:
   - The elements meet at a single point that is a vertex for both elements.
   - The elements meet along a common edge and the endpoints of the edge are vertices of the two elements.
   - The elements meet at a common face and the vertices of the face are vertices of both elements.

We also need the elements to be "regular" or non-degenerate by which we mean that if ρ_k denotes the diameter of the largest inscribed sphere in the kth element then there is a constant $\gamma > 0$ such that

$$\frac{h_k}{\rho_k} < \gamma \text{ for } 1 \leq k \leq N_h, \text{ and for all } h > 0.$$

This condition prevents elements from becoming too flat as the mesh is refined. These conditions are from [43] where a more detailed discussion of meshes can be found.

Some of these requirements can be relaxed for discontinuous Galerkin methods, but in our case these conditions rule out meshes of mixed tetrahedral and cube elements. To have such mixed meshes requires a third element: pyramidal elements having one rectangular face and four triangular faces. Finite elements of this type have been constructed in [67, 108] and we direct the reader there for more details.

We shall now give a somewhat brief description of finite elements from the classical point of view. This is elaborated in [104]. For a more sophisticated view from the point of view of finite element exterior calculus see [10, 11, 29, 81, 82]. This theory motivates some of the choices we shall make and has resulted in several technical advances that greatly improve the error estimates for finite element methods. For a presentation related to computational electromagnetism see [42]. Because of the amount of overhead needed to lay the foundations for this approach, we shall adopt the classical approach.

The simplest family of elements are subspaces of $H^1(\Omega)$. These scalar elements are described in all textbooks on finite element methods. Let P_p denote the set of all polynomials of total degree at most p (we denote by $P_p(f)$ the space of polynomials of degree at most p two variables in the plane of f, similarly $P_p(e)$ for an edge). Then the scalar subspace $\tilde{S}_h \subset H^1(\Omega)$ of continuous piecewise p degree polynomials is defined for tetrahedra by

$$\tilde{S}_h = \{u_h \in H^1(\Omega) \mid u_h|_{\Omega_k} \in P_p \text{ for all tetrahedra } \Omega_k \in \mathscr{T}_h\}.$$

Functions in $\tilde{S}_h$ are continuous because continuity is required for any piecewise polynomial to be in $H^1(\Omega)$ [43].

On each element, the finite element can be described by giving $\dim(P_p)$ scalar numbers called degrees of freedom of the element. For tetrahedron Ω_k these are given by functionals of $u_h|_{\Omega_k}$ as follows: Any function $u_h \in P_p$ is uniquely specified by giving values for:

1. Vertex values $u_h(a_{k,j}) 1 \leq j \leq 4$ where $a_{k,j}$ is the jth vertex of Ω_k.
2. The integrals $\int_e u_h q \, ds$ for all $q \in P_{p-2}(e)$ where $P_{p-2}(e)$ is the set of polynomials in arc length s along each of the six edges e of Ω_k.
3. The integrals $\int_f u_h q \, dA$ for all $q \in P_{p-3}(f)$ for each of the four faces f of Ω_k.
4. The integrals $\int_{\Omega_k} u_h q \, dV$ for all $q \in P_{p-4}$.

In this definition we understand that $P_\ell = \emptyset$ if $\ell < 0$. By specifying these degrees of freedom on all vertices (and edges, faces and volumes as needed), we uniquely specify $u_h \in \tilde{S}_h$ and the degrees of freedom shared between neighboring elements guarantee the continuity of u_h. For a given polynomial degree, these degrees of freedom can then be used to construct a basis of $\tilde{S}_h$ (if p is to vary from element to element the situation is much more complex, see for example [55, 56]).

For parallelepipeds with edges parallel to the coordinate axes, the situation is similar. Now the fundamental polynomial space is $Q_{r,s,t}$, the tensor product space of polynomials of degree at most r in x_1, s in x_2 and t in x_3. For parallelepiped elements with edges parallel to the coordinate axis we now define

$$\tilde{S}_h = \{u_h \in H^1(\Omega) \mid u_h|_{\Omega_h} \in Q_{p,p,p} \quad \text{for all parallelepipeds in } \mathscr{T}_h\}.$$

The degrees of freedom are similar to the tetrahedral case:

1. Vertex degrees: $u_h(a_{k,j})$ for the eight vertices of Ω_k.
2. Edge degrees: $\int_e u_h q\,ds$ for all $q \in P_{k-2}(e)$ and each edge of Ω_k.
3. $\int_f u_{h,q}\,dA \quad \forall\, q \in Q_{p-2,p-2}(f)$ for each of the 6 faces f of Ω_k.
4. $\int_{\Omega_k} u_h q\,dV \quad \forall\, q \in Q_{p-3,p-3,p-3}(\Omega_k)$.

Again these degrees are H^1 conforming (i.e. specifying all degrees on vertices, edges, faces and volumes, specifies a function $u_h \in \tilde{S}_h \subset H^1(\Omega)$) and unisolvent ($u_h$ is uniquely specified by giving the degrees of freedom).

For tetrahedral or hexahedral elements the degrees of freedom define an interpolation operator $\pi_S : H^2(\Omega) \to \tilde{S}_h$.

Continuous piecewise linear elements have an important part to play in the numerical solution of Maxwell's equations by discretizing the scalar space S defined in (50):

$$S_h = \{u_h \in \tilde{S}_h \mid u_h = 0 \text{ on } \Gamma, \ u_h = \text{ constant on } \Sigma\}.$$

If we wish to mimic the argument leading to the conservation of charge result in (52) we need to use a finite element subspace $X_h \subset X$ such that $\nabla S_h \subset X_h$ and furthermore ensure that if $\boldsymbol{u}_h \in X_h$ with $\boldsymbol{u}_h \in \nabla S_h^\perp$ then $\nabla \times \boldsymbol{u}_h = 0$ implies $\boldsymbol{u}_h = 0$. All these desirable properties are provided by the edge elements of Nédélec [106, 107].

To define the Nédélec elements on tetrahedra we need some extra auxiliary spaces. Let $\tilde{P}_p$ denote the space of polynomials of total degree exactly p and define

$$S_p = \{\boldsymbol{q} \in (\tilde{P}_p)^3 \mid \boldsymbol{x} \cdot \boldsymbol{q} = 0\}$$

then define

$$R_p = (P_{p-1})^3 \oplus S_p.$$

The space R_p can now be used to construct the first family of curl conforming elements on tetrahedra. In particular we may define $\tilde{X}_h \subset H(\text{curl}; \Omega)$ for $p \geq 0$ by

$$\tilde{X}_h = \{\boldsymbol{u}_h \in H(\text{curl}; \Omega) \mid \boldsymbol{u}_h|_{\Omega_k} \in R_p \quad \text{for} \quad 1 \leq k \leq N_h\}.$$

The degrees of freedom on Ω_k are:

1. $\int_e \boldsymbol{u}_h \cdot \boldsymbol{\tau} q\,ds$ for all $q \in P_{p-1}(e)$ and all edges e of Ω_k where $\boldsymbol{\tau}$ is the unit tangent along that edge.
2. $\int_f \boldsymbol{u} \cdot \boldsymbol{q}\,dA$ for all $\boldsymbol{q} \in (P_{p-2}(f))^3$ with $\boldsymbol{q}$ tangential to f for all faces f of Ω_k.
3. $\int_{\Omega_k} \boldsymbol{u} \cdot \boldsymbol{q}\,dV$ for all $\boldsymbol{q} \in (P_{p-3}(\Omega_k))^3$.

The first set of degrees of freedom motivate the name for these elements: edge elements. In fact, the above degrees of freedom may not be most convenient in practice and there has been considerable effort expended to provide convenient bases for these spaces [6, 126].

We can also define edge elements on parallelepipeds with edges parallel to the coordinate axes [106]. On such an element we have we have

$$\tilde{X}_h = \left\{ \boldsymbol{u}_h \in H(\text{curl}; \ \Omega) | \boldsymbol{u}_h|_{\Omega_k} \in Q_{p-1,p,p} \times Q_{p,p-1,p} \times Q_{p,p,p-1} \ \forall \Omega_k \in \tau_h \right\}$$

with degrees of freedom.

1. $\int_e \boldsymbol{u} \cdot \boldsymbol{\tau} \, q \, ds$ for $q \in P_{p-1}(e)$ and each edge e of Ω_k where τ is the unit tangent along e.
2. $\int_f \boldsymbol{u} \cdot \boldsymbol{q} \, dA$ for all $q \in Q_{p-2,p-1} \times Q_{p-1,p-2}$ tangential to the face f for each face of Ω_k.
3. $\int_{\Omega_k} \boldsymbol{u} \cdot \boldsymbol{q} \, dV$ for all $q \in Q_{p-1,p-2,p-2} \times Q_{p-2,p-1,p-2} \times Q_{p-2,p-2,p-1}$

A diagram showing the degrees of freedom for hexahedral elements for $p = 1, 2$ is given in Fig. 4.

Particularly if one is interested in varying p from element to element, the above description of edge elements needs to become a good deal more complex (see for example [55, 56]). However, variable p is not common in the time domain yet (but see [85] for a Discontinuous Galerkin method for the wave equation with variable p).

For either tetrahedral or parallelepiped grids, we can now define

$$X_h = \left\{ \boldsymbol{u}_h \in \tilde{X}_h \mid \boldsymbol{u}_h \times \boldsymbol{n} = 0 \quad \text{on } \Gamma \right\}.$$

Note that imposing the zero perfect conducting boundary condition is easy: we simply set the degrees of freedom associated with edges or faces on Γ to zero. Of course the requirement that $\boldsymbol{u}_{h,T} \in L^2_Z(\Sigma)$ is always satisfied for piecewise polynomial vector functions.

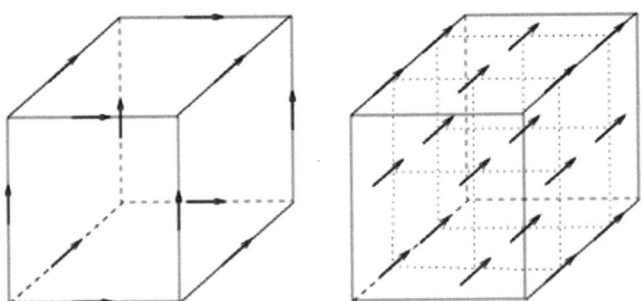

Fig. 4 *Left: $p = 1$*; the average value of tangential component of the finite element vector field is given on each edge. *Right: $p = 2$*; only the degrees of freedom for the second component u_2 of the field are shown. By permission of Oxford University Press from [104]

Both the tetrahedral and hexahedral edge elements have the same relationship with the corresponding H^1 vertex elements summarized in the following fragment of the discrete de Rham diagram

$$
\begin{array}{ccc}
S \cap C^\infty & \xrightarrow{\nabla} & X \cap C^\infty(\Omega) \\
\pi_S \downarrow & & \pi_X \downarrow \\
S_h & \xrightarrow{\nabla} & X_h
\end{array}
\tag{70}
$$

To complete the discrete de Rham diagram, the picture is extended to the right to include divergence conforming and L^2 conforming elements in the full theory of compatible discretizations [10, 11, 29, 81, 82, 104]. However, the fragment shown here is most relevant to us in this paper.

Hidden in this diagram is the fact that if $u_h \in X_h$ and $\nabla \times u_h = 0$ then $u_h = \nabla p_h$ for some $p_h \in S_h$. Moreover, we can show that if $u_h \in (\nabla S_h)^\perp \subset X_h$ then there is a constant C such that [11, 104]

$$
\| \nabla \times u_h \|_{L^2_{\mu-1}(\Omega)} \geq C \| u_h \|_{L^2_\epsilon(\Omega)}.
$$

This observation can be used to construct a projection that would allow us to avoid the curl terms on the right hand side of our estimate in (55). This in turn improves the error estimates, but we shall not pursue this point here.

Using the degrees of freedom we can define an interpolant π_X which is well defined on functions in

$$
H^r(\text{curl} \,; \Omega) = \{ u \in H^r(\Omega) \mid \nabla \times u \in H^r(\Omega) \}, \quad r > 1/2,
$$

(see for example [104]). We have the following error estimate (many other possible interpolation error estimates appear in the literature [104], see also [82]).

Lemma 2. *Suppose* $u \in H^r(\text{curl}; \Omega)$, $1 \leq r \leq p$, *then*

$$
\| u - \pi_X u \|_{L^2(\Omega)} + \| \nabla \times (u - \pi_X u) \|_{L^2(\Omega)}
$$
$$
+ \| (u - \pi_X u)_T \|_{L^2(\Sigma)} \leq C h^r \| u \|_{H^r(\text{curl} \,; \Omega)}.
$$

Using this error estimate we can now provide estimates for the error in our previous theory. By using the definition of $\mathscr{E}$ in (54) and choosing $\eta_h = \pi_X E$ then using the previous lemma we obtain:

Lemma 3. *Suppose* X_h *is constructed from degree* p *edge elements on regular tetrahedral or cube meshes and suppose*

$$
E, \dot{E} \in L^\infty(0, T; H^r(\text{curl} \,; \Omega)), \quad \frac{1}{2} < r \leq p,
$$

*then there is a constant C depending on r, **E** and T but independent of h such that*

$$\inf_{\boldsymbol{\eta}_h \in X_h} \mathscr{E}(\boldsymbol{E} - \boldsymbol{\eta}_h) \leq C h^r.$$

Remark 6. Using a more sophisticated approximation operator, the "bounded cochain projection" from [11], in place of π_X we could decrease the minimum regularity for this theorem to $r = 0$.

Similarly, using the same choice of $\boldsymbol{\eta}_h$ in (55) we can prove:

Lemma 4. *If X_h is constructed from p degree edge elements then provided*

$$\ddot{\boldsymbol{E}}, \dot{\boldsymbol{E}}, \boldsymbol{E} \in H^r(curl\,; \Omega), \quad 1/2 < r \leq p,$$

*there is a constant C depending on T, **E** and r but independent of h such that*

$$\inf_{\eta_h \in X_h} \mathscr{R}(\boldsymbol{E} - \boldsymbol{\eta}_h) \leq C h^r.$$

So we can obtain up to an $O(h^p)$ convergence rate for the semi-discrete approximation if the solution is smooth enough.

Then we obtain a corollary to Theorem 6:

Corollary 3. *Given a final time T, suppose we use a qth order A-stable multistep method (satisfying (66)), and p-degree edge elements, then if the solution **E** of the Maxwell system is sufficiently smooth*

$$\|\boldsymbol{E}_h^n - \boldsymbol{E}(\cdot, n\,\Delta\, t)\|_{L_\epsilon^2(\Omega)} = O(\Delta\, t^q + h^p)$$

for $0 \leq n\Delta t \leq T$, where $\boldsymbol{E}_h^n$ is the fully discrete solution of (65).

This theorem begs the question: when is the solution "sufficiently smooth". This is a difficult question, see for example [135], involving a priori estimates for hyperbolic problems.

3.5 *The Use of Continuous Elements*

We have presented the FETD method using edge elements, but there are arguments for considering other elements. An obvious choice of finite element space consists of continuous piecewise polynomial elements. These have the potential attraction over edge elements that they are well understood, in many finite element packages, and easier to use in graphics software by giving a smoother field. But as we shall see the use of these elements faces several obstacles.

We could try to build X_h by taking three copies of $\tilde{S}_h$ and adjusting for boundary conditions:

$$X_h = \{ u_h \in \tilde{S}_h \times \tilde{S}_h \times \tilde{S}_h \mid u_h \times n = 0 \text{ on } \Gamma \}.$$

There is an immediate objection to this space: it is difficult for us to implement the boundary condition since it imposes a face by face constraint on the elements (the Nitsche method can be used to alleviate this problem [15]).

Leaving aside that difficulty we see a second difficulty: suppose ϵ is discontinuous across a surface S in the mesh having normal n_S and suppose F and σ vanish in the neighborhood of S, then charge conservation $\nabla \cdot (\epsilon E) = 0$ implies $n_S \cdot \epsilon E$ is continuous across S so $\epsilon_1 n_S \cdot E_1 = \epsilon_2 n_S \cdot E_2$ where ϵ_1 and ϵ_1 are the values of ϵ on either side of S and E_1 and E_2 are the fields on either side. Thus, the field E is discontinuous across S and so will be poorly approximated by continuous finite elements. This difficulty can be handled using Lagrange multipliers [14], discontinuous Galerkin techniques or the Nitsche method [15].

Lastly suppose we are solving in a metallic air filled cavity so that ϵ and μ are constant, σ vanishes, $\Sigma = \phi$, and $F = 0$. As in our derivation of (61) the semi-discrete in time problem is to solve

$$M^\epsilon \frac{d^2 \mathbf{E}}{dt^2} + S\mathbf{E} = 0$$

where $\mathbf{E}$ is the vector of degrees of freedom. This problem has solutions $\mathbf{E}_n e^{-i\omega_n t}$ where ω_n and $\mathbf{E}_n \neq 0$ satisfy the generalized eigenvalue problem

$$- \omega_n^2 M^\epsilon \mathbf{E}_n + S\mathbf{E}_n = 0, \quad n = 1, 2, \cdots. \tag{71}$$

Here S is symmetric and M^ϵ is positive definite and symmetric so the eigenvalues are real and the discrete solution $\mathbf{E}$ is a superposition of modes $\mathbf{E}_n$ with frequency ω_n where $\mathbf{E}_n \neq 0$ that satisfy the eigenvalue equation (71) so

$$\mathbf{E} = \sum_n a_n \mathbf{E}_n e^{-i\omega_n t} \tag{72}$$

where $\{a_n\}$ are coefficients determined by the initial data. The solution given by (72) will be accurate if the eigenmodes of (71) are a sufficiently accurate approximation to the eigenvalues ω and eigenvectors $E \in X$ such that $E_0 \neq 0$ of

$$\omega^2 E - \nabla \times \nabla \times E = 0 \text{ in } \Omega. \tag{73}$$

Unfortunately, Boffi et al [28] (for more details see his paper in this volume) show that piecewise linear elements result in discrete eigenvalues ω_n amongst the lowest non-zero eigenvalues of (73) but having no physical counterparts. In addition, some eigenvalues approximating physical eigenvalues have eigenspaces of the wrong multiplicity.

Confusingly, our error estimates show that provided the exact solution can be approximated well by finite elements, then for fixed T and for h small enough we will get convergence of the time domain method. But for a particular T and particular h the initial data may be such as to excite a spurious mode. This may then corrupt the solution (usually as an unexpected oscillatory feature in the solution). This problem can be remedied by adding a carefully designed divergence stabilization term [44].

So we see that to use continuous elements in a general purpose solver requires several modifications to the basic FETD scheme. Since edge elements give a robust discretization compatible with the conservation properties of the problem, we prefer to base our conforming FETD solver on these elements.

3.6   Extensions to the Basic Solver

The physical parameters in Maxwell's equations are not truly constant. In fact, they depend on the frequency of the radiation. As we have seen this dependence is often neglected either because we are interested in propagation in the air where propagation is weakly dependent on frequency or because we are interested in propagation in a relatively narrow band of frequencies. If the parameters ϵ, μ and σ in (11) depend on frequency, terms such as $\epsilon \ddot{E}$ in (14) need to be reinterpreted as $\partial/\partial t (\epsilon * \partial E/\partial t)$ where $*$ denotes convolution in time. For general frequency dependent coefficient this requires storing all past history of E (causality requires that future fields cannot be involved, so limiting the possible choices of the coefficients).

More commonly, a model is usually assumed for the frequency dependence, and this allows us to solve in the time domain without excessive storage demands. To explain this in more detail, consider a generalization of the frequency domain problem (11) with $J^i = 0$ where we seek $\hat{E}$ (subject to suitable boundary conditions) such that

$$\omega^2 \epsilon_0 \left(\hat{\epsilon}_R(\omega) + \frac{i\sigma}{\omega} \right) \hat{E} - \nabla \times \mu_0^{-1} \nabla \times \hat{E} = 0$$

where the relative electric permittivity $\hat{\epsilon}_R$ depends on frequency via ω. Formally taking the inverse Fourier transform gives

$$\epsilon_0 \frac{\partial}{\partial t} \left(\epsilon_R * \frac{\partial E}{\partial t} \right) + \sigma \dot{E} + \nabla \times \mu_0^{-1} \nabla \times E = 0$$

where ϵ_R is the inverse Fourier transform of $\hat{\epsilon}_R$.

There are several models for $\hat{\epsilon}_R$ in common use: Debye, Drude or Lorentz media being three examples [92, 94]. Water, and therefore biological tissue, is often modeled by a Debye model [88] where

$$\hat{\epsilon}_R(\omega) = \left(\epsilon_\infty + \frac{(\epsilon_s - \epsilon_\infty)}{1 - i\omega t_0}\right).$$

Here ϵ_∞ is the relative permittivity at infinite frequency ($\hat{\epsilon}_R \to \epsilon_\infty$ as $\omega \to \infty$), ϵ_s is the static relative permittivity ($\hat{\epsilon}_R(0) = \epsilon_s$) and t_0 is a relaxation time. It is assumed that $\epsilon_s > \epsilon_\infty$ [94].

The idea is to introduce additional variables such that, after inverse Fourier transformation, we obtain a system of differential equations. In fact, we need only introduce one auxiliary field in this case: define $\hat{P}$ by

$$\hat{P} = \epsilon_0 \frac{(\epsilon_s - \epsilon_\infty)}{1 - i\omega t_0} \hat{E}$$

so that we have

$$(1 - i\omega t_0)\hat{P} = \epsilon_0(\epsilon_s - \epsilon_\infty)\hat{E}$$

and taking the inverse Fourier transform

$$t_0 \dot{P} + P = \epsilon_0(\epsilon_s - \epsilon_\infty)E.$$

We obtain

$$\omega^2 \epsilon_0 \epsilon_\infty \hat{E} + i\omega\sigma\hat{E} + \nabla\times \mu_0^{-1} \nabla\times \hat{E} + \omega^2 \hat{P} = 0. \tag{74}$$

Following [94], note that

$$-i\omega\hat{P} = \frac{1}{t_0}(\epsilon_0(\epsilon_s - \epsilon_\infty)\hat{E} - \hat{P})$$

so (74) can be written, in the time domain, as

$$\epsilon_0 \epsilon_\infty \ddot{E} + \sigma\dot{E} + \nabla\times \mu_0^{-1} \nabla\times E + \frac{1}{t_0}(\epsilon_0(\epsilon_s - \epsilon_\infty)\dot{E} - \dot{P}) = 0 \tag{75}$$

$$t_0 \dot{P} + P = \epsilon_0(\epsilon_s - \epsilon_\infty)E. \tag{76}$$

This is a second order version of the mixed system analyzed by Li and Zhang [94]. Let us suppose that we use only the Perfect Electrically Conducting boundary condition

$$E \times n = 0 \text{ on } \Gamma = \partial\Omega.$$

To obtain an energy equality we follow [94], suitably modified for the second order system considered here, and multiply (75) by $\dot{E}$ and integrate over Ω to obtain

$$\frac{1}{2}\frac{d}{dt}\left[\epsilon_\infty \epsilon_0 \|\dot{E}\|^2_{L^2(\Omega)} + \mu_0^{-1}\|\nabla\times E\|^2_{L^2(\Omega)}\right] + \|\dot{E}\|^2_{L^2_\sigma(\Omega)}$$

$$+ \frac{\epsilon_0}{t_0}(\epsilon_s - \epsilon_\infty)\|\dot{E}\|^2_{L^2(\Omega)} - \frac{1}{t_0}(\dot{P}, \dot{E}) = 0.$$

Taking the time derivative of (76) and multiplying by $\dot{P}$, then integrating over Ω we get

$$\frac{1}{\epsilon_0(\epsilon_s - \epsilon_\infty)}\left[(\ddot{P}, \dot{P}) + \frac{1}{t_0}(\dot{P}, \dot{P})\right] = \frac{1}{t_0}(\dot{E}, \dot{P}).$$

Adding these equations gives

$$\frac{1}{2}\frac{d}{dt}\left[\epsilon_\infty\epsilon_0\|\dot{E}\|^2_{L^2(\Omega)} + \mu_0^{-1}\|\nabla\times E\|^2_{L^2(\Omega)} + \frac{1}{\epsilon_0(\epsilon_s - \epsilon_\infty)}\|\dot{P}\|^2_{L^2(\Omega)}\right] \quad (77)$$

$$+\|\dot{E}\|^2_{L^2_\sigma(\Omega)} + \frac{\epsilon_0}{t_0}(\epsilon_s - \epsilon_\infty)\|\dot{E}\|^2_{L^2(\Omega)} + \frac{1}{\epsilon_0(\epsilon_s - \epsilon_\infty)t_0}\|\dot{P}\|^2_{L^2(\Omega)} - \frac{2}{t_0}(\dot{E}, \dot{P}) = 0.$$

The last three terms can be combined to give

$$\frac{1}{2}\frac{d}{dt}\left[\epsilon_\infty\epsilon_0\|\dot{E}\|^2_{L^2(\Omega)} + \mu_0^{-1}\|\nabla\times E\|^2_{L^2(\Omega)} + \frac{1}{\epsilon_0(\epsilon_s - \epsilon_\infty)}\|\dot{P}\|^2_{L^2(\Omega)}\right] \quad (78)$$

$$+\frac{1}{t_0}\left\|\frac{1}{\sqrt{\epsilon_0(\epsilon_s - \epsilon_\infty)}t_0}\dot{P} - \sqrt{\frac{\epsilon_0(\epsilon_s - \epsilon_\infty)}{t_0}}\dot{E}\right\|^2_{L^2(\Omega)} + \|\dot{E}\|^2_{L^2_\sigma(\Omega)} = 0.$$

This argument, and in particular the final step, follows directly the proof of Lemma 2.1 in [94] with appropriate modifications for the second order equation. It could now be used as the basis of an error analysis for the second order approximation of this problem. Li and Zhang [94] take the alternative route of analyzing the trapezoidal rule and edge elements applied to the first order system governing (E, B, P). The reader should consult their paper for error estimates, as well as extensions to Lorentz media and cold plasma.

An elegant alternative approach, using history integrals, can be found in [120].

3.7 Linear System Solver

If we use an implicit time stepping scheme, we need to solve a linear system at each timestep. Let us suppose, for simplicity, that $\sigma = 0$ and $\Sigma = \emptyset$ then to implement (65) we need to solve the problem of finding $E_h^{n+1} \in X_h$ such that

$$\left(\frac{a_0}{b_0\,\Delta t}\right)^2(\epsilon E_h^{n+1}, \xi_h) + (\mu^{-1}\nabla\times E_h^{n+1}, \nabla\times\xi_h) = (G^n, \xi_h) \quad (79)$$

for all $\xi_h \in X_h$. Here G^n can be computed from E_h^m and F^m for $m \leq n$ so is a known function. Obviously solving (79) is an extremely important task when developing a practical implicit FETD code. Fortunately this is a problem with a coercive bilinear form with a positive definite matrix on the left hand side and so is solvable at each timestep.

Typically the matrix for (79) is too large to be factored in memory and so an iterative scheme needs to be applied. There are several possible choices already in the literature including Hiptmair [80] or Arnold-Falk-Winther [12] multigrid methods or alternating Schwarz methods [66, 123].

3.8 Mass Lumping

An alternative approach, applicable when explicit timestepping like the leapfrog method is used, is mass lumping to speed the solution algorithm. The approach we shall take is from [47,49] (see also [42,62]). For this to work we need to assume that the mesh consists of parallelepipeds with edges parallel to the coordinate axes. In particular let us suppose we are using degree $p = 1$ edge elements (for extensions to higher order elements, see [47]). Let us also assume that $\Sigma = \emptyset$, and $\sigma = 0$. The finite element space X_h has a basis $\{\xi_j^h\}_{j=1}^{N_h} \subset X_h$ of shape functions (basis functions) where N_h is the number of interior edges in the mesh. The basis function ξ_j^h has vanishing tangential component on each edge except on the jth interior edge of the mesh where the degree of freedom is unity. Of course the support of this basis function is just the elements sharing this edge. Now we can expand the semidiscrete solution as

$$E_h(x,t) = \sum_{j=1}^{N_h} E_j(t)\xi_j^h(x)$$

then if $\mathbf{E} = (E_1, E_2, \ldots)^T$ we have

$$M^\epsilon \frac{d^2}{dt^2}\mathbf{E} + S\mathbf{E} = \mathbf{F}.$$

Here the sparse $N_h \times N_h$ symmetric matrices M^ϵ and S are given by

$$M_{\ell,m}^\epsilon = \int_\Omega \epsilon \xi_\ell^h \cdot \xi_m^h \, dV, \text{ and } S_{\ell,m} = \int_\Omega \mu^{-1}\nabla \times \xi_\ell^h \cdot \nabla \times \xi_m^h \, dV.$$

To discretize in time we can use standard centered differences

$$M^\epsilon \frac{(E^{n+1} - 2E^n + E^{n-1})}{\Delta t^2} + SE^n = F^n.$$

Since S is symmetric positive semi-definite an energy analysis shows that the method is stable.

But as we have seen there is a problem: at each timestep we must solve a linear system with matrix M^ϵ (e.g. by diagonally preconditioned conjugate gradients [105]). If we wish to avoid this cost, one further approximation saves the day: We can approximate the integrals defining M^ϵ by a suitable quadrature. On an element

$\Omega_k = [0, h] \times [0, h] \times [0, h]$ (we can assume one corner is at the origin by translation) we can use the quadrature

$$\int_{\Omega_k} \boldsymbol{u} \cdot \boldsymbol{v} \, dV \approx Q_{\Omega_k}(\boldsymbol{u}, \boldsymbol{v})$$

$$= \frac{h^3}{12} \left[\sum_{j=1}^{4} u_1(\boldsymbol{a}_j) \cdot v_1(\boldsymbol{a}_j) + \sum_{j=1}^{4} u_2(\boldsymbol{b}_j) \cdot v_2(\boldsymbol{b}_j) + \sum_{j=1}^{4} u_3(\boldsymbol{c}_j) \cdot v_3(\boldsymbol{c}_j) \right]$$

where $\boldsymbol{a}_1 = (h/2, 0, 0)$, $\boldsymbol{a}_2 = (h/2, h, 0)$, $\boldsymbol{a}_3 = (h/2, 0, h)$, $\boldsymbol{a}_4 = (h/2, h, h)$ and similarly for mid-points $\boldsymbol{b}_j$ on edges parallel to the y axis, and $\boldsymbol{c}_j$ on edges parallel to the z-axis.

The mass lumped FETD method is then to approximate M^ϵ by a diagonal matrix $M^{\epsilon, Q}$ computed using this special quadrature. In particular

$$M^{\epsilon, Q}_{\ell, m} = \sum_{k=1}^{N_h} Q_{\Omega_k}(\boldsymbol{\xi}_\ell^h, \boldsymbol{\xi}_m^h), \quad F^{Q, n}_m = \sum_{k=1}^{N_h} Q_{\Omega_k}(\boldsymbol{F}^n, \boldsymbol{\xi}_m^h),$$

and $M^{\epsilon, Q}$ is diagonal. We can time-step

$$M^{\epsilon, Q} \frac{(\boldsymbol{E}^{n+1} - 2\boldsymbol{E}^n + \boldsymbol{E}^{n-1})}{\Delta t^2} + S\boldsymbol{E}^n = \boldsymbol{F}^{Q, n}$$

without inverting a matrix. This is called the Finite Difference Time Domain method (FDTD) [121, 132]: we can show that the method converges with error $O(h^2 + \Delta t^2)$ provided we satisfy the CFL condition $\Delta t \leq h/\sqrt{3}c$. Obviously this is not the most direct derivation of FDTD!

We also have discrete conservation of charge: if $\boldsymbol{F} = 0$

$$\sum_{K \in \mathcal{T}_h} Q_K(\boldsymbol{E}_h^{n+1} - 2\boldsymbol{E}_h^n + \boldsymbol{E}_h^{n-1}, \nabla p_h) = 0 \qquad \forall p_h \in S_h$$

at each timestep $n \geq 1$. Hence, if the initial data is chosen so that

$$\sum_{K \in \mathcal{T}_h} Q_K(\boldsymbol{E}_h^0, \nabla p_h) = \sum_{K \in \mathcal{T}_h} Q_K(\boldsymbol{E}_h^1, \nabla p_h) = 0 \qquad \forall p_h \in S_h$$

we have

$$\sum_{K \in \mathcal{T}_h} Q_K(\boldsymbol{E}_h^0, \nabla p_h) = 0 \qquad \forall p_h \in S_h$$

for each n. We say that the fields are "discrete divergence free". On a uniform grid, taking p_h to be unity at one vertex and zero at all the rest we have that at that vertex

$$(E_z^+ - E_z^-) + (E_y^+ - E_y^-) + (E_x^+ - E_x^-) = 0$$

Fig. 5 Notation for the
discrete divergence at a node
in the mesh

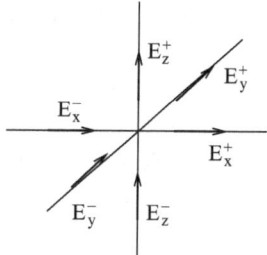

where the $\pm$ superscripts represent fields to the left/right, and above/below the point at which p_h is non-zero as shown in Fig. 5. This is just a finite difference approximation of the divergence at each node.

FDTD is the workhorse of CEM and results are to be found in every issue of the journal IEEE Transactions on Antennas and Propagation. The MAFIA code, which is based on the integral formulation of Maxwell's equations but results in the same FDTD when applied to the simple case here, is an example of a sophisticated implementation [100, 127, 128].

Unfortunately, if this method is used with hexahedral elements that are not rectangular parallelepipeds it does not produce a lumped (i.e. diagonal) matrix. The same technique can also be tried on lowest order edge elements on a tetrahedral mesh [77, 78] but unfortunately the quadrature weights needed for zeroth order accuracy of the quadrature may be negative (depending on the shape of the element). In tests we have not been able to obtain an automatically generated mesh for which all the weights are positive (necessary for stability). However, it is possible to obtain a mass lumped tetrahedral method by using an extended family of edge elements having extra basis function resulting in more degrees of freedom that give rise to positive quadrature weights [58].

Convergence of the mass lumped scheme can be proved using the Strang Lemma [42] and for a Fourier-Laplace domain based argument for the mass lumped wave equations see [48].

3.9 Dispersion Analysis

It is difficult to carry out a general dispersion analysis for edge elements on tetrahedra (see [105]), but for rectangular parallelepipeds an analysis is possible and has been given for all orders of elements by Ainsworth [5].

The novel part of this review is the suggestion to use backward Euler or BDF2 for timestepping the Maxwell system. So here we will just consider the dispersion error due to timestepping. Let $\delta_t^{\Delta t}$ denote the relevant backward difference operator and consider the time discrete Maxwell system

$$\frac{1}{c^2}(\delta_t^{\Delta t})^2 \boldsymbol{E}^{\Delta t} + \nabla \times \nabla \times \boldsymbol{E}^{\Delta t} = 0$$

in all space. Assuming that the solution is divergence free, this can be rewritten as

$$\frac{1}{c^2}(\delta_t^{\Delta t})^2 \boldsymbol{E}^{\Delta t} = \Delta\, \boldsymbol{E}^{\Delta t}$$

and as in Sect. 2 it suffices to consider the time discrete scalar wave equation

$$\frac{1}{c^2}(\delta_t^{\Delta t})^2 u^n = \Delta u^n.$$

Using the BE method

$$\frac{1}{c^2}\frac{u^n - 2u^{n-1} + u^{n-2}}{(\Delta t)^2} = \Delta u^n.$$

We then seek solutions of the form $u^n = \exp(i(\boldsymbol{k} \cdot \boldsymbol{x} - \omega_{\Delta t} t_n))$ and arrive at

$$\frac{4\exp(i\omega_{\Delta t}\Delta t)}{(\Delta t)^2} \sin^2(\omega_{\Delta t}\Delta t/2) = c^2|\boldsymbol{k}|^2.$$

If the method has no dispersion error $\omega = \pm c|\boldsymbol{k}|$ (see the one dimensional analysis in Sect. 2), however for backward Euler we find

$$\omega_{\Delta t} = c|\boldsymbol{k}| - \frac{1}{2}ic|\boldsymbol{k}|\Delta t - \frac{1}{2}c^3|\boldsymbol{k}|^3(\Delta t)^2 + \cdots.$$

The method is in general dispersive at second order and dissipative of first order. It will tend to damp out wave propagation. For BDF2, BDF3 and the trapezoidal rule we find [40]

$$\text{BDF2}: \quad \omega_{\Delta t} = \pm|c\boldsymbol{k}| \mp \frac{1}{3}|c\boldsymbol{k}|^3 \Delta t^2 + \frac{i}{4}|c\boldsymbol{k}|^4 \Delta t^3 + \cdots,$$

$$\text{BDF3}: \quad \omega_{\Delta t} = \pm|c\boldsymbol{k}| + \frac{i}{4}|c\boldsymbol{k}|^4 \Delta t^3 \mp \frac{3}{10}|c\boldsymbol{k}|^5 \Delta t^4 + \cdots,$$

$$\text{Trapezoidal}: \quad \omega_{\Delta t} = \pm|c\boldsymbol{k}| \mp \frac{1}{12}|c\boldsymbol{k}|^3 \Delta t^2 \pm \frac{1}{80}|c\boldsymbol{k}|^5 \Delta t^4 + \cdots,$$

Thus, these methods are much less dissipative (now the dissipation is third order, fourth order or non-existent). BDF2 and the trapezoidal rule have the same order of dispersion error. Neither BDF3 nor the trapezoidal rule are covered by the theory here.

To obtain better dispersion and dissipation error would require us to use an implicit Runge-Kutta scheme if we want to stay within a straight forward application of Lubich's convergence theory [97].

Finally, we note that if one is willing to work with a modified mass matrix and special quadrature, it is possible to improve the dispersion error from the space discretization very considerably [8] without ruining the overall convergence rate. This does not make the mass matrix diagonal, but it does make higher order edge elements on cubes particularly attractive.

3.10   A Discontinuous Galerkin Method

As we have seen, mass lumping is only possible for rectangular parallelepiped edge elements in general. To avoid the problem of mass matrices many researchers have turned to discontinuous Galerkin methods. We shall describe just one version: the Symmetric Interior Penalty Galerkin method (SIPG). This method was first analyzed in the time domain in [69–72, 79] and for the related eigenvalue problem in [36, 37].

For simplicity let us assume that $\partial\Omega = \Gamma$, so that the field satisfies the Perfectly Electrically Conducting boundary condition on the boundary, and that $\sigma = 0$. We cover Ω by a mesh $\mathcal{T}_h$ of regular tetrahedra of maximum diameter h (we could use cubes equally well). To develop the method in the classical way we proceed as follows. On an element $K = \Omega_k$ in the mesh, we multiply the interior Maxwell system by a smooth test function $\boldsymbol{\xi}^K$ and integrate over the domain. The key term is the spatial derivative term:

$$\int_K \nabla \times \mu^{-1}\nabla \times \boldsymbol{E} \cdot \boldsymbol{\xi}^K dV = \int_K \mu^{-1}\nabla \times \boldsymbol{E} \cdot \nabla \times \boldsymbol{\xi}^K dV$$

$$+ \int_{\partial K} \mu^{-1}\nabla \times \boldsymbol{E} \cdot \boldsymbol{\xi}^K \times \boldsymbol{n}^K dA$$

where $\boldsymbol{n}^K$ is the unit outward normal to K. Adding over all elements and letting $\boldsymbol{\xi}$ denote the piecewise defined function $\boldsymbol{\xi}|_K = \boldsymbol{\xi}^K$ and defining $\nabla_h \times \boldsymbol{\xi}|_K = \nabla \times \boldsymbol{\xi}^K$ we have

$$\int_\Omega \nabla \times \mu^{-1}\nabla \times \boldsymbol{E} \cdot \boldsymbol{\xi} dV = \int_\Omega \mu^{-1}\nabla_h \times \boldsymbol{E} \cdot \nabla_h \times \boldsymbol{\xi} dV + \sum_{f \in \mathcal{E}_h} \int_f \mu^{-1}\nabla \times \boldsymbol{E} \cdot [\![\boldsymbol{\xi}]\!] dA$$

where $\mathcal{E}_h$ denotes the set of all faces in the mesh and the jump $[\![\cdot]\!]$ is defined by

$$[\![\boldsymbol{\xi}]\!] = \begin{cases} \boldsymbol{\xi}^K \times \boldsymbol{n}^K + \boldsymbol{\xi}^{K'} \times \boldsymbol{n}^{K'} & \text{if } f = \overline{K} \cap \overline{K}' \\ \boldsymbol{\xi}^K \times \boldsymbol{n}^K & \text{if } \quad f \in \Gamma. \end{cases}$$

We will allow the discrete approximation to $\boldsymbol{E}$ to jump so define the average value operator $\{\cdot\}$ by

$$\{\boldsymbol{\xi}\} = \begin{cases} (\boldsymbol{\xi}^K + \boldsymbol{\xi}^{K'})/2 & \text{if } f = \overline{K} \cap \overline{K}' \\ \boldsymbol{\xi}^K & \text{if } \quad f \in \Gamma. \end{cases}$$

We are also motivated to define the "broken" space

$$H(\text{curl}; \mathcal{T}_h) = \{v \in L^2(\Omega)^3 \mid \nabla \times v|_{\Omega_k} \in L^2(\Omega_k)^3 \quad \forall \Omega_k \in \mathcal{T}_h\}.$$

Then for all $\xi \in H(\text{curl}; \mathcal{T}_h)$ and sufficiently smooth E we have

$$\int_\Omega \nabla \times \mu^{-1} \nabla \times E \cdot \xi \, dV = \int_\Omega \mu^{-1} \nabla_h \times E \cdot \nabla_h \times \xi \, dV$$

$$+ \sum_{f \in \mathscr{E}_h} \int_f \{\mu^{-1} \nabla \times E\} \cdot [\![\xi]\!] \, dA.$$

The right hand side is not symmetric (which we want so we can use an energy analysis) so we add a consistent symmetrizing term

$$\sum_{f \in \mathscr{E}_h} \int_f \{\mu^{-1} \nabla \times \xi\} \cdot [\![E]\!] \, dA$$

which vanishes for the exact solution.

So if E is smooth enough we have

$$\int_\Omega \nabla \times \mu^{-1} \nabla \times E \cdot \xi \, dV = \int_\Omega \mu^{-1} \nabla_h \times E \cdot \nabla_h \times \xi \, dV$$

$$+ \sum_{f \in \mathscr{E}_h} \int_f \{\mu^{-1} \nabla \times E\} \cdot [\![\xi]\!] + \{\mu^{-1} \nabla \times \xi\} \cdot [\![E]\!] \, dA.$$

The right hand side turns out not to be positive semidefinite (again needed for an energy analysis) so we add a consistent stabilizing term

$$\sum_{f \in \mathscr{E}_h} \int_f a[\![E]\!][\![\xi]\!] dA$$

where $a = \alpha/h$ and

$$h = \begin{cases} \min(h_K, h_{K'}) & f \in \overline{K} \cap \overline{K'}, \\ h_K & f \in \Gamma \cap \overline{K}, \end{cases}$$

The coefficient $\alpha > 0$ is a penalty parameter which will need to be chosen large enough. We now use the discontinuous Galerkin space

$$V_h = \{v_h \in L^2(\Omega) \mid v_h|_K \in (P_p)^3 \quad \forall K \in \mathcal{T}_h\}$$

where P_p is the set of all polynomials of degree at most p. The semi-discrete solution is $E_h(t) \in V_h$ such that

$$\int_\Omega \epsilon \ddot{E}_h \cdot \xi_h dV + a_{DG}(E_h, \xi_h) = \int F \cdot \xi_h dV \quad \forall \xi_h \in V_h, \tag{80}$$

where

$$a_{DG}(E_h, \xi_h) = \int_\Omega \mu^{-1} \nabla_h \times E_h \cdot \nabla_h \times \xi_h dV$$

$$+ \sum_{f \in \mathscr{E}_h} \int_f \{\mu^{-1} \nabla \times E_h\} \cdot [\![\xi_h]\!] + \{\xi_h\} \cdot [\![\mu^{-1} \nabla \times E_h]\!]$$

$$+ \sum_{f \in \mathscr{E}_h} \int a [\![E_h]\!] \cdot [\![\xi_h]\!] dA.$$

To show the stability of this method, a key finite element result is the local inverse inequality (see Brenner's paper in this volume): if $v_h \in V_h$ then

$$\|v_h\|^2_{L^2(\partial K)} \leq C h_K^{-1} \|v_h\|^2_{L^2(K)}.$$

So we can estimate the flux terms in a_{DG} as follows: if $v \in V_h, z \in V_h$ then

$$\sum_{f \in \mathscr{E}_h} \int_f [\![v]\!]\{z\} dA \leq \sum_{f \in \mathscr{E}_h} \|a^{1/2}[\![v]\!]\|_{L^2(f)} \|a^{-1/2}\{z\}\|_{L^2(f)}$$

$$\leq C\alpha^{-1/2} \left(\sum_{f \in E_h} \|a^{1/2}[\![v]\!]\|_{L^2(f)} \right) \cdot \|z\|_{L^2(\Omega)}.$$

We can then show if α is large enough, for all $u_h \in V_h$ there is a constant $C > 0$ independent of h and u_h such that

$$|a_{DG}(u_h, u_h)| \geq C \|u_h\|^2_{V_h}$$

where

$$\|u\|^2_{V_h} = \|\mu^{-1} \nabla_h \times u\|^2_{L^2(\Omega)} + \sum_{f \in E_h} \|a^{1/2}[\![u]\!]\|^2_{L^2(f)}.$$

So we will have a stable time marching scheme if we use centered differences in time and a suitable CFL condition (decreasing in α).

Convergence can then be proved by Laplace transforms and a Strang Lemma argument. By more classical arguments the following theorem is proved in [70]:

Theorem 7. *Suppose for $r > 1/2$, and*

$$E, \dot{E} \in L^\infty(0, T; H^{1+r}(\Omega)^3), \quad \ddot{E} \in L^1(0, T; H^r(\Omega)^3)$$

then if α is large enough, and $e_h = E - E_h$ where E_h is the solution of the semi-discrete SIPG method (80), we have the error estimate

$$\|\dot{e}_h\|_{L^\infty(0,T;L^2_\epsilon(D)^3)} + \|e_h\|_{L^\infty(0,T;V(h))} \leq C_1(\|\dot{e}_h(0)\|_{L^2_\epsilon(\Omega)} + \|e_h(0)\|_{V_h}) + C_2 h^{\min(r,p)}$$

where C_2 depends on the indicated norms of E, and the $V(h)$ norm is

$$\|e_h\|^2_{V(h)} = \|e_h\|^2_{L^2_\epsilon(\Omega)} + \|e_h\|^2_{V_h}.$$

Note that the matrix resulting from the time derivative term in the DG formulation is block diagonal and so it can be inverted easily. Often explicit Runge-Kutta is used for discontinuous Galerkin formulations [46], but in [69] a new flexible alternative based on local time steps is formulated. This allows short timesteps where needed and longer timesteps elsewhere up to the standard CFL condition.

Fully discrete error estimates, for the wave equation, using DG in space and leapfrog in time can be found in [72]. Because the method uses discontinuous piecewise polynomial elements, it has more degrees of freedom than the standard conforming edge element scheme. So it does not seem to us that it is a good candidate for implicit timestepping. On the other hand, when explicit timestepping is applicable, it can be an extremely efficient and flexible method [64, 65, 116].

For another approach to building a DG method based on upwind fluxes, see [79] and for central fluxes see [61].

4 · Background on Time Domain Integral Equations

Integral equations reduce the solution of homogeneous boundary value problems to equations on the boundary of the scatterer. This reduction in dimension carries with it the lure of fewer degrees of freedom, and hence considerable effort has been expended to develop solvers based on this formulation. In the time domain it is only quite recently that stable and accurate integral equation formulations have appeared. In some applications they are now the state of the art solver (see for example, [1, 17, 51, 53, 54, 60, 73, 74, 118, 119, 122, 133, 134]).

The approach to formulating the Time Domain Boundary Integral Equation (TDBIE) that we shall present here is from the work of Terrasse [122] and Pujols [17] who developed a stable discretization of the time domain Electric Field Integral Equation (EFIE). Their work was based on earlier work in [19] nicely summarized in [73]. For a general survey of TDBIEs and methods of discretization for the wave equation, Maxwell system and linear elasticity see [52].

We are now going to outline the derivation of a time domain boundary integral equation for Maxwell's equation. Suppose we want to solve

$$\left.\begin{array}{l} \epsilon_0 \dot{E} - \nabla \times H = 0 \\ \mu_0 \dot{H} + \nabla \times E = 0 \end{array}\right\} \text{ in } D' = \mathbb{R}^3 \backslash \overline{D} \qquad (81)$$

subject to the perfect conducting boundary condition

$$E \times n = g := -E^i \times n,$$

on Γ where (E^i, H^i) is a solution of the Maxwell system in the absence of D (see Chap. 1), and we take n to be the unit outward normal to D. We assume $E^i = 0$ in a neighborhood of D of $t = 0$ and select

$$E(x, 0) = H(x, 0) = 0 \quad \text{at} \quad t = 0.$$

We will now show how to reduce this problem to a time domain integral equation.

The first point is that integral equations effectively solve the interior and exterior problem simultaneously. So we now assume that (81) is satisfied both in D and in $D' = \mathbb{R}^3 \setminus \overline{D}$. However, the field is in general discontinuous across the boundary Γ of D.

Boundary integral equations are based on integral representations of the solution. We are now going to outline how to derive a Fourier-Laplace domain "Stratton-Chu" formula. This plays the role of Green's representation formula for the Helmholtz equation. We shall consider D and D' separately. Suppose $(\hat{E}, \hat{H})$ is a smooth solution of the Fourier-Laplace domain Maxwell's equations:

$$\left. \begin{array}{l} s\epsilon_0 \hat{E} - \nabla \times \hat{H} = 0 \\ s\mu_0 \hat{H} + \nabla \times \hat{E} = 0 \end{array} \right\} \text{ in } D \text{ or } D'. \tag{82}$$

Taking the divergence of the first equation and recalling that $\Re(s) > 0$, we see also that $\nabla \cdot \hat{E} = 0$.

We shall make use of the fundamental solution of the Fourier-Laplace domain Helmholtz equation. We need to find the solution $\hat{\Phi}_0$ of

$$\frac{s^2}{c^2} \hat{\Phi}_0 - \Delta \hat{\Phi}_0 = \delta_0,$$

with $\hat{\Phi}_0$ bounded at infinity, where $s = \varphi + i\omega$ with $\varphi > \varphi_0 > 0$, $c = 1/\sqrt{\epsilon_0 \mu_0}$, and δ_0 is the Dirac delta at the origin. Separating variables and noting that the solution is invariant with respect to rotation about the origin we see that the solution is

$$\hat{\Phi}_0(x) = \frac{\exp(-s|x|/c)}{4\pi |x|}.$$

More generally

$$\hat{\Phi}(x, y) := \hat{\Phi}_0(|x - y|) \tag{83}$$

gives the fundamental solution for a source point at y evaluated at x.

Let p be a fixed vector and suppose that $x \in D$, then using the definition of $\hat{\Phi}$ and the standard Laplacian vector identity

$$p \cdot \hat{E}(x) = \int_D \hat{E}(y) \cdot p \left(-\Delta_y \hat{\Phi} + \frac{s^2}{c^2} \hat{\Phi} \right)(x, y) \, dV(y)$$

$$= \int_D \left(\hat{E} \cdot (\nabla_y \times \nabla_y \times (p\hat{\Phi}) - \nabla_y \nabla_y \cdot (\hat{\Phi} p)) + \frac{s^2}{c^2} \hat{E} \cdot p\Phi \right) dV(y).$$

Here ∇_y denotes derivatives with respect y.

Integrating by parts, adding and subtracting terms and then performing a further integration by parts shows that

$$p \cdot \hat{E}(x) = \int_D \nabla_y \times \hat{E} \cdot \nabla_y \times (p\hat{\Phi}) + \nabla_y \cdot (\hat{\Phi} p) \nabla_y \cdot \hat{E} + \frac{s^2}{c^2} \hat{E} \cdot p\hat{\Phi} \, dV(y)$$

$$+ \int_\Gamma \hat{E} \cdot n \times \nabla_y \times (p\hat{\Phi}) - \nabla_y \cdot (\hat{\Phi} p) \hat{E} \cdot n \, dA(y)$$

$$= \int_D (\nabla_y \times \hat{E} + s\mu_0 \hat{H}) \cdot \nabla_y \times (p\hat{\Phi}) \, dV(y)$$

$$+ \int_D (\frac{s^2}{c^2} \hat{E} \cdot p\hat{\Phi} - s\mu_0 \hat{H} \cdot \nabla_y \times (p\hat{\Phi})) \, dV(y)$$

$$+ \int_\Gamma (\hat{E} \times n) \cdot \nabla_y \times (p\hat{\Phi}) - \hat{E} \cdot n \, \nabla_y \cdot (\hat{\Phi} p) \, dA(y).$$

Now using Maxwell's equations (82) this simplifies to

$$p \cdot \hat{E}(x) = \int_\Gamma (\hat{E} \times n) \cdot \nabla_y \times (p\hat{\Phi}) - \hat{E} \cdot n \, \nabla_y \cdot (\hat{\Phi} p) - s\mu_0 \hat{H} \times n \cdot p\hat{\Phi} \, dA(y).$$

Using vector calculus we can show that

$$\nabla_y \times (p\Phi) \cdot \xi(y) = p \cdot \nabla_x \times (\Phi \xi),$$

$$\nabla_y \cdot (p\Phi) = -p \cdot \nabla_x \Phi.$$

In addition, since the above equality holds for all p, we arrive at

$$\hat{E}(x) = \int_\Gamma \nabla_x \times [(\hat{E} \times n)\hat{\Phi}] + \hat{E} \cdot n \, \nabla_x \hat{\Phi} - s\mu_0 (\hat{H} \times n)\hat{\Phi} \, dA(y).$$

But $\hat{\Phi} = -c^2 \Delta_x \hat{\Phi}/s^2$ since $x \neq y$ and so, using the expansion for the Laplacian in terms of curl-curl and grad-div,

$$\hat{E}(x) = \int_\Gamma \nabla_x \times ((\hat{E} \times n)\hat{\Phi}) + \hat{E} \cdot n \nabla_x \hat{\Phi}$$

$$+ s\mu_0 \frac{c^2}{s^2} [\nabla_x \times \nabla_x \times (\hat{H} \times n)\hat{\Phi} - \nabla_x \nabla_x \cdot (H \times n)\hat{\Phi}] \, dA.$$

Now we need to integrate by parts again to show that

$$\int_{\Gamma} \nabla_x \cdot ((\hat{H} \times n)\hat{\Phi}) \, dA(y) = - \int_{\Gamma} \hat{H} \times n \cdot \nabla_y \hat{\Phi} \, dA(y)$$

$$= \int_D \nabla \times \hat{H} \cdot \nabla_y \hat{\Phi} \, dV(y) = \int_D s\epsilon_0 \hat{E} \cdot \nabla_y \hat{\Phi} \, dV(y) = \int_{\Gamma} s\epsilon_0 \hat{E} \cdot n \, \hat{\Phi} \, dA(y)$$

where we have used the fact that $\nabla \cdot \hat{E} = 0$.

By slightly more careful reasoning (see [50, 104]) we can fully justify the following result:

Theorem 8 (Stratton-Chu). *Suppose D is a bounded Lipschitz domain and $(\hat{E}, \hat{H})$ in $H(\mathrm{curl}; D) \times H(\mathrm{curl}; D)$ satisfies (81) then $\hat{E}$ can be expressed as follows*

$$\hat{E}(x) = \nabla \times \int_{\Gamma} (\hat{E} \times n)(y) \, \hat{\Phi}(x, y) \, dA(y)$$

$$+ \frac{\mu_0 c^2}{s} \nabla \times \nabla \times \int_{\Gamma} (\hat{H} \times n)(y) \, \hat{\Phi}(x, y) \, dA(y).$$

Remark 7. A similar representation can easily be derived for H by taking the curl of this expression.

It is convenient to rearrange terms yet more. Using the curl-curl vector identity and the equation for $\hat{\Phi}$ as well as denoting by $\nabla_{\Gamma}\cdot$, the surface divergence, we have

$$\hat{E}(x) = \nabla \times \int_{\Gamma} \hat{E} \times n\hat{\Phi} \, dA(y) + \frac{1}{\epsilon_0 s}(-\Delta + \nabla\nabla\cdot) \int_{\Gamma} \hat{H} \times n\hat{\Phi} \, dA(y)$$

$$= \nabla \times \int_{\Gamma} \hat{E} \times n\hat{\Phi} \, dA(y) - \frac{1}{\epsilon_0 s} \frac{s^2}{c^2} \int_{\Gamma} \hat{H} \times n \, \hat{\Phi} \, dA(y)$$

$$+ \frac{1}{\epsilon_0 s} \nabla \int_{\Gamma} \nabla_{\Gamma} \cdot (\hat{H} \times n)\hat{\Phi} \, dA(y)$$

$$= \nabla \times \int_{\Gamma} \hat{E} \times n\hat{\Phi} \, dA(y) - \mu_0 s \int_{\Gamma} \hat{H} \times n\hat{\Phi} \, dA(y)$$

$$+ \frac{1}{\epsilon_0 s} \nabla \int_{\Gamma} \nabla_{\Gamma} \cdot (\hat{H} \times n) \, \hat{\Phi} \, dA(y).$$

Note that repeating the same calculation for $x \in D'$ so that $\hat{\Phi}$ is smooth in D shows that if the evaluation point $\hat{x}$ is outside $\overline{D}$, the right hand side above vanishes.

We conclude at last that, if $x \in D$, using the traces $\hat{E} \times n$ and $\hat{H} \times n$ taken from the field inside D, for $(\hat{E}, \hat{H}) \in H(\mathrm{curl}; D) \times H(\mathrm{curl}; D)$,

$$-\frac{s}{c^2}\,\widehat{\mathrm{SL}}(\hat{H}\times n)+\frac{1}{s}\nabla\widehat{\mathrm{SL}}(\nabla_\Gamma\cdot(\hat{H}\times n))+\epsilon_0\nabla\times\widehat{\mathrm{SL}}(\hat{E}\times n)=\begin{cases}0 & \text{if } x\in D',\\ \epsilon_0\hat{E}(x) & \text{if } x\in D,\end{cases}$$

where $c=1/\sqrt{\epsilon_0\mu_0}$ and $\widehat{\mathrm{SL}}$ is the single layer potential operator given by

$$\widehat{\mathrm{SL}}(\hat{u})=\frac{1}{4\pi}\int_\Gamma\frac{e^{-s|x-y|/c}}{|x-y|}\hat{u}(y)\,dA(y).$$

We can perform similar calculations for $x\in D'$ for the field outside D to obtain for $x\in\mathbb{R}^3\setminus\Gamma$ the same result (with sign changes due to the assumed outward pointing normal). Before quoting this result, we note that the appropriate space for traces of fields where $u\in H(\mathrm{curl};\Omega)$ is $u\times n\in H^{-1/2}(\mathrm{Div};\Gamma)$ where, for a smooth domain,

$$H^{-1/2}(\mathrm{Div};\Gamma)=\left\{u\in H^{-1/2}(\Gamma)^3\mid u\cdot n=0\text{ a.e. on }\Gamma,\nabla_\Gamma\cdot u\in H^{-1/2}(\Gamma)\right\}.$$

For less smooth domains, this space is a good deal more technically difficult to define, see [34], and for the use of these spaces to analyze time-harmonic integral equations for Maxwell's equations see [32, 33, 35]. We will give no details but assume here that such a space is well defined. The dual space of $H^{-1/2}(\mathrm{Div};\Gamma)$ is (again for smooth domains)

$$H^{-1/2}(\mathrm{Curl};\Gamma)=\left\{u\in H^{-1/2}(\Gamma)^3\mid u\cdot n=0\text{ a.e. on }\Gamma,\nabla_\Gamma\times u\in H^{-1/2}(\Gamma)\right\},$$

where $\nabla_\Gamma\times u=\nabla_\Gamma\cdot(u\times n)$. This space can also be defined for polyhedral domains, we again refer to [32, 33, 35].

These observations can then be combined to give (see [35] for the time-harmonic case):

Theorem 9. *Let Γ be a smooth boundary. Suppose $(\hat{E},\hat{H})\in H(\mathrm{curl};D\cup D')\times H(\mathrm{curl};D\cup D')$ are solutions of the Fourier-Laplace Maxwell system (82) in D' and D, then*

$$\epsilon_0\hat{E}(x)=-\frac{s}{c^2}\widehat{\mathrm{SL}}(\hat{j})+\frac{1}{s}\nabla\widehat{\mathrm{SL}}(\nabla_\Gamma\cdot(\hat{j}))-\epsilon_0\nabla\times\widehat{\mathrm{SL}}(\hat{m})$$

where $\hat{j}=[\![\hat{H}\times n]\!]$ and $\hat{m}=[\![\hat{E}\times n]\!]$ where $[\![\cdot]\!]$ denotes the jump across Γ from outside to inside D so that for $x\in\Gamma$

$$[\![\hat{H}\times n]\!](x)=\lim_{\epsilon\to0}(\hat{H}(x-\epsilon\nu(x))\times n(x)-\hat{H}(x+\epsilon\nu(x))\times n(x)).$$

and $\hat{j},\hat{m}\in H^{-1/2}(\mathrm{Div};\Gamma)$.

We can derive two distinct integral equation formulations by choosing $\hat{j}$ and $\hat{m}$ appropriately (the field inside Γ is an arbitrary solution of the Maxwell system and so can be chosen to yield a desirable system). The one we will consider is the "indirect" single layer representation where we choose $\hat{m} = 0$ (the "direct" formulation is to choose $\hat{E} = \hat{H} = 0$ inside Γ, and then use the boundary condition to write $\hat{m} = -\hat{E}^i \times n$). The indirect representation gives

$$\epsilon_0 \hat{E}(x) = -\frac{s}{c^2}\widehat{\mathrm{SL}}\left(\hat{j}\right) + \frac{1}{s}\nabla\widehat{\mathrm{SL}}\left(\nabla_\Gamma \cdot \hat{j}\right)$$

where $\hat{j} \in H^{-1/2}(\mathrm{Div}; \Gamma)$ is a tangential vector field on Γ. Note that

$$s\epsilon_0 \hat{E}(x) = -\frac{s^2}{c^2}\widehat{\mathrm{SL}}\left(\hat{j}\right) + \nabla\widehat{\mathrm{SL}}(\nabla_\Gamma \cdot \hat{j}). \tag{84}$$

To obtain an integral equation we now let $x \to \Gamma$ and define the tangential projection

$$\Pi_T(a) = n \times (a \times n).$$

The tangential component of the single layer potential operator is continuous across Γ and we obtain the problem of finding $\hat{j} \in H^{-1/2}(\mathrm{Div}; \Gamma)$ such that, on Γ,

$$\epsilon_0 s n \times \hat{g} = -\frac{s^2}{c^2}\Pi_T\hat{S}\left(\hat{j}\right) + \nabla_\Gamma \hat{S}(\nabla_\Gamma \cdot \hat{j}), \tag{85}$$

where, for $x \in \Gamma$, the single layer operator is given by

$$\hat{S}(\hat{u}) = \frac{1}{4\pi}\int_\Gamma \frac{e^{-s|x-y|/c}}{|x-y|}\hat{u}(y)\,dA(y)$$

(see [52, 99] for mapping properties of the single layer operator). Equation (85) is the Fourier-Laplace domain Electric Field Integral Equation.

We now define $\hat{V}(s) : H^{-1/2}(\mathrm{Div}; \Gamma) \to H^{-1/2}(\mathrm{Curl}; \Gamma)$ by

$$\hat{V}(s)\hat{j} = -\frac{s^2}{c^2}\Pi_T\hat{S}\left(\hat{j}\right) + \nabla_\Gamma\hat{S}(\nabla_\Gamma \cdot \hat{j}). \tag{86}$$

Suppose $n \times \hat{g} \in H^{-1/2}(\mathrm{Curl}; \Gamma)$. To determine a Galerkin method for the Fourier-Laplace EFIE, we multiply (85) by a test function and integrate over Γ, and then integrate the gradient term by parts to obtain the problem of finding $\hat{j} \in H^{-1/2}(\mathrm{Div}; \Gamma)$ such that

$$\frac{s^2}{c^2}\int_\Gamma \hat{S}(\hat{j}) \cdot \xi\,dA + \int_\Gamma \hat{S}(\nabla_\Gamma \cdot \hat{j})\nabla_\Gamma \cdot \xi\,dA = -\int_\Gamma \epsilon_0 n \times s\hat{g} \cdot \xi\,dA, \tag{87}$$

for all $\xi \in H^{-1/2}(\mathrm{Div}; \Gamma)$. We shall show in the next section that (87) has a solution for any s with $s = \varphi + i\omega$ with $\varphi \geq \varphi_0 > 0$.

We shall now proceed formally to derive the time domain electric field integral equation. Consider first the problem of computing the time domain fundamental solution of the wave equation denoted $\Phi(x, y, t, \tau)$ due to a point source positioned at y and active at time τ. Taking the inverse Laplace transform of the fundamental solution given by (83)

$$\Phi(x, y, t, \tau) = \frac{\delta(t - |x - y|/c - \tau)}{4\pi|x - y|}. \tag{88}$$

Applying the inverse Laplace transform to $\hat{S}$, we obtain the retarded potential single layer operator [52]

$$S(u)(x, t) = \frac{1}{4\pi} \int_{-\infty}^{\infty} \int_{\Gamma} \frac{\delta(t - |x - y|/c - \tau)}{|x - y|} u(y, \tau)\, dA(y)d\tau$$

$$= \frac{1}{4\pi} \int_{\Gamma} \frac{u(y, t - |x - y|/c)}{|x - y|}\, dA(y),$$

and we obtain the problem of finding j such that

$$-\frac{1}{c^2}\Pi_T S\left(\frac{\partial^2 j}{\partial t^2}\right) + \nabla_\Gamma S(\nabla_\Gamma \cdot j) = -\epsilon_0 n \times \dot{g} \text{ on } \Gamma \text{ for } t > 0. \tag{89}$$

This is the time domain Electric Field Integral Equation. Note that since $\dot{j}$ is difficult to read we spell out the time derivatives of j explicitly. We now define the time domain operator

$$V(\partial_t)j = -\frac{1}{c^2}\Pi_T S\left(\frac{\partial^2 j}{\partial t^2}\right) + \nabla_\Gamma S(\nabla_\Gamma \cdot j)$$

using the left hand side of (89) where the argument ∂_t reminds us of the connection to $\hat{V}(s)$.

Once we have determined a solution j to (89) we can use the time domain version of (84)

$$\epsilon_0\dot{E} = -\frac{1}{c^2}S\left(\frac{\partial^2 j}{\partial t^2}\right) + \nabla S(\nabla_\Gamma \cdot j)$$

and obtain E by a further integration. Note that for $x \in \Gamma$, we have $\epsilon_0\dot{E}_T = V(\partial_t)j$.

To directly analyze the time-doman EFIE we could try to use an energy argument. Let the electromagnetic energy $\mathscr{E}$ be given by

$$\mathscr{E}(t) = \frac{1}{2}\int_{\mathbb{R}^3}\left(\frac{1}{c^2}\left|\dot{E}\right|^2 + |\nabla \times E|^2\right)dV,$$

Note that in writing this energy we have used the assumption that $m = [\![E \times n]\!] = 0$ so the electric field is globally in $H(\mathrm{curl}; \mathbb{R}^3)$. Then, using integration by parts,

$$\frac{d\mathcal{E}}{dt} = \int_{\mathbb{R}^3} \left(\frac{1}{c^2} \dot{E} \cdot \ddot{E} + \nabla \times E \cdot \nabla \times \dot{E} \right) dV \tag{90}$$

$$= \int_{D \cup D'} \left(\frac{1}{c^2} \ddot{E} + \nabla \times \nabla \times E \right) \cdot \dot{E} \, dV - \int_{\Gamma} [\![\nabla \times E \times n]\!] \cdot \dot{E} \, dA.$$

Now recall that

$$[\![\nabla \times E \times n]\!] = -\mu_0 [\![\dot{H} \times n]\!] = -\mu_0 \frac{\partial j}{\partial t},$$

Thus, using the fact that $\epsilon_0 \dot{E}_T = V(j)$, we obtain (recalling that the impedance of free space is $Z = \sqrt{\mu_0/\epsilon_0}$)

$$\frac{d\mathcal{E}}{dt} = Z^2 \int_{\Gamma} \frac{\partial j}{\partial t} \cdot V(\partial_t) j \, dA.$$

Integrating both sides, for any $\eta > 0$, assuming the various terms are well defined

$$\int_0^\infty e^{-2\eta t} \, \mathcal{E}(t) \, dt = \mu_0 \int_0^\infty \frac{e^{-2\eta t}}{2\eta} \int_{\Gamma} \frac{\partial j}{\partial t} \cdot V(\partial_t) j \, dA \, dt. \tag{91}$$

Thus, if $V(\partial_t)j = 0$ then $\mathcal{E}(t) = 0$, so $E = 0$ and thus $j = 0$, and so V is injective. To obtain explicit norm estimates, we would need a trace inequality for some norm of j in terms of $\mathcal{E}$. This is the approach we will take in the next section, but using the Fourier-Laplace transform.

For the wave equation, a fully time domain approach to TDIEs is considered in [3].

The above energy analysis shows that:

1. The space of test functions for a space-time discretization of (87) needs to include $\partial j / \partial t$.
2. "Coercivity" is found in unusual norms defined by the energy integral.

5 Numerical Analysis of the Time Domain EFIE

We now look in more detail at the Time Domain EFIE. Completing this analysis is highly technical and we shall only sketch a few arguments in the next section (see [17, 122] for details). Given a bounded domain D with connected complement, we want to find the solution $E = E(x, t)$ of the model time domain exterior scattering problem of finding E such that

$$\frac{1}{c^2}\ddot{E} + \nabla \times \nabla \times E = 0 \text{ in } D' = \mathbb{R}^3 \setminus \overline{D} \text{ for } t \in (0, T),$$

$$E \times n = g \text{ on } \Gamma := \partial D',$$

$$E = \dot{E} = 0 \text{ at } t = 0 \text{ in } D',$$

where g is a tangential vector field ($g = -E^i \times n$ where E^i is a smooth solution of Maxwell's equations vanishing for $t < 0$ in the neighborhood of Γ). In addition, n is the unit outward normal to Γ.

In the previous section we outlined how to show that if $x \notin \Gamma$ then E has the representation

$$\epsilon_0 \dot{E}(x, t) = -\frac{1}{c^2} \int_0^t \int_\Gamma \Phi(x, y, t, \tau)\ddot{j}(y, \tau) \, dA(y) \, d\tau$$

$$+ \nabla \int_0^t \int_\Gamma \Phi(x, y, t, \tau)(\nabla_\Gamma \cdot j)(y, \tau) \, dA(y) \, d\tau \quad (92)$$

for some surface current j where the kernel Φ is given by (88).

Recalling that $\Pi_T u = (n \times u) \times n$ on Γ and letting ∇_Γ denote the surface gradient and $\nabla_\Gamma \cdot$ denote the surface divergence, we then gave a heuristic argument that j satisfies the time domain Electric Field Integral Equation:

$$-\frac{1}{c^2}\Pi_T \int_0^t \int_\Gamma \Phi(x, y, t, \tau)\ddot{j}(y, \tau) \, dA(y) \, d\tau$$

$$+ \nabla_\Gamma \int_0^t \int_\Gamma \Phi(x, y, t, \tau)(\nabla_\Gamma \cdot j)(y, \tau) \, dA(y) \, d\tau = \epsilon_0 n \times \dot{g} \quad (93)$$

for all $x \in \Gamma$ and for $0 < t < T$. We now want to discretize and solve the time domain EFIE.

5.1 Space-Time Petrov-Galerkin Methods

Historically time discretization of (93) has been a challenge due to stability problems. There have been several successful solutions including Band Limited Interpolation and Extrapolation (BLIFS) [129] and Space-Time Petrov-Galerkin methods [1,17,51,60,73,118,119,122,133,134]. BLIFS have yet to be analyzed. The most widely used method is the Petrov-Galerkin method, or a related collocation method, and this is probably the current method of choice for scattering problems.

Motivated by (91) we see that a reasonable space time bilinear form for determining j is to require that

$$\int_0^\infty e^{-2\eta t} \int_\Gamma \xi \cdot V(j) \, dA \, dt = \int_0^\infty e^{-2\eta t} \int_\Gamma \xi \cdot n \times \dot{g} \, dA \, dt, \quad (94)$$

for some suitable choice of test and trial functions. In practice, on finite time intervals, the term $e^{-2\eta t}$ is dropped [73]. For a space-time Galerkin method, we could use space-time function spaces U and W and seek $j \in U$ such that

$$\int_0^T \int_\Gamma \xi \cdot V(j) \, dA \, dt = \int_0^T \int_\Gamma \xi \cdot n \times \dot{g} \, dA \, dt \qquad \forall \xi \in W.$$

For example, in [51], j is approximated by

$$j_h(x,t) = \sum_{i=1}^{N_T} \sum_{n=1}^{N_S} j_{i,n} T_i^{\Delta t}(t) \xi_n^h(x)$$

where the basis functions ξ_n^h are the usual lowest order Raviart-Thomas (RT) divergence conforming elements on a triangular mesh of elements of maximum diameter h on Γ that we shall describe shortly. For time discretization, the authors of [51] use continuous piecewise $k > 1$ degree polynomial functions $T_j^{\Delta t}$. Denoting by $\mathbf{J}_n$ the vector of degrees of freedom at $t_n = n\Delta t$, and collocating the equation at the timesteps while using the usual Galerkin approach in space, a convolution type equation is obtained where the convolution appears due to the retarded potential.

In particular, suppose we know $\mathbf{J}_1, \cdots, \mathbf{J}_{n-1}$ then $\mathbf{J}_n$ satisfies a discrete system of the form

$$\sum_{k=0}^n T_k \mathbf{J}_{n-k} = \mathbf{F}_n$$

where T_k are matrices arising from the spatial Galerkin scheme and retarded potential integrals, and $\mathbf{F}_n$ is a suitable data vector. Since T_0 is invertible, we can solve

$$T_0 \mathbf{J}_n = \mathbf{F}_n - \sum_{k=1}^n T_k \mathbf{J}_{n-k}$$

to obtain $\mathbf{J}_n$. This process is termed "Marching On in Time" in the engineering literature. At this stage algorithmic aspects, such as fast operator evaluation and preconditioning, become very important (for example see [31, 51, 60, 118] for more details).

The method we have sketched is used very successfully in practical applications [51] (see e.g. Terrasse [122] for other suggestions). Some difficulties remain however:

- Due to the retarded potential, basis functions are evaluated at delayed times via the argument $(t - |x - y|/c)$ and so the region of integration may only cover part of an integration element. This makes Gaussian integration rules inaccurate and can lead to instability unless great care is exercised. These integrals must be performed with special rules that respect the light cone. See Fig. 6.
- The need to handle integrals accurately on complicated domains suggests a difficulty in handling curved patches or higher order elements.

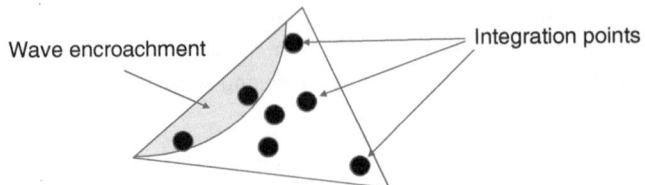

Fig. 6 Because of the shift in the retarded potential, integrals must be performed over domains where light cones intersect spatial elements. This gives rise to complicated integration regions that are not accurately integrated by standard Gaussian quadrature (marked with dots above). Special accurate integration rules are used for the various possible intersections (thanks to Prof. Daniel Weile for this graphic)

5.2 Discrete Convolution

Given the difficulties of implementing higher order versions of the standard Petrov-Galerkin approach, it is reasonable to ask if other approaches could be used. We shall take the rather non-standard approach of discretizing the retarded potential using a method called Convolution Quadrature (CQ) due to Lubich [95,96]. This has been used successfully in elastodynamics [115] and has been extensively developed for the acoustic wave equation [20–22, 75, 84, 86]. For Maxwell's equations see [124, 125], while the analysis presented here is from [40].

The main problem is that we need to discretize integrals of the form

$$\int_0^t \delta(t - R/c - \tau)\xi(\tau)\,d\tau$$

where $R = |x - y|$. This integral has a highly singular kernel that requires ξ at arbitrary past times. Introducing a time step $\Delta t = T/N$, $N > 0$ and $t_n = n\Delta t$, we seek an approximation of the form

$$\int_0^{t_n} \delta(t_n - R/c - \tau)\xi(\tau)\,d\tau \approx \sum_{j=1}^{n} w_{n-j}(R/c)\xi_j$$

where $\xi_j = \xi(t_j)$ and the kernels $w_j(R/c)$ are to be computed. CQ is targeted at integrals of this type. It is based on the Fourier-Laplace transform, and before proceeding it is useful to recall two facts about the Fourier-Laplace transform:

- $\widehat{\delta(\cdot - R/c)} = \exp(-sR/c)$.
- Multiplication by s corresponds to time differentiation.

To see how to proceed, consider the discrete "ideally sampled" function

$$f(t) = \sum_{n=0}^{\infty} f_n \delta(t - n\Delta t)$$

with $\sum_{n=0}^{\infty} |f_n| < \infty$. Then, setting $z = \exp(s\Delta t)$ in the Fourier-Laplace transform $\hat{f}$ of f, we see that

$$(\hat{f})(\ln(z)/\Delta t) = \sum_{n=0}^{\infty} f_n z^{-n}$$

for $|z| > 1$. So we can define, for an absolutely summable sequence $\mathbf{f} = \{f_n\}_{n=0}^{\infty}$, the z-transform by

$$\mathscr{Z}(\mathbf{f})(z) = \sum_{n=0}^{\infty} f_n z^{-n}$$

for all z with $|z|$ large enough. The inversion formula is

$$f_n = \frac{1}{2\pi i} \oint_C \mathscr{Z}(\mathbf{f})(z) z^{n-1} \, dz$$

where C encloses the origin, sufficiently far away. The z-transform (discrete Laplace transform) is a standard control engineering technique for converting analogue filters to digital filters [109].

An important property of the z-transform is as follows. Suppose $f_0 = 0$ and we define the delay operator by $(D\mathbf{f})_n = f_{n-1}$ then

$$\mathscr{Z}(D\mathbf{f}) = z^{-1} \mathscr{Z}(\mathbf{f}).$$

Notice also the product property of the z-transform and discrete convolution: if $(\mathbf{w} * \xi)_n = \sum_{j=0}^{n} w_{n-j}(R/c)\xi_j$, $n = 0, \cdots$ then

$$\mathscr{Z}(\mathbf{w} * \xi) = \sum_{n=0}^{\infty} \sum_{j=0}^{n} w_{n-j}(R/c)\xi_j z^{-n} = \sum_{n=0}^{\infty} \sum_{j=0}^{n} w_{n-j} z^{n-j}(R/c)\xi_j z^{-j}$$

$$= \mathscr{Z}(\mathbf{w})\mathscr{Z}(\xi).$$

To transform from an analogue filter (a function $w(s)$) to a digital filter (a sequence $\mathbf{w}$), we need to replace s (related to ∂_t) by a discrete counterpart. It is plausible to use a difference scheme to approximate the time derivative, and because of the delay property of the z-transform a finite difference operator corresponds to a polynomial in z^{-1} in z-space. More generally, a multistep difference operator corresponds to a rational function in z-space and in particular, it is usual to replace s by $\delta(z)/\Delta t$ where $\delta(z)$ is defined by (63). For an alternative argument following Lubich see [84].

As in Sect. 3, it turns out, see [96], that a sub-class of A-stable schemes, as given in Theorem 5, provide the way to construct δ for Maxwell's equations. This class includes BE and BDF2. As an example we use BE. We are interested in $w(t) = \delta(t - R/c)$, the kernel we wish to discretize, then following the above prescription replacing s and expanding the resulting function we have

$$\mathscr{L}(\mathbf{w})(z) = \exp\left(\frac{-R(1-z^{-1})}{c\,\Delta t}\right) = \exp(-R/c\,\Delta t)\sum_{n=0}^{\infty}\frac{1}{n!}\left(\frac{R}{c\,\Delta t}\right)^n z^{-n}.$$

So for BE, the weights corresponding to the shifted delta function are

$$w_n^{(0)} = \frac{\exp(-R/c\,\Delta t)}{n!}\left(\frac{R}{c\,\Delta t}\right)^n.$$

where the super-script (0) records that these weights are for the zeroth derivative of the delta function. Similarly for $\hat{w}(s) = s^2\exp(-Rs/c)$ we can derive weights $w_n^{(2)}$ corresponding to the second derivative term in the time domain EFIE (these weights turn out to be divided differences of $w_n^{(0)}$, see [125] for more details). The weights $w_n^{(0)}$ for BE are shown graphically in Fig. 7.

The strong form of the semi-discrete time domain EFIE is to find the $\boldsymbol{j}_n$ in $H^{-1/2}(\mathrm{Div};\Gamma)$, $n = 0, 1, 2, \cdots$, such that

$$\sum_{j=0}^{n}\left\{\Pi_T\int_{\Gamma}-\frac{w_{n-j}^{(2)}(|\boldsymbol{x}-\boldsymbol{y}|/c)}{4\pi c^2|\boldsymbol{x}-\boldsymbol{y}|}\boldsymbol{j}_j(\boldsymbol{y})\,dA(\boldsymbol{y})\right.$$

$$\left.+\nabla_{\Gamma}\int_{\Gamma}\frac{w_{n-j}^{(0)}(|\boldsymbol{x}-\boldsymbol{y}|/c)}{4\pi|\boldsymbol{x}-\boldsymbol{y}|}(\nabla_{\Gamma}\cdot\boldsymbol{j}_j)(\boldsymbol{y})\,dA(\boldsymbol{y})\right\} = \epsilon_0 \boldsymbol{n}\times\dot{\boldsymbol{g}}_n$$

for $n = 0, 1, 2, \cdots, N$ (of course $\boldsymbol{j}_n = 0$ for $n \leq 0$).

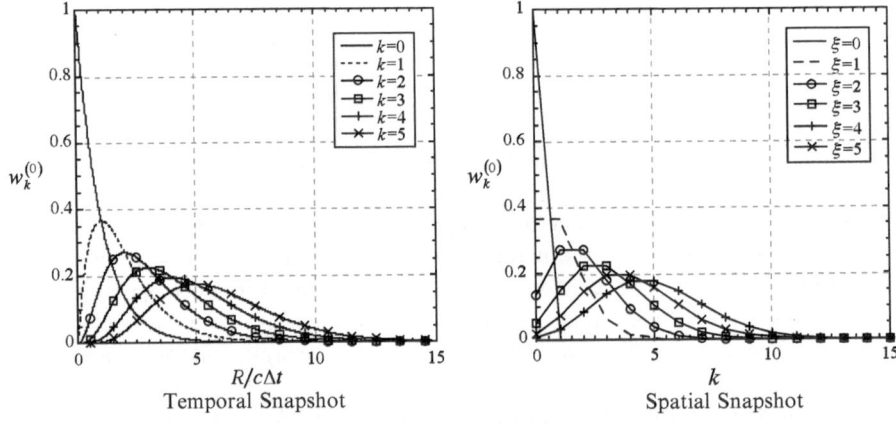

Fig. 7 Here we show plots of $w_k^{(0)}(R/c\,\Delta t)$ for the BE scheme. The left panel shows temporal snapshots fixing k and plotting as a function of scaled distance $R/c\,\Delta t$. In the right panel we show spatial snapshots fixing the scaled distance $\xi = R/c\,\Delta t$ and plotting as a discrete function of time. Note that the weights smear out back in time. Reproduced from [125] with permission ©2008 IEEE

To allow spatial discretization we use the following weak form: for each n we seek $\boldsymbol{j}_n \in H^{-1/2}(\mathrm{Div};\Gamma)$ such that

$$
\sum_{j=0}^{n} \int_{\Gamma} \int_{\Gamma} \left\{ \frac{w_{n-j}^{(2)}(|\boldsymbol{x} - \boldsymbol{y}|/c)}{4\pi c^2 |\boldsymbol{x} - \boldsymbol{y}|} \boldsymbol{j}_j(\boldsymbol{y}) \cdot \boldsymbol{\xi}(\boldsymbol{x}) + \frac{w_{n-j}^{(0)}(|\boldsymbol{x} - \boldsymbol{y}|/c)}{4\pi |\boldsymbol{x} - \boldsymbol{y}|} \right. \tag{95}
$$

$$
\left. (\nabla_\Gamma \cdot \boldsymbol{j}_j)(\boldsymbol{y})(\nabla_\Gamma \cdot \boldsymbol{\xi})(\boldsymbol{x}) \right\} \, dA(\boldsymbol{y}) \, dA(\boldsymbol{x}) = -\epsilon_0 \int_\Gamma \boldsymbol{n} \times \dot{\boldsymbol{g}}_n \cdot \boldsymbol{\xi} \, dA,
$$

for all $\boldsymbol{\xi} \in H^{-1/2}(\mathrm{Div};\Gamma)$ and $n = 0, 1, 2, \cdots, N$.

We now need to discretize in space. As usual we start with a mesh $\mathscr{T}_h$ of elements of maximum diameter h. We have assumed a polyhedral domain so each mesh patch on the boundary (we will not describe rectangular elements here) is a triangle. We of course assume that the mesh is regular and satisfies the meshing constraints in Sect. 3.4 reinterpreted for a surface mesh.

We use pth order Raviart-Thomas elements in $H(\mathrm{Div};\Gamma)$ using the basis from [68]. To describe the pth order Raviart-Thomas (RT) element [30, 111] we first need a new polynomial space. Let us assume that the triangle lies in the (x_1, x_2) plane (otherwise we need to use a local Cartesian coordinate system in the plane of the triangle) and let

$$
\mathscr{P}_p = (P_p)^2 + \boldsymbol{x} P_p
$$

where here we understand P_p to be the space of polynomials in two variables $\boldsymbol{x} = (x_1, x_2)$ of maximum degree p. Note that $p = 0$ is the lowest order element unlike for the Nédélec elements where $p = 1$ is the lowest order element.

Then, for $p \geq 0$, the RT subspace $Y_h \subset H(\mathrm{Div};\Gamma)$ is given by

$$
Y_h = \left\{ \boldsymbol{u} \in H(\mathrm{Div};\Gamma) \mid \boldsymbol{u} \in \mathscr{P}_p \text{ for all elements } K \in \mathscr{T}_h \right\}.
$$

Of course to be useful we need a set of degrees of freedom. Each triangle $K \in \mathscr{T}_h$ defines a plane, and in that plane the edges e of the triangle have normal $\boldsymbol{n}_e^K$. This is a tangent vector to $K \subset \Gamma$. In the local coordinates of the element K we can thus view the normal as having two components. Then the degrees of freedom of a function $\boldsymbol{u} \in \mathscr{P}_p$ on an element K are:

1. The edge integrals $\int_e \boldsymbol{u} \cdot \boldsymbol{n}_e^K q \, ds$ for each $q \in P_p(e)$ and all three edges e of K.
2. The area integrals $\int_K \boldsymbol{u} \cdot \boldsymbol{q} \, dA$ for all $\boldsymbol{q} \in (P_{p-1})^2$.

These degrees are unisolvent, and provided they allow for incoming and outgoing normals, the element is $H(\mathrm{Div};\Gamma)$ conforming. For a smooth domain, the elements have to be defined using a mapping that from a planar triangle to a curvilinear triangle in the mesh [24]. Note that higher order elements on curved surfaces require curved patches to preserve accuracy. The fully discrete time domain EFIE is obtained by replacing $H^{-1/2}(\mathrm{Div};\Gamma)$ in (95) by Y_h.

For theoretical aspects of the discretization of the EFIE in the Fourier frequency domain, including finite element error estimates and comments on meshing curved boundaries, see [24, 35]. In the electrical engineering literature the lowest order elements in the RT family are known as Rao-Wilton-Glisson (RWG) elements [110].

5.3 Convergence

The analysis of CQ rests on understanding the Fourier-Laplace domain EFIE and spatially discretized EFIE. In the Fourier-Laplace domain the EFIE is to find $\hat{\boldsymbol{j}} \in H^{-1/2}(\text{Div}; \Gamma)$ such that

$$b_s(\hat{\boldsymbol{j}}, \boldsymbol{\xi}) = -\int_\Gamma \boldsymbol{n} \times s\hat{\boldsymbol{g}} \cdot \overline{\boldsymbol{\xi}} \, dA, \qquad \forall \boldsymbol{\xi} \in H^{-1/2}(\text{Div}; \Gamma)$$

where

$$b_s(\hat{\boldsymbol{j}}, \boldsymbol{\xi}) = \int_\Gamma \int_\Gamma \frac{\exp(-s|\boldsymbol{x} - \boldsymbol{y}|/c)}{4\pi c^2 |\boldsymbol{x} - \boldsymbol{y}|} \left(s^2 \hat{\boldsymbol{j}}(\boldsymbol{y}) \cdot \overline{\boldsymbol{\xi}}(\boldsymbol{x}) + \right.$$
$$\left. (\nabla_\Gamma \cdot \hat{\boldsymbol{j}})(\boldsymbol{y})(\nabla_\Gamma \cdot \overline{\boldsymbol{\xi}})(\boldsymbol{x}) \right) dA(\boldsymbol{y}) \, dA(\boldsymbol{x})$$

for $s = \varphi + i\omega$ where $\varphi \geq \varphi_0 > 0$.

Often s dependent norms are used to analyze the Fourier-Laplace frequency domain problem. But, as we have seen, Theorem 5 needs s independent norms so we use standard $H^{-1/2}(\text{Div}; \Gamma)$ and $H^{-1/2}(\text{Curl}; \Gamma)$ norms. The integral operator $\hat{V}(s) : H^{-1/2}(\text{Div}; \Gamma) \rightarrow H^{-1/2}(\text{Curl}; \Gamma)$ defined in (86) is related to the sesquilinear form b_s by

$$\left\langle \hat{V}(s)\boldsymbol{\eta}, \boldsymbol{\xi} \right\rangle = b_s(\boldsymbol{\eta}, \boldsymbol{\xi}) \quad \forall \boldsymbol{\xi} \in H^{-1/2}(\text{Div}; \Gamma)$$

for any $\boldsymbol{\eta} \in H^{-1/2}(\text{Div}; \Gamma)$. We want to show that $\hat{V}(s)$ is invertible. Given $\hat{\boldsymbol{j}} \in H^{-1/2}(\text{Div}; \Gamma)$, let $\hat{\boldsymbol{u}} \in H(\text{curl}; \mathbb{R}^3)$ denote the field due to $\hat{\boldsymbol{j}}$ computed via (84). Then

$$0 = \int_{D \cup D'} \frac{s^2}{c^2} \hat{\boldsymbol{u}} \cdot \overline{s\hat{\boldsymbol{u}}} + \nabla \times \nabla \times \hat{\boldsymbol{u}} \cdot \overline{s\hat{\boldsymbol{u}}} \, dV$$
$$= \frac{s|s|^2}{c^2} \|\hat{\boldsymbol{u}}\|^2_{L^2(\mathbb{R}^3)} + \overline{s} \|\nabla \times \hat{\boldsymbol{u}}\|^2_{L^2(\mathbb{R}^3)} - \int_\Gamma [\![\nabla \times \hat{\boldsymbol{u}} \times \boldsymbol{n}]\!] \cdot \overline{s\hat{\boldsymbol{u}}}_T \, dA.$$

This is the Fourier-Laplace domain analogue of (90). Noting that $\epsilon_0 s\hat{\boldsymbol{u}} = \hat{V}(s)\hat{\boldsymbol{j}}$ and $[\![\nabla \times \hat{\boldsymbol{u}} \times \boldsymbol{n}]\!] = \mu_0 s\hat{\boldsymbol{j}}$ on Γ we have shown that

$$\Re \left(Z^2 \int_\Gamma s\hat{\boldsymbol{j}} \cdot \overline{\hat{V}(s)\hat{\boldsymbol{j}}} \, dA \right) = \varphi \left(\frac{1}{c^2} \|s\hat{\boldsymbol{u}}\|^2_{L^2(\mathbb{R}^3)} + \|\nabla \times \hat{\boldsymbol{u}}\|^2_{L^2(\mathbb{R}^3)} \right).$$

This shows that $\hat{V}(s)$ is coercive. Let us define $s\hat{v} = -\nabla \times \hat{u}$ and recall that $\hat{j} = [\![\hat{v} \times n]\!]$. Then $\frac{1}{c^2} s\hat{u} = \nabla \times \hat{v}$ and rewriting the above equality in terms of $\hat{v}$ we have

$$\Re\left(Z^2 \int_\Gamma s\hat{j} \cdot \overline{\hat{V}(s)\hat{j}}\, dA\right) = \varphi\left(c^2 \|\nabla \times \hat{v}\|^2_{L^2(D \cup D')} + \|s\hat{v}\|^2_{L^2(\mathbb{R}^3)}\right).$$

so estimating crudely

$$\Re\left(Z^2 \int_\Gamma s\hat{j} \cdot \overline{\hat{V}(s)\hat{j}}\, dA\right) \geq \min(\varphi, \varphi^3)\left(\|\nabla \times \hat{v}\|^2_{L^2(D \cup D')} + \|\hat{v}\|^2_{L^2(\mathbb{R}^3)}\right),$$

and using the trace inequality for functions in $H(\text{curl}, D \cup D')$ we have

$$\Re\left(Z^2 \int_\Gamma s\hat{j} \cdot \overline{\hat{V}(s)\hat{j}}\, dA\right) \geq C \min(\varphi, \varphi^3) \|[\![\hat{v} \times n]\!]\|^2_{H^{-1/2}(\text{Div};\Gamma)}$$

$$= C \min(\varphi, \varphi^3) \|\hat{j}\|^2_{H^{-1/2}(\text{Div};\Gamma)}.$$

For a more rigorous and careful analysis see [122]. We obtain the following theorem (a slight modification of one in [122]):

Theorem 10. *The sesquilinear form b_s is continuous and coercive on*

$$H^{-1/2}(\text{Div}; \Gamma) \times H^{-1/2}(\text{Div}; \Gamma).$$

In particular

$$|\langle \hat{V}(s)\xi, s\xi\rangle| \geq C(\varphi_0)\|\xi\|^2_{H^{-1/2}(\text{Div};\Gamma)}.$$

for all $\xi \in H^{-1/2}(\text{Div}; \Gamma)$ where we recall that $s = \varphi + i\omega$ with $\varphi \geq \varphi_0 > 0$.

Note that this theorem could now be used to give a precise description of the space-time function spaces in which the solution j is to be found, see [122].

From this theorem we can conclude that

$$\|\hat{V}(s)^{-1}\| \leq C|s|.$$

Applying Theorem 5, assuming g and sufficiently many of its time derivatives vanish at $t = 0$, we have:

Theorem 11. *Using BDF2 we have the semi-discrete error estimate that for $0 \leq n\Delta t \leq T$,*

$$\|j_n - j(\cdot, n\Delta t)\|_{H^{-1/2}(\text{Div};\Gamma)} \leq C\,\Delta t^2 \int_0^t \|n \times g^{(6)}(\tau)\|_{H^{-1/2}(\text{Curl};\Gamma)}\, d\tau$$

for any g such that $\dot{g}(0) = \ddot{g}(0) = \cdots = g^{(5)}(0) = 0$.

Similarly, we obtain first order convergence for BE with

$$n \times g \in H_{\varphi}^{5}(\mathbb{R}_{+}, H^{-1/2}(\mathrm{Curl};\, \Gamma)).$$

We can then prove convergence of the fully discrete scheme using RT elements in space by verifying that Theorem 5 can be applied to the fully discrete integral operator as in [96]. For details see [40].

More interestingly, let $\delta_t^{\Delta t}$ denote the relevant backward difference operator. Following similar arguments to those in Lubich [96] for the wave equation we have:

Theorem 12. *[40] For smooth compatible data (see Theorem 11), the solution of the boundary integral equation is equivalent to solving*

$$\frac{1}{c^2}(\delta_t^{\Delta t})^2 E^n + \nabla \times \nabla \times E^n = 0 \quad \text{in } D',$$

$$E^n \times n = g^n \text{ on } \Gamma,$$

for $n = 0, 1, 2, \cdots$ *where* $E^n = 0$ *for* $n \le 0$ *and where* $(\delta_t^{\Delta t})^2$ *is the BDF2 difference operator (the same theorem also holds for BE).*

This theorem connects the CQ EFIE method to the time stepping methods considered in Sect. 3 and implies that the CQ method is dispersive and dissipative. This is unusual for an integral equation based method, and a potential drawback. Dispersion and dissipation can be controlled to some extent by using implicit Runge-Kutta methods that give, for example, fifth order in time convergence and very low phase error [20, 124]. For dispersion error results, see Sect. 3.9 of this paper.

5.4 Numerical Examples

In our numerical examples, which are reproduced from [40, 125], the incident electric field is taken to be

$$E^{\mathrm{inc}}(r, t) = \hat{p}\exp\left[\frac{1}{2\sigma^2}\left(t - \frac{r \cdot \hat{k}}{c} - t_p\right)^2\right]\cos\left[2\pi f_0\left(t - \frac{r \cdot \hat{k}}{c}\right)\right]$$

where $\sigma = \frac{6}{2\pi B}$, $t_p = 8\sigma$, and $\hat{p} \cdot \hat{k} = 0$, B is a nominal bandwidth, and f_0 is a center frequency.

The time step is related to the frequency parameters and oversampling rate ψ according to

$$\psi = \frac{1}{2(f_0 + B)\Delta t}.$$

Here ψ is defined to give a measure of how the timestep compares to the minimum sampling rate of two points per temporal wavelength at the upper end of the nominal frequency band.

We start with results from [40] showing scattering from a sphere of diameter 1m which is convenient because we can compute the solution at each frequency by an infinite Mie series (see e.g. [104]). The center frequency $f_0 = 120$MHz, band width $B = 40$MHz and we use $p = 1$ RT elements on curvilinear patches. In each case we solve the time domain problem, then Fourier transform the results to get the solution at several desired frequency (e.g. f_0). We can compare the resulting computed radar cross-section (RCS, see [18]) from the time domain code with one computed by a frequency domain integral equation using the same spatial elements and spatial grid at several frequencies. The difference between these solutions gives an indication of time discretization error.

Figures 8 and 9 show the average magnitude of the surface current j as the wave passes over the sphere. Figure 8 shows a slight long term instability in the solution which we believe is related to the well-known low frequency instability of the EFIE (this is allowed in our estimates). The low frequency instability can be removed by several techniques and we used the loop tree decomposition [131]. Once this is done, no growth in the solution is seen. This is even true for BDF3 which is not covered by our theory (although an overshoot is visible for BDF3 in Fig. 9). All remaining results for EFIE are computed with this stabilization.

Usually the RCS is important for antenna calculations and in Fig. 10 we show the RCS of the sphere computed from the surface current found by CQ EFIE with BE, BDF2 and BDF3 using the oversampling factor $\psi = 5$ and $\psi = 10$. We compared to the results from a frequency domain EFIE scheme with the same spatial elements on the same spatial mesh called the Method-of-Moments (MoM).

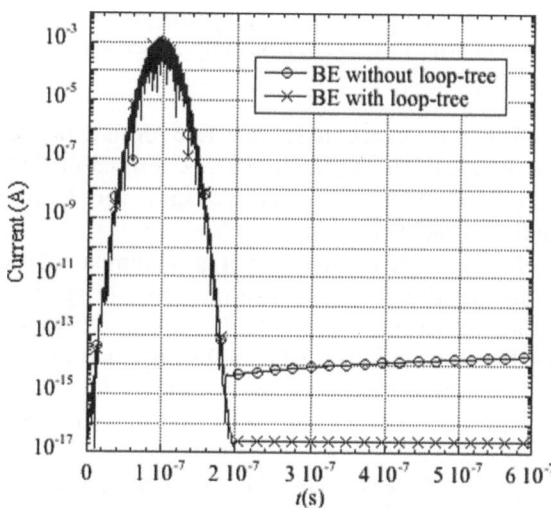

Fig. 8 The average of the surface current j computed by the CQ EFIE using $p = 1$ RT elements on curvilinear patches of the sphere with and without the stabilization. Reproduced from [40] with permission

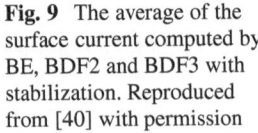

Fig. 9 The average of the
surface current computed by
BE, BDF2 and BDF3 with
stabilization. Reproduced
from [40] with permission

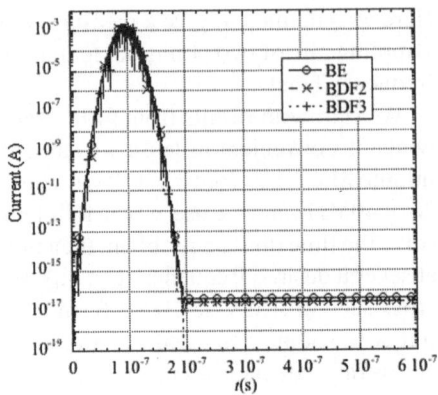

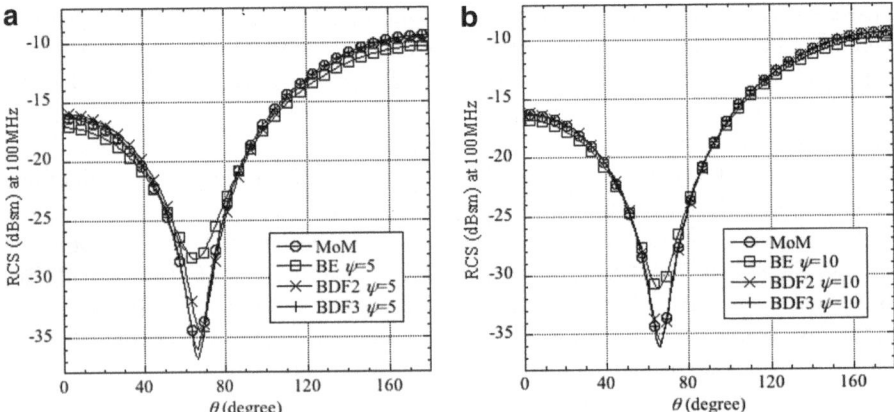

Fig. 10 The RCS with oversampling factor $\psi = 5$ and $\psi = 10$ for the sphere as a function of
elevation angle. Reproduced from [40] with permission

The theoretical convergence rate for time discretization for BE is first order and
for BDF2 is second order. This is verified in Fig. 11. The range of oversampling
factor for which convergence is seen is very large and reflects the stability inherent
in the A-stable marching methods.

In electrical engineering it is common to use the Combined Field Integral
Equation (CFIE) on closed surfaces like the sphere. Next we reproduce some results
from [125] using the CQ CFIE. In Fig. 12 we show the results for BDF2 using
analytically computed CQ coefficients, and coefficients computed by the discrete
inverse z-transform. Results are shown for 360 spatial degrees of freedom with
polarization p along the z-axis and direction of propagation d along the x-axis
(Fig. 13).

In Fig. 14 we show the time step convergence rate of the CQ CFIE by comparing
the CFIE RCS to a frequency domain RCS on the same grid. We see convergence

Fig. 11 Relative least square RCS error as a function of the oversampling factor. The convergence rates agree with our theory for BE and BDF2. Reproduced from [40] with permission

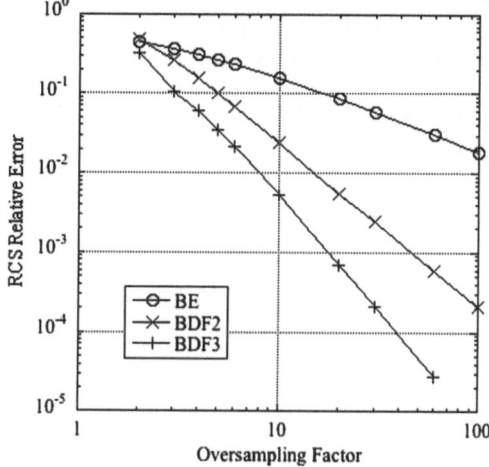

Fig. 12 Results for 1m conducting sphere using CFIE. Reproduced from [125] with permission, ©2008 IEEE

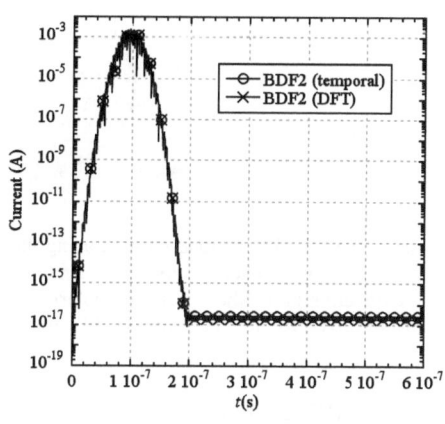

Fig. 13 RCS computed via the CFIE by several methods with oversampling factor $\psi = 10$. Reproduced from [125] with permission, ©2008 IEEE

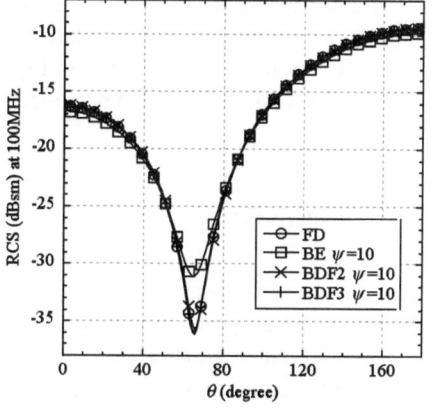

Fig. 14 Convergence of the
RCS for the CFIE. We see the
expected convergence rate for
all the methods shown over a
wide range of oversampling
factor. Reproduced from
[125] with permission,
©2008 IEEE

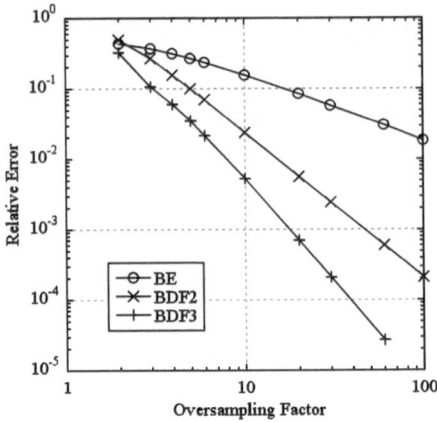

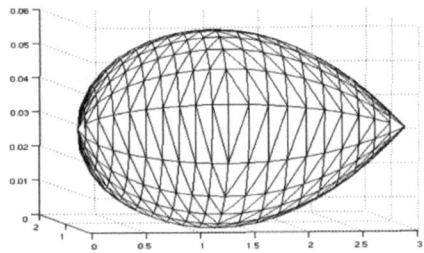

Fig. 15 The NASA Almond meshed using flat patches (note that the drawing is not to scale) and
lowest order divergence conforming elements. Here we have 1140 spatial unknowns. We use a
wave traveling along the x-axis and polarized along the z-axis. The center frequency $f_0 = 0$ and
$B = 400$MHz. Reproduced from [125] with permission, ©2008 IEEE

even for BDF3, though this is not predicted by our theory (but we have also seen
cases where BDF3 did not converge).

Finally, we show a more challenging example: the NASA almond shown in
Fig. 15. This geometry is a standard test case for CEM codes. Results are compared
to an RCS computed by FDTD in Fig. 16. The convolution quadrature CFIE gives
comparable results to FDTD provided the oversampling factor is large enough.

6   Conclusion

We have tried to describe some issues and methods of analysis for finite element
methods in computational electromagnetics. We emphasized analysis based on
energy estimates, and we took the rather non-standard approach of concentrating
on Fourier-Laplace methods, with implicit timestepping because we wanted to draw
parallels between the boundary integral equation and volume based methods.

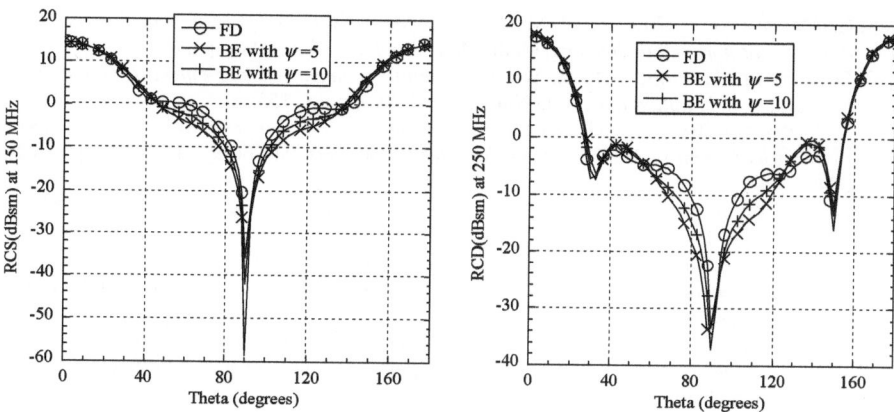

Fig. 16 Comparison of RCS for the almond at 150MHz (left) and 250MHz (right). Reproduced from [125] with permission, ©2008 IEEE

Whatever method is used, CEM usually results in very large computational problems with vast memory requirements. This survey has barely touched on solvers or implementational aspects.

Future work in CEM will probably focus on more exotic media, and the inclusion of small features (such as wires) in simulations.

Acknowledgements Our research is supported in part by a grant from NSF (DMS-0811104). PM would like to thank BICOM at Brunel University, UK for a visiting position during the writing of most of this paper.

References

1. T. ABBOUD, J.-C. NÉDÉLEC, AND J. VOLAKIS, *Stable solution of the retarded potential equations*, in Proc. 17th Ann. Rev. Progress in Appl. Comp. Electromagnetics, Monterey, CA, 2001, pp. 146–151.
2. J. ADAM, A. SERVENIERE, J. NÉDÉLEC, AND P. RAVIART, *Study of an implicit scheme for integrating Maxwell's equations*, Comput. Meth. Appl. Mech. Eng., 22 (1980), pp. 327–46.
3. A. AIMI, M. DILIGENTI, C. GUARDASONI, I. MAZZIERI, AND S. PANIZZI, *An energy approach to spacetime Galerkin BEM for wave propagation problems*, Int. J. Numer. Meth. Eng., DOI:10.1002/nme.2660 (2009).
4. M. AINSWORTH, *Dispersive and dissipative behaviour of high order discontinuous Galerkin finite element methods*, J. Comp. Phys., 198 (2004), pp. 106–130.
5. M. AINSWORTH, *Dispersive properties of high-order Nedelec/edge element approximation of the time-harmonic Maxwell equations*, Phil. Trans. Roy. Soc. A, 362 (2004), pp. 471–91.
6. M. AINSWORTH AND J. COYLE, *Hierarchic finite element bases on unstructured tetrahedral meshes*, Int. J. Numer. Meth. Eng., 58 (2003), pp. 2103–30.
7. M. AINSWORTH, P. MONK, AND W. MUNIZ, *Dispersive and dissipative properties of discontinuous Galerkin methods for the wave equation*, J. Sci. Comput., 27 (2006), pp. 5–40.
8. M. AINSWORTH AND H. WAJID, *Optimally blended spectral-finite element scheme for wave propagation and nonstandard reduced integration*, SIAM J. Numer. Anal., 48 (2010), pp. 346–71.

9. C. AMROUCHE, C. BERNARDI, M. DAUGE, AND V. GIRAULT, *Vector potentials in three-dimensional nonsmooth domains*, Math. Meth. Appl. Sci., 21 (1998), pp. 823–64.

10. D. ARNOLD, R. FALK, AND R. WINTHER, *Finite element exterior calculus, homological techniques, and applications*, Acta Numer., 15 (2006), pp. 1–155.

11. D. ARNOLD, R. FALK, AND R. WINTHER, *Finite element exterior calculus from Hodge theory to numerical stability*, Bulletin of the American Mathematical Society, 47 (2010), pp. 281–354.

12. D. ARNOLD, R. FALK, AND R. WINTHUR, *Multigrid in $H(div)$ and $H(curl)$*, Numer. Math., 85 (2000), pp. 197–217.

13. F. ASSOUS, P. DEGOND, E. HEINTZE, P. RAVIART, AND J. SEGREÉ, *On a finite-element method for solving the three-dimensional Maxwell equations*, J. Comput. Phys., 109 (1993), pp. 222–37open.

14. F. ASSOUS, P. DEGOND, AND J. SEGRÉ, *Numerical approximation of the Maxwell equations in inhomogeneous media by a p^1 conforming finite element method*, J. Comput. Phys., 128 (1996), pp. 363–80.

15. F. ASSOUS AND M. MIKHAELI, *Nitsche type method for approximating boundary conditions in the static maxwell equations*, in Proceedings of the 26th IASTED International Conference on Modelling, Identification, and Control, MIC '07, Anaheim, CA, USA, 2007, ACTA Press, pp. 402–407.

16. I. BABUŠKA AND S. SAUTER, *Is the pollution effect of the FEM avoidable for the Helmholtz equation considering high wave numbers?*, SIAM J. Numer. Anal., 34 (1997), pp. 2392–423.

17. A. BACHELOT AND A. PUJOLS, *Equations intégrales espace-temps pour le systeème de Maxwell*, C.R. Acad. Sc. Paris, Série I, 314 (1992), pp. 639–44.

18. C. BALANIS, *Advanced Engineering Electromagnetics*, Wiley, 1989.

19. A. BAMBERGER AND T. HA-DUONG, *Formulation variationnelle espace-temps pour le calcul par potentiel retarde de la diffraction d'une onde acoustique(i)*, Math. Mech. in the Appl. Sci, 8 (1986), pp. 405–435.

20. L. BANJAI, *Multistep and multistage convolution quadrature for the wave equation: Algorithms and experiments*, SIAM J. Sci. Comput., 32 (2010), pp. 2964–2994.

21. L. BANJAI AND W. HACKBUSCH, *Hierarchical matrix techniques for low- and high-frequency Helmholtz problems*, IMA J. Numer. Anal., 28 (2008), pp. 46–79.

22. L. BANJAI AND S. SAUTER, *Rapid solution of the wave equation in unbounded domains*, SIAM J. Numer. Anal., 47 (2008), pp. 227–49.

23. R. BARTHELMÉ, P. CIARLET, AND E. SONNENDRÜCKER, *Generalized formulations of Maxwell's equations for numerical Vlasov-Maxwell simulations*, Math. Meth. Appl. Sci., 17 (2007), pp. 659–80.

24. A. BENDALI, *Numerical analysis of the exterior boundary value problem for the time harmonic Maxwell equations by a boundary finite element method. Part ii: The discrete problem*, Math. Comput., 43 (1984), pp. 47–68.

25. J. BÉRENGER, *A perfectly matched layer for the absorption of electromagnetics waves*, J. Comput. Phys., 114 (1994), pp. 185–200.

26. H. BERTRAM, *Theory of Magnetic Recording*, Cambridge University Press, ber94.

27. D. BISKAMP, *An Introduction to Magnetohydrodynamics*, Cambridge University Press, 2001.

28. D. BOFFI, L. GASTALDI, AND A. BUFFA, *Convergence analysis for hyperbolic evolution problems in mixed form*, Tech. Rep. 17-PV, I.M.A.T.I.-C.N.R., University of Pavia, Italy, 2005.

29. A. BOSSAVIT, *Computational Electromagnetism*, Academic Press, San Diego, 1998.

30. F. BREZZI AND M. FORTIN, *Mixed and Hybrid Finite Element Methods*, Springer, New York, 1991.

31. A. BUFFA AND S. CHRISTIANSEN, *A dual finite element complex on the barycentric refinement*, Math. Comput., 76 (2007), pp. 1743–69.

32. A. BUFFA AND P. CIARLET JR., *On traces for functional spaces related to Maxwell's equations. I. An integration by parts formula in Lipschitz polyhedra*, Math. Meth. Appl. Sci., 24 (2001), pp. 9–30.

33. A. BUFFA AND P. CIARLET JR., *On traces for functional spaces related to Maxwell's equations. II. Hodge decompositions on the boundary of Lipschitz polyhedra*, Math. Meth. Appl. Sci., 24 (2001), pp. 31–48.

34. A. BUFFA, M. COSTABEL, AND D. SHEEN, *On the traces of* $\mathbf{H}(\mathbf{curl}, \Omega)$ *in Lipschitz domains*, J. Math. Anal. Appl., 276 (2003), pp. 845–67.

35. A. BUFFA, R. HIPTMAIR, T. VON PETERSDORFF, AND C. SCHWAB, *Boundary element methods for Maxwell transmission problems in Lipschitz domains*, Numer. Math., 95 (2003), pp. 459–85.

36. A. BUFFA, P. HOUSTON, AND I. PERUGIA, *Discontinuous Galerkin computation of the Maxwell eigenvalues on simplicial meshes*, J. Comput. Appl. Math., 204 (2007), pp. 317–33.

37. A. BUFFA AND I. PERUGIA, *Discontinuous galerkin approximation of the Maxwell eigen-problem*, SIAM J. Numer. Anal., 44 (2006), pp. 2198–226.

38. A. BUFFA, G. SANGALLI, AND R. VAZQUEZ, *Isogeometric analysis in electromagnetics: B-splines approximation*, Comput. Meth. Appl. Mech. Eng., 199 (2010), pp. 1143–52.

39. A. CAMPBELL, *An introduction to numerical methods in superconductors*, J. Supercond. Nov. Magn., (2010). DOI 10.1007/s10948-010-0895-5.

40. Q. CHEN, P. MONK, D. WEILE, AND X. WANG, *Analysis of convolution quadrature applied to the time-domain electric field integral equation*. to appear in Communications in Computational Physics.

41. W. C. CHEW AND W. H. WEEDON, *A 3D perfectly matched medium from modified Maxwell's equations with stretched coordinates*, Microwave Opt. Technol. Lett., 7 (1994), pp. 599–604.

42. S. CHRISTIANSEN, *Foundations of finite element methods for wave equations of Maxwell type*, in Applied Wave Mathematics, E. Quak and T. Soomere, eds., Springer-Verlag, Berlin Heidelberg, 2009, pp. 335–93.

43. P. CIARLET, *The Finite Element Method for Elliptic Problems*, vol. 4 of Studies In Mathematics and Its Applications, North-Holland, New York, 1978.

44. P. CIARLET AND E. JAMELOT, *Continuous Galerkin methods for solving the time-dependent Maxwell equations in 3D geometries*, J. Comput. Phys., 226 (2007), pp. 1122–35.

45. P. CIARLET AND J. ZOU, *Fully discrete finite element approaches for time-dependent Maxwell's equations*, Numer. Math., 82 (1999), pp. 193–219.

46. B. COCKBURN AND C.-W. SHU, *The Runge-Kutta discontinuous Galerkin method for conservation laws V: Multidimensional systems*, J. Comput. Phys., 141 (1998), pp. 199–224.

47. G. COHEN, *Higher-order numerical methods for transient wave equations*, Springer, Berlin, 2002.

48. G. COHEN, P. JOLY, J. ROBERTS, AND N. TORDJMAN, *Higher order triangular finite elements with mass lumping for the wave equation*, SIAM J. Numer. Anal., 38 (2001), pp. 2047–78.

49. G. COHEN AND P. MONK, *Gauss point mass lumping schemes for Maxwell's equations*, Numer. Meth. Partial Diff. Eqns., 14 (1998), pp. 63–88.

50. D. COLTON AND R. KRESS, *Inverse Acoustic and Electromagnetic Scattering Theory*, Springer-Verlag, New York, 2nd ed., 1998.

51. K. COOLS, F. ANDRIULLI, F. OLYSLAGER, AND E. MICHIELSSEN, *Time domain Calderón identities and their application to the integral equation analysis of scattering by PEC objects Part I: Preconditioning*, IEEE Trans. Antennas Propagat., 57 (2009), pp. 2352–64.

52. M. COSTABEL, *Fundamentals*, vol. 1 of Encyclopedia of Computational Mechanics, John Wiley & Sons, 2004, ch. Time-dependent Problems with the Boundary Integral Equation Method.

53. P. DAVIES, *Numerical stability and convergence of approximations of retarded potential integral equations*, SIAM J. Numer. Anal., 31 (1994), pp. 856–75.

54. P. DAVIES AND D. DUNCAN, *Averaging techniques for time-marching schemes for retarded potential integral equations*, Applied Numerical Mathematics, 23 (1997), pp. 291–310.

55. L. DEMKOWICZ, *hp-adaptive finite elements for time-harmonic Maxwell equations*, in Topics in Computational Wave Propagation: Direct and Inverse Problems, M. Ainsworth, P. Davies, D. Duncan, P. Martin, and B. Rynne, eds., vol. 31 of Lecture Notes in Computational Science and Engineering, Springer, 2003.

56. L. DEMKOWICZ, P. MONK, AND L. VARDAPETYAN, *de Rham diagram for hp finite element spaces*, Comput. Math. Appl., 39 (2000), pp. 29–38.
57. G. DUVAUT AND J.-L. LIONS, *Inequalities in Mechanics and Physics*, Springer, New York, 1976.
58. A. ELMKIES AND P. JOLY, *Elements finis d'arete et condensation de masse pour les equations de Maxwell: le cas 3D*, C. R. Acad. Sci. Paris, Série 1, 325 (1997), pp. 1217–22.
59. B. ENGQUIST AND A. MAJDA, *Absorbing boundary conditions for the numerical simulation of waves*, Math. Comput., 31 (1977), pp. 629–51.
60. A. ERGIN, B. SHANKER, AND E. MICHIELSSEN, *The plane-wave time-domain algorithm for fast analysis of transient phenomena*, IEEE Antennas and Propagation Magazine, 41 (1999), pp. 39–52.
61. L. FEZOUI, S. LANTERI, S. LOHRENGEL, AND S. PIPERNO, *Convergence and stability of a discontinuous Galerkin time-domain method for the 3D heterogeneous Maxwell equations on unstructured meshes*, ESAIM: Mathematical Modeling and Numerical Analysis, 39 (2005), pp. 1149–76.
62. A. FISHER, R. RIEBEN, G. RODRIGUE, AND D. WHITE, *A generalized mass lumping technique for vector finite-element solutions of the time-dependent maxwell equations*, IEEE Trans. Antennas Propagat., 53 (2005), pp. 2900–10.
63. M. GANDER, F. MAGOULÈS, AND F. NATAF, *Optimized Schwarz methods without overlap for the Helmholtz equation*, SIAM J. Sci. Comput., 24 (2002), pp. 36–80.
64. N. GEODEL, S. SCHOMANN, AND T. W. M. CLEMENS, *GPU accelerated Adams-Bashforth multirate discontinuous Galerkin FEM simulation of high-frequency electromagnetic fields*, IEEE Trans. Mag., 46 (2010), pp. 2735–8.
65. N. GOEDEL, N. N. T. WARBURTON, AND M. CLEMENS, *Scalability of higher-order discontinuous Galerkin FEM computations for solving electromagnetic wave propagation problems on GPU clusters*, IEEE Trans. Mag., 46 (2010), pp. 3469–72.
66. J. GOPALAKRISHNAN AND J. PASCIAK, *Overlapping Schwarz preconditioners for indefinite time harmonic Maxwell equations*, Math. Comput., 72 (2003), pp. 1–15.
67. V. GRADINARU AND R. HIPTMAIR, *Whitney elements on pyramids*, ETNA, 8 (1999), pp. 154–68.
68. R. GRAGLIA, D. WILTON, AND A. PETERSON, *Higher order interpolatory vector bases for computational electromagnetics*, IEEE Trans. Antennas Propagat., 45 (1997), pp. 329–342.
69. M. GROTE AND T. MITKOVA, *Explicit local time-stepping methods for Maxwell's equations*, J. Comput. Appl. Math., 234 (2010), pp. 3283–302.
70. M. GROTE, A. SCHNEEBELI, AND D. SCHÖTZAU, *Interior penalty discontinuous Galerkin method for Maxwell's equations:Energy norm error estimates*, IMA J. Numer. Anal., 204 (2007), pp. 375–86.
71. M. GROTE, A. SCHNEEBELI, AND D. SCHÖTZAU, *Interior penalty discontinuous Galerkin method for Maxwell's equations: Optimal L^2-norm error estimates*, IMA J. Numer. Anal., 28 (2008), pp. 440–68.
72. M. GROTE AND D. SCHOETZAU, *Optimal Error Estimates for the Fully Discrete Interior Penalty DG Method for the Wave Equation*, JOURNAL OF SCIENTIFIC COMPUTING, 40 (2009), pp. 257–272.
73. T. HA-DUONG, *On retarded potential boundary integral equations and their discretizations*, in Topics in Computational Wave Propagation: Direct and Inverse Problems, M. Ainsworth, ed., Springer, 2003, pp. 301–36.
74. T. HA-DUONG, B. LUDWIG, AND I. TERRASSE, *A galerkin BEM for transient acoustic scattering by an absorbing obstacle*, Int. J. Numer. Meth. Engng., 57 (2003), pp. 1845–82.
75. W. HACKBUSCH, W. KRESS, AND S. SAUTER, *Sparse convolution quadrature for time domain boundary integral formulations of the wave equation*, IMA J. Numer. Anal., 29 (2009), pp. 158–79.
76. T. HAGSTROM, WARBURTON, AND D. GIVOLI, *Radiation boundary conditions for time-dependent waves based on complete plane wave expansions*, J. Comput. Appl. Math., 234 (2010), pp. 1988–95.

77. Y. HAUGAZEAU AND P. LACOSTE, *Condenstaion de la matrice masse pour les éléments finis mixtes de h(rot)*, Comptes Rendu, Series 1, 316 (1993), pp. 509–12.

78. Y. HEAUGAZEAU AND P. LACOSTE, *Lumping of the mass matrix for 1st-order mixed finite elements in h(curl)*, in Mathematical and Numerical Aspects of Wave Propagation, R. Kleinman, T. Angell, D. Colton, F. Santosa, and I. Stakgold, eds., SIAM, Philadelphia, 1993, pp. 259–68.

79. J. HESTHAVEN AND T. WARBURTON, *Nodal high-order methods on unstructured grids - I. Time-domain solution of Maxwell's equations*, J. Comput. Phys., 181 (2002), pp. 186–221.

80. R. HIPTMAIR, *Multigrid method for Maxwell's equations*, SIAM J. Numer. Anal., 36 (1998), pp. 204–25.

81. R. HIPTMAIR, *Discrete Hodge-operators: An algebraic perspective*, Journal of Electromagnetic Waves and Applications, 15 (2001), pp. 343–4.

82. R. HIPTMAIR, *Finite elements in computational electromagnetism*, Acta Numerica, 11 (2002), pp. 237–339.

83. H. HOLTER, *Some experiences from FDTD analysis of infinite and finite multi-octave phased arrays*, IEEE Trans. Antennas Propagat., 50 (2002), pp. 1725–31.

84. W. KRESS AND S. SAUTER, *Numerical treatment of retarded boundary integral equations by sparse panel clustering*, IMA J. Numer. Anal., 28 (2008), pp. 162–85.

85. T. LAHIVAARA AND T. HUTTUNEN, *A non-uniform basis order for the discontinuous Galerkin method of the acoustic and elastic wave equations*, Applied Numerical Mathematics, 61 (2011), pp. 473–86.

86. A. LALIENA AND F. SAYAS, *Theoretical aspects of the application of convolution quadrature to scattering of acoustic waves*, Numer. Math., 112 (2009), pp. 637–78.

87. P. LAX, *Functional Analysis*, Wiley, New York, 2002.

88. M. LAZEBNIK, M. OKONIEWSKI, J. BOOSKE, AND S. HAGNESS, *Highly accurate Debye models for normal and malignant breast tissue dielectric properties at microwave frequencies*, IEEE Microwave and Wireless Components Letters, 17 (2007), pp. 822–4.

89. J. LEE AND B. FORNBERG, *Some unconditionally stable time stepping methods for the 3D Maxwell's equations*, J. Comput. Appl. Math., 166 (2004), pp. 497–523.

90. J.-F. LEE, *WETD - A finite element time domain approach for solving Maxwell's equations*, IEEE Microwave and Guided Wave Lett., 4 (1994), pp. 11–3.

91. J.-F. LEE, R. LEE, AND A. CANGELLARIS, *Time-domain finite-element methods*, IEEE Trans. Antennas Propagat., 45 (1997), pp. 430–42.

92. J. LI, *Finite element study of the Lorentz model in metamaterials*, Comput. Meth. Appl. Mech. Eng., 200 (2011), pp. 626–37.

93. J. LI, Y. CHEN, AND V. ELANDER, *Mathematical and numerical study of wave propagation in negative-index materials*, Comput. Meth. Appl. Mech. Eng., 197 (2008), pp. 45–8.

94. J. LI AND Z. ZHANG, *Unified analysis of time domain mixed finite element methods for Maxwell's equations in dispersive media*, Journal of Computational Mathematics, 28 (2010), pp. 693–710.

95. C. LUBICH, *Convolution quadrature and discretized operational calculus. I and II*, Numer. Math., 52 (1988), pp. 129–45 and 413–425.

96. C. LUBICH, *On the multistep time discretization of linear initial-boundary value problems and their boundary integral equations*, Numer. Math., 67 (1994), pp. 365–89.

97. C. LUBICH AND A. OSTERMANN, *Runge-Kutta methods for parabolic equations and convolution quadratur*, Math. Comput., 60 (1993), pp. 105–31.

98. C. MAKRIDAKIS AND P. MONK, *Time-discrete finite element schemes for Maxwell's equations*, RAIRO - Math. Model. Numer. Anal., 29 (1995), pp. 171–97.

99. W. MCLEAN, *Strongly Elliptic Systems and Boundary Integral Equations*, Cambridge University Press, Cambridge, 2000.

100. T. W. M.CLEMENS, *Transient eddy-current calculation with the FI-method*, IEEE Trans. Mag., 35 (1999), pp. 1163–6.

101. J. MELENK AND S. SAUTER, *Convergence analysis for finite element discretizations of the Helmholtz equation with Dirichlet-to-Neumann boundary conditions*, Math. Comput., 79 (2010), pp. 1871–914.

102. P. MONK, *Analysis of a finite element method for Maxwell's equations*, SIAM J. Numer. Anal., 29 (1992), pp. 714–29.

103. P. MONK, *An analysis of Nédélec's method for the spatial discretization of Maxwell's equations*, J. Comput. Appl. Math., 47 (1993), pp. 101–21.

104. P. MONK, *Finite Element Methods for Maxwell's Equations*, Oxford University Press, Oxford, 2003.

105. P. MONK AND A. PARROTT, *Phase-accuracy comparisons and improved far-field estimates for 3-D edge elements on tetrahedral meshes*, J. Comput. Phys., 170 (2001), pp. 614–41.

106. J. NÉDÉLEC, *Mixed finite elements in* $\mathbb{R}^3$, Numer. Math., 35 (1980), pp. 315–41.

107. J. NÉDÉLEC, *A new family of mixed finite elements in* $\mathbb{R}^3$, Numer. Math., 50 (1986), pp. 57–81.

108. N.NIGAM AND J. PHILLIPS, *High-order finite elements on pyramids*. accepted, IMA J. Numer.Anal., 2011.

109. J. PROAKIS AND D. MANOLAKIS, *Digital Signal Processing: Principles, Algorithms, and Applications*, Prentice Hall, Upper Saddle River, NJ, third edition ed., 1996.

110. S. RAO, D. WILTON, AND A. GLISSON, *Electromagnetic scattering by surfaces of arbitrary shap*, IEEE Trans. Antennas Propagat., 30 (1982), pp. 409–18.

111. P. RAVIART AND J. THOMAS, *Primal hybrid finite element methods for 2nd order elliptic equations*, Math. Comput., 31 (1977), pp. 391–413.

112. R. RIEBEN, G. RODRIGUE, AND D. WHITE, *A high order mixed vector finite element method for solving the time dependent Maxwell equations on unstructured grids*, J. Comput. Phys., 204 (2005), pp. 490–519.

113. D. RILEY AND J. JIN, *Finite-element time-domain analysis of electrically and magnetically dispersive periodic structures*, IEEE Trans. Antennas Propagat., (2008), pp. 3501–9.

114. T. RYLANDER AND A. BONDESON, *Stability of explicit-implicit hybrid time-stepping schemes for Maxwell's equations*, J. Comput. Phys., 179 (2002), pp. 426–438.

115. M. SCHANZ AND H. ANTES, *A new visco- and elastodynamic time domain boundary element formulation*, Computational Mechanics, 20 (1997), pp. 452–9.

116. S. SCHOMANN, N. GOEDEL, T. WARBURTON, AND M. CLEMENS, *Local timestepping techniques using Taylor expansion for modeling electromagnetic wave propagation with discontinuous Galerkin-FEM*, IEEE Trans. Mag., 46 (2010), pp. 3504–7.

117. W. SHA, X. WU, Z. HUANG, AND M. CHEN, *Waveguide simulation using the high-order symplectic finite-difference time-domain scheme*, Progress In Electromagnetics Research B, 13 (2009), pp. 217–56.

118. B. SHANKER, A. ERGIN, M. LU, AND E. MICHIELSSEN, *Fast analysis of transient electromagnetic scattering phenomena using the multilevel plane wave time domain algorithm*, IEEE Trans. Antennas Propagat., 51 (2003), pp. 628–41.

119. B. SHANKER, A. A. ERGIN, K. AYGUN, AND E. MICHIELSSEN, *Analysis of transient electromagnetic scattering phenomena using a two-level plane wave time-domain algorithm*, IEEE Trans. Antennas Propagat., 48 (2000), pp. 510–23.

120. S. SHAW, *Finite element approximation of Maxwell's equations with Debye memory*. to appear in Advances in Numerical Analysis.

121. A. TAFLOVE, *Computational Electrodynamics*, Artech House, Boston, 1995.

122. I. TERRASSE, *Résolution mathématique et numérique des équations de Maxwell instationnaires par une méthode de potentiels retardés*, Spécialité: Mathématiques Appliquées, Ecole Polytechnique, Paris, France, 1993.

123. A. TOSELLI, *Overlapping Schwarz methods for Maxwell's equations in three dimensions*, Numer. Math., 86 (2000), pp. 733–52.

124. X. WANG AND D. WEILE, *Implicit Runge-Kutta methods for the discretization of time domain integral equations*. IEEE Trans. Antennas Propagat., (2011), pp. 4651–63.

125. X. WANG, R. WILDMAN, D. WEILE, AND P. MONK, *A finite difference delay modeling approach to the discretization of the time domain integral equations of electromagnetism*, IEEE Trans. Antennas Propagat., 56 (2008), pp. 2442–52.

126. J. WEBB AND B. FORGHANI, *Hierarchal scalar and vector tetrahedra*, IEEE Trans. Mag., 29 (1993), pp. 1495–8.

127. T. WEILAND, *A discretization method for the solution of Maxwell's equations for six-component fields*, Electronics and Communications AEUE, 31 (1977), pp. 116–120.
128. T. WEILAND, *Numerical solution of Maxwell's equation for static, resonant and transient problems*, in Studies in Electrical and Electronic Engineering 28B, T. Berceli, ed., URSI International Symposium on Electromagnetic Theory Part B, Elsevier, New York, 1986, pp. 537–42.
129. D. WEILE, G. PISHARODY, N.-W. CHEN, B. SHANKER, AND E. MICHIELSSEN, *A novel scheme for the solution of the time-domain integral equations of electromagnetics*, IEEE Trans. Antennas Propagat., 52 (2004), pp. 283–95.
130. *Wikipedia: Finite-difference time-domain method.* http://en.wikipedia.org/wiki/Finite-difference-time-domain_method, 2010.
131. R. WILDMAN AND D. WEILE, *An accurate broad-band method of moments using higher order basis functions and tree-loop decomposition*, IEEE Trans. Antennas Propagat., 52 (2004), pp. 3005–11.
132. K. YEE, *Numerical solution of initial boundary value problems involving Maxwell's equations in isotropic media*, IEEE Trans. Antennas Propagat., 16 (1966), pp. 302–7.
133. A. YILMAZ, D. W. E. MICHIELSSEN, AND J.-M. JIN, *A fast Fourier transform accelerated marching-on-in-time algorithm for electromagnetic analysis*, Electromagnetics, 21 (2001), pp. 181–97.
134. A. YILMAZ, D. WEILE, E. MICHIELSSEN, AND J.-M. JIN, *A hierarchical FFT algorithm (HIL-FFT) for the fast analysis of transient electromagnetic scattering phenomena*, IEEE Trans. Antennas Propagat., (2002).
135. J. ZHAO, *Analysis of the finite element method for time-dependent Maxwell problems*, Math. Comput., 247 (2003), pp. 1089–105.

Numerical Approximation of Large Contrast Problems with the Unfitted Nitsche Method

Erik Burman and Paolo Zunino

Abstract These notes are concerned with the numerical treatment of the coupling between second order elliptic problems that feature large contrast between their characteristic coefficients. In particular, we study the application of Nitsche's method to set up a robust approximation of interface conditions in the framework of the finite element method. The notes are subdivided in three parts. Firstly, we review the weak enforcement of Dirichlet boundary conditions with particular attention to Nitsche's method and we discuss the extension of such technique to the coupling of Poisson equations. Secondly, we review the application of Nitsche's method to large contrast problems, discretised on computational meshes that capture the interface of discontinuity between coefficients. Finally, we extend the previous schemes to the case of unfitted meshes, which occurs when the computational mesh does not conform with the interface between subproblems.

1 A Review of Nitsche's Method

1.1 Weak Enforcement of Boundary Conditions for Poisson's Problem

The aim of this section is to review some well known techniques to enforce boundary conditions of Dirichlet type for second order problems. In particular, we

E. Burman
Department of Mathematics, University of Sussex, Falmer, Brighton, BN1 9RF, United Kingdom
e-mail: E.N.Burman@sussex.ac.uk

P. Zunino (✉)
MOX - Department of Mathematics, Politecnico di Milano, P.zza Leonardo da Vinci 32, 20133 Milano, Italy
e-mail: paolo.zunino@polimi.it

J. Blowey and M. Jensen (eds.), *Frontiers in Numerical Analysis – Durham 2010*,
Lecture Notes in Computational Science and Engineering 85,
DOI 10.1007/978-3-642-23914-4_4, © Springer-Verlag Berlin Heidelberg 2012

will focus on the techniques that allow to enforce such boundary conditions within
the definition of the bilinear form associated to the variational formulation of the
problem at hand, rather than enforcing the constraints at the boundary in the search
space for the solution. We refer to such schemes as those using weak enforcement
of Dirichlet boundary conditions, in contrast to the case where Dirichlet boundary
conditions appear in the definition of the trial space, often addressed as strong
enforcement of boundary constraints. Concerning Neumann or mixed type boundary
conditions we observe that they are naturally embedded in the set up of the problem
bilinear form. Some alternatives for the treatment of natural boundary conditions
have been recently addressed in [34].

We start from the simplest model problem, that is Poisson's problem with Dirichlet boundary conditions, which can be straightforwardly formulated as follows. Let
Ω be a convex polygonal domain in $\mathbb{R}^d$. Given $f \in L^2(\Omega)$ and $g \in H^{\frac{1}{2}}(\partial\Omega)$, find
$\hat{u} \in H^1(\Omega)$ a weak solution of

$$\begin{cases} -\Delta u = f, & \text{in } \Omega, \\ u = g, & \text{on } \partial\Omega. \end{cases} \tag{1}$$

The most straightforward way to enforce Dirichlet type constraints at the
boundary is to embed the variational formulation into an Hilbert space whose
functions satisfy the boundary constraints. Given $\mathscr{R}g \in H^1(\Omega)$, a lifting of g on
the entire Ω, we aim to find $u \in H_0^1(\Omega)$ such that

$$a(u,v) = F(v) - a(\mathscr{R}g,v) \quad \forall v \in H_0^1(\Omega), \tag{2}$$
$$a(u,v) := (\nabla u, \nabla v)_\Omega ,$$
$$F(v) := (f,v)_\Omega ,$$

where $(\cdot,\cdot)_\Omega$ denotes the L^2 inner product on Ω. In the framework of the finite
element method, the enforcement of Dirichlet boundary conditions in the trial space
is also easily translated to the discrete level. Thus, for the approximation of classical
second order problems there is no need to consider alternatives. However, the
continuous expansion of computational analysis in several engineering disciplines
often requires to consider non standard problem formulations. Among many other
examples we mention problems that feature multiple domains, accounting for the
contact between different materials or fluids, problems with moving boundaries,
such as the ones arising from fluid-structure interaction analysis, problems set
on domains with very complex dendritic shapes, which are often encountered
in the application of computational analysis to life sciences. In these cases, the
strong enforcement of Dirichlet boundary or interface conditions may turn out to
be cumbersome when applied at the discrete level, while the weak treatment of
Dirichlet constraints, which allows to relax their satisfaction, may lead to numerical
schemes that are more efficient or easily implemented.

An effective technique for weak enforcement of Dirichlet boundary constraints is the application of Lagrange multipliers. The original idea, due to Babuška [3], is based on the fact that the weak formulation of the Poisson's problem is equivalent to the minimisation among all functions $v \in H_0^1(\Omega)$ of the energy functional

$$J(u) = \min_{v \in H_0^1(\Omega)} J(v) \tag{3}$$

$$J(v) := a(v, v) - 2F(v) \quad \forall v \in H^1(\Omega). \tag{4}$$

The problem of finding the minimum $u \in H_0^1(\Omega)$ can be seen as a constrained minimisation problem, because the solution is sought in a subspace $H_0^1(\Omega)$ of the natural space $H^1(\Omega)$ where the functional is well defined. This convex constrained minimisation problem can be translated into an unconstrained problem by resorting to the Lagrangian functional accounting for the constraint $u = 0$ on $\partial\Omega$. Let $H^{-\frac{1}{2}}(\partial\Omega)$ be the dual space of $H^{\frac{1}{2}}(\partial\Omega)$ with the duality pairing $\langle \cdot, \cdot \rangle_{\partial\Omega}$, then

$$L(v, \mu) := J(v) + \langle \mu, v \rangle, \quad \forall v \in H^1(\Omega), \ \mu \in H^{-\frac{1}{2}}(\partial\Omega)$$

is the Lagrangian functional and we look for a couple (u, λ), where the additional unknown λ is called Lagrange multiplier such that,

$$L(u, \lambda) = \inf_{v \in H^1(\Omega)} \sup_{\mu \in H^{-1/2}(\partial\Omega)} L(v, \mu).$$

This is an instance of a saddle point problem, involving minimisation with respect to one unknown and maximisation with respect to the other. Owing to fundamental results of convex analysis, this constrained minimisation problem admits the following equivalent formulation: setting $b(\lambda, v) := \langle \lambda, v \rangle_{\partial\Omega}$ and given $f \in L^2(\Omega)$, find $u \in H^1(\Omega)$, $\lambda \in H^{-\frac{1}{2}}(\partial\Omega)$ such that

$$\begin{cases} a(u, v) + b(\lambda, v) = F(v) & \forall v \in H^1(\Omega), \\ b(\mu, u) = b(\mu, g) & \forall \mu \in H^{-\frac{1}{2}}(\partial\Omega). \end{cases} \tag{5}$$

We notice that the new formulation with Lagrange multipliers involves an additional unknown that at the discrete level increases the computational cost of the problem. However, this is not only a drawback, because the unknown λ has a relevant physical meaning,

$$\lambda + \partial_n u = 0 \text{ in } H^{-\frac{1}{2}}(\partial\Omega).$$

Anyway, the most relevant remark concerning the weak enforcement of Dirichlet boundary condition with Lagrange multipliers is the fact that the corresponding variational problem does not conform with the assumptions of Lax-Milgram's Lemma, which ensures well posedness of the usual weak formulation of Poisson's problem. The crucial point is that the introduction of Lagrange multipliers breaks the coercivity of the entire weak problem, whose well posedness holds true under the following set of conditions,

$$a(\cdot, \cdot) \text{ and } b(\cdot, \cdot) \text{ are bilinear and continuous,}$$

$$\text{coercivity: } \exists \, \alpha > 0 \text{ s.t. } a(v, v) \geq \alpha \|v\|_{1,\Omega}^2$$

$$\forall v \in Z := \{v \in H^1(\Omega) : b(\lambda, v) = 0, \ \forall \lambda \in H^{-\frac{1}{2}}(\partial\Omega)\},$$

$$\text{``inf-sup'': } \exists \, \beta > 0 \text{ s.t. } \forall \lambda \in H^{-\frac{1}{2}}(\partial\Omega), \quad \sup_{v \in H^1(\Omega) \setminus \{0\}} \frac{b(\lambda, v)}{\|v\|_{1,\Omega}} \geq \beta.$$

It is immediately evident that the verification of such conditions is a more challenging task than the check of Lax-Milgram's assumptions. In the case of the variational formulation of problem (1) with weak enforcement of boundary conditions they are satisfied, see for instance [35].

However, a fundamental problem appears when we look at the discretisation by means of finite elements. With a conforming finite element discretisation, the classical Lax-Milgram's coercivity is automatically inherited at the discrete level, but this is not the case for the aforementioned "inf-sup"condition. According to the particular choices for the approximation spaces of $H^1(\Omega)$ and $H^{-\frac{1}{2}}(\partial\Omega)$ such condition may not be verified at the discrete level. The correct formalisation of this difficulty and the constructive development of suitable couples of discrete spaces for the approximation of saddle point problems have been an important milestone of finite element analysis in the last decades, see [9, 39, 40] among many others.

More precisely, given the finite element spaces $V_h \subset H^1(\Omega)$, $\Lambda_h \subset H^{-\frac{1}{2}}(\partial\Omega)$, the application of Galerkin method to (5) consists in finding $u_h \in V_h$ and $\lambda_h \in \Lambda_h$ such that

$$\begin{cases} a(u_h, v_h) + b(\lambda_h, v_h) = F(v_h) & \forall v_h \in V_h, \\ b(\mu_h, u_h) = b(\mu_h, g) & \forall \mu_h \in \Lambda_h \end{cases} \qquad (6)$$

and proceeding similarly to the infinite-dimensional case, it has been proved that, see [3], the discrete problem is well posed provided that

$$\exists \, \alpha_h > 0 \text{ s.t. } a(v_h, v_h) \geq \alpha_h \|v_h\|_{1,\Omega}^2 \qquad (7)$$

$$\forall v_h \in Z_h := \{v_h \in V_h \text{ s.t. } b(v_h, \mu_h) = 0, \ \forall \mu_h \in \Lambda_h\},$$

$$\exists \, \beta_h > 0 \text{ uniformly independent of } h \text{ s.t.} \qquad (8)$$

$$\forall \, \lambda_h \in \Lambda_h, \quad \sup_{v_h \in V_h \setminus \{0\}} \frac{b(\lambda_h, v_h)}{\|v_h\|_{1,\Omega}} \geq \beta_h.$$

Note that, since the search space for the solution u_h has been extended, by removing the strong enforcement of the constraints at the boundary, coercivity of

$a(\cdot, \cdot)$ is lost in $V_h \subset H^1(\Omega)$. For this reason, the Lax-Milgram's theory does not apply any more. Furthermore, the satisfaction of the discrete "inf-sup" condition is not straightforward, for instance intuitive choices of discrete spaces such as linear finite elements in Ω for u_h and linear finite elements on $\partial\Omega$ for λ_h lead to an unstable discrete problem. For standard H^1-conforming affine finite elements for the approximation of u, a piecewise constant approximation for λ is stable provided that the multiplier space is defined on a boundary mesh with size $3h$, with h being the characteristic element size. For Crouzeix-Raviart approximation of u, piecewise constant multipliers on the unrestricted boundary mesh are stable. Recalling the equation $\lambda + \partial_n u = 0$ one can expect that the regularity for the Lagrange multiplier space should be lower than the one for the primal unknown u_h. Such rule of thumb is also confirmed by observing that a generalisation of the previous stable couple of elements is given by k-order H^1-conforming finite elements on Ω combined with fully discontinuous $(k - 1)$-order finite elements on $\partial\Omega$. We refer the interested reader to [39–41] for a detailed analysis.

The relaxation of the strong enforcement of Dirichlet boundary conditions by means of Lagrange multipliers leads to an accurate but expensive problem at the discrete level. For this reason, some alternatives have been developed, with the aim to perform the weak approximation of boundary conditions using a numerical method that can still be cast in the framework of Lax-Milgram's lemma.

Starting from the minimisation problem (3), the most straightforward strategy consists in the application of a penalty method. The idea is to enrich the energy functional $J(v)$ with an additional quadratic term that takes its minimum when the Dirichlet boundary conditions are exactly satisfied. The magnitude of the additional functional should be modulated by means of a constant factor that ensures that the minimum of the augmented functional accurately, but not exactly, satisfies the prescribed boundary conditions. Given $\varepsilon > 0$ the penalty method consists in finding $u_\varepsilon \in H^1(\Omega)$ such that

$$J_\varepsilon(u_\varepsilon) = \min_{v \in H^1(\Omega)} J_\varepsilon(v), \tag{9}$$

$$J_\varepsilon(v) := J(v) + \frac{1}{2}\varepsilon^{-1}\|v - g\|_{0,\partial\Omega}^2, \quad \forall v \in H^1(\Omega), \tag{10}$$

whose Euler equations require to find $u_\varepsilon \in H^1(\Omega)$ such that

$$a(u, v) + \varepsilon^{-1}(u - g, v)_{\partial\Omega} = F(v), \quad \forall v \in H^1(\Omega), \tag{11}$$

which seem to share all the good properties of (2) with the additional advantage that the natural search and test spaces are the entire $H^1(\Omega)$. The application of Galerkin method to (11) consists in finding $u_{h,\varepsilon} \in V_h \subset H^1(\Omega)$ such that

$$a(u_{h,\varepsilon}, v_h) + \varepsilon^{-1}(u_{h,\varepsilon} - g, v_h)_{\partial\Omega} = F(v_h), \quad \forall v_h \in V_h, \tag{12}$$

where V_h could be any H^1-conformal finite element space on Ω. However, to analyze the efficiency of the penalty method, we remind that (12) has been developed to approximate (2). In this respect, the first property to be considered is the consistency of such an approximation scheme. Starting from (12) and performing integration by parts on Ω we obtain a residual,

$$\mathscr{R}(u_{h,\varepsilon}) := (-\Delta u_{h,\varepsilon} - f, v_h)_\Omega + (\partial_n u_{h,\varepsilon}, v_h)_{\partial\Omega} + \varepsilon^{-1}(u_{h,\varepsilon} - g, v_h)_{\partial\Omega} \quad \forall v_h \in V_h.$$

Replacing $u_{h,\varepsilon}$ with $u \in H_0^1(\Omega)$ such that $-\Delta u - f = 0$ weakly in Ω and $u = g$ on $\partial\Omega$ we observe that the residual does not vanish, i.e.

$$\mathscr{R}(u) = (\partial_n u, v_h)_{\partial\Omega} \neq 0.$$

This proves that the penalty method is not strongly consistent with the original weak Poisson's problem. Then, the fundamental question is how to choose the penalty parameter ε with respect to the characteristic mesh size h and the finite element polynomial order k so that $u_{h,\varepsilon}$ converges to u with possibly optimal rate as h becomes infinitesimal. We refer to [4, 7] for a thorough discussion and error analysis of the penalty method, which will be briefly summarized later on. Anyway, the penalty method has received a considerable attention in literature, in particular for the approximation of problems where the computational mesh is not fitted to the boundary, because the penalty term can be easily implemented also in this setting.

Among several interpretations, Nitsche's method can be seen as a variant to override the major drawback of the penalty method, restoring the strong consistency of the discrete scheme with respect to (2). More precisely, we aim to find $u_{h,\varepsilon} \in V_h \subset H^1(\Omega)$ such that

$$a_\varepsilon(u_{h,\varepsilon}, v_h) = F_\varepsilon(v_h) \quad \forall v_h \in V_h, \tag{13}$$

with

$$a_\varepsilon(u_{h,\varepsilon}, v_h) := a(u_{h,\varepsilon}, v_h) - (\partial_n u_{h,\varepsilon}, v_h)_{\partial\Omega} - s(\partial_n v_h, u_{h,\varepsilon})_{\partial\Omega} + \varepsilon^{-1}(u_{h,\varepsilon}, v_h)_{\partial\Omega},$$

$$F_\varepsilon(v_h) := F(v_h) + \varepsilon^{-1}(g, v_h)_{\partial\Omega} - s(\partial_n v_h, g)_{\partial\Omega},$$

where ε plays the role of penalty parameter and $s(\partial_n v_h, u_{h,\varepsilon} - g)_{\partial\Omega}$ with $s \in \{-1, 0, 1\}$ is an additional term that if $s = 1$ restores the symmetry of $a_\varepsilon(u_{h,\varepsilon}, v_h)$, according to the fact that $a(u, v)$ is supposed to be a symmetric bilinear form. However, all choices $s = \pm 1$ and $s = 0$ are admissible and will be discussed later on. Another fundamental part of the scheme is the selection of the penalty parameter that will clearly emerge from the error analysis of the scheme.

Quoting R. Stenberg 1995, [45], *"In view of our analysis it seems that the Nitsche method is the most straightforward method to use. Unfortunately, this method seems to be quite unknown. We think, however, that it would be worthwhile to explore it in applications such as contact problems, for fictitious domain methods and for domain decomposition*". Indeed, Nitsche's method has been recently applied to all of these cases with success and the scope of the present work is to review those studies, developing and discussing further extensions.

1.2   Analysis of Nitsche's Method

Let $\mathscr{T}_h$ be a family of shape regular and quasi uniform triangulations of Ω. Let K be a generic element of $\mathscr{T}_h$ and let h_K be its diameter (the radius of the smallest ball containing this set) and the characteristic mesh size is $h := \max_{K \in \mathscr{T}_h} h_K$. Without loss of generality, we refer with our notation and choice of symbols to the case of two space dimensions. In particular, we apply the subscript E to denote element edges (or faces in three dimensions). Let $\mathscr{B}_h$ be the collection of mesh edges lying on the boundary $\partial\Omega$. On each mesh $\mathscr{T}_h$ we set up a Lagrangian finite element space of order k denoted as

$$V_h := \{v_h \in C^0(\Omega) : v_h|_K \in \mathbb{P}^k(K) \ \forall K \in \mathscr{T}_h\}.$$

We endow the finite element space with the following norms that are adapted to the analysis of the scheme

$$\|v\|_{\pm\varepsilon,\partial\Omega}^2 := \sum_{E \in \mathscr{B}_h} \varepsilon^{\mp 1} \|v\|_{0,E}^2, \quad \forall v \in L^2(\partial\Omega),$$

$$\|v\|_{1,\varepsilon,\Omega}^2 := |v|_{1,\Omega}^2 + \|v\|_{\varepsilon,\partial\Omega}^2, \quad \forall v \in H^1(\Omega).$$

For the forthcoming analysis we remind of the following basic inequalities, for which we refer to [10]. For simplicity of notation, we write $a \lesssim b$ if there exists a positive constant C independent of h such that $a \leq Cb$. The standard L^2 Cauchy-Schwarz inequality can be straightforwardly extended to,

$$(v,w)_{\partial\Omega} \leq \|v\|_{\pm\varepsilon,\partial\Omega} \|w\|_{\mp\varepsilon,\partial\Omega}, \quad \forall v,w \in L^2(\partial\Omega).$$

We will also make use of a generalised Poincaré inequality, also known as Poincaré-Friedrichs inequality, which holds in $H^1(\Omega)$ provided that an additional term is introduced to enrich the H^1-seminorm in order to account for constant functions,

$$\|v\|_{1,\Omega} \lesssim |v|_{1,\Omega}^2 + \|v\|_{\varepsilon,\partial\Omega}^2, \quad \forall v \in H^1(\Omega).$$

Finally, the following discrete inequalities will be fundamental for the analysis of Nitsche's method,

$$h_E^{\frac{1}{2}} \|v_h\|_{0,E} \lesssim \|v_h\|_{0,K}, \quad h_K \|\nabla v_h\|_{0,K} \lesssim \|v_h\|_{0,K}, \quad \forall v_h \in V_h. \tag{14}$$

The first inequality implies that there exists a positive constant C_I such that

$$\sum_{E\in\mathscr{B}_h} h_E \|v_h\|_{0,E}^2 \leq C_I \sum_{K\in\mathscr{T}_h} \|v_h\|_{0,K}^2. \tag{15}$$

We notice that problem (13) consists of a standard Galerkin method using an H^1-conformal approximation space. Then, owing to Lax-Milgram's lemma its well posedness is ensured by consistency, stability and boundedness of $a_\varepsilon(\cdot,\cdot)$ together with linearity and boundedness of the right hand side.

Recalling that Nitsche's method can be seen as a correction of a simple penalty method in order to recover consistency, it is easy to verify that, given $u \in H^2(\Omega) \cap H_0^1(\Omega)$ the weak solution of $-\Delta u = f$ in Ω with $u = g$ on $\partial\Omega$, then problem (13) satisfies $a_\varepsilon(u - u_{h,\varepsilon}, v_h) = 0$ for any $v_h \in V_h$, which states that Nitsche's method is strongly consistent for any admissible value of ε and s.

In the framework of Lax-Milgram's lemma, stability is equivalent to coercivity of $a_\varepsilon(\cdot,\cdot)$ that holds true if there exists $\alpha > 0$, uniformly independent of h, such that

$$a_\varepsilon(v_h, v_h) \geq \alpha \|v_h\|_{1,\varepsilon,\Omega}^2, \quad \forall v_h \in V_h.$$

To investigate the validity of such property in the particular case $s = 1$, we proceed as follows

$$\begin{aligned}
a_\varepsilon(v_h, v_h) &= |v_h|_{1,\Omega}^2 + \|v_h\|_{\varepsilon,\partial\Omega}^2 - 2\,(v_h, \partial_n v_h) \\
&\geq |v_h|_{1,\Omega}^2 + \|v_h\|_{\varepsilon,\partial\Omega}^2 - 2\|v_h\|_{\varepsilon,\partial\Omega}\|\partial_n v_h\|_{-\varepsilon,\partial\Omega} \\
&\geq |v_h|_{1,\Omega}^2 + \|v_h\|_{\varepsilon,\partial\Omega}^2 - \delta^{-1}\|v_h\|_{\varepsilon,\partial\Omega}^2 - \delta\|\partial_n v_h\|_{-\varepsilon,\partial\Omega}^2 \\
&\leq |v_h|_{1,\Omega}^2 + \|v_h\|_{\varepsilon,\partial\Omega}^2 - \delta^{-1}\|v_h\|_{\varepsilon,\partial\Omega}^2 - \delta \sum_{E\in\mathscr{B}_h} \varepsilon|v_h|_{1,E}^2. \tag{16}
\end{aligned}$$

In order to combine the first with the fourth term of previous inequality, it is convenient to select ε such that it is ε is directly proportional to h_E on $\mathscr{B}_h$. As a result of that, the norm $\|v\|_{\mp\varepsilon,\partial\Omega}$ is equivalent to

$$\|v_h\|_{\pm\frac{1}{2},h,\partial\Omega}^2 := \sum_{E\in\mathscr{B}_h} h_E^{\mp 1}\|v\|_{0,E}^2, \quad \forall v \in L^2(\partial\Omega)$$

and we denote $\|v\|_{1,h,\Omega}^2 := |v|_{1,\Omega}^2 + \|v\|_{\frac{1}{2},h,\partial\Omega}^2$ accordingly. Owing to inverse inequality (14), we notice that it holds

$$\|v_h\|_{-\frac{1}{2},h,\partial\Omega}^2 = \sum_{E\in\mathscr{B}_h} h_E\|v_h\|_E^2 \lesssim \sum_{K\in\mathscr{T}_h} \|v_h\|_K^2 \lesssim \|v_h\|_{0,\Omega}^2, \quad \forall\, v_h \in V_h.$$

Given a positive constant γ we select $\varepsilon = h_E/\gamma$ for notational convenience. Then, the bilinear form $a_\varepsilon(\cdot,\cdot)$ and the right hand side $F_\varepsilon(\cdot)$ should be modified as follows,

$$a_h(u_h, v_h) := a(u_h, v_h) - (\partial_n u_h, v_h)_{\partial\Omega} - s\,(\partial_n v_h, u_h)_{\partial\Omega} + \gamma \sum_{E \in \mathscr{B}_h} h_E^{-1}\,(u_h, v_h)_E,$$

$$F_h(v_h) := F(v_h) + \gamma \sum_{E \in \mathscr{B}_h} h_E^{-1}\,(g, v_h)_E - s\,(\partial_n v_h, g)_{\partial\Omega},$$

and we aim to find $u_h \in V_h$ such that $a_h(u_h, v_h) = F_h(v_h)$ for any $v_h \in V_h$, which precisely define the Nitsche's method, except from the constant γ. To conclude the analysis of coercivity of $a_h(\cdot, \cdot)$, we mimic the reasoning of (16) for the case $s = 1$ and exploiting (15) we obtain

$$a_h(v_h, v_h) \gtrsim (1 - \delta C_I)|v_h|_{1,\Omega}^2 + (\gamma - \delta^{-1})\|v_h\|_{\frac{1}{2}, h, \partial\Omega}^2$$

such that the coercivity of $a_h(\cdot, \cdot)$ holds true for any $C_I^{-1} > \delta > 0$ provided that the penalty parameter γ is such that $\gamma > \delta^{-1} > C_I$. An estimate of constant C_I for piecewise affine approximation is provided in [31].

Boundedness of Nitsche's method is equivalent to continuity of $a_h(\cdot, \cdot)$. In view of the forthcoming error analysis, we introduce the augmented norm

$$|||v|||_{1,h,\Omega}^2 := |v|_{1,\Omega}^2 + \|v\|_{+\frac{1}{2}, h, \partial\Omega}^2 + \|\partial_n v\|_{-\frac{1}{2}, h, \partial\Omega}^2, \quad \forall v \in (H^2(\Omega) + V_h).$$

Then, there exists $M > 0$ uniformly independent of h such that

$$a_h(u, v) \leq M|||u|||_{1,h,\Omega}\|v\|_{1,h,\Omega}, \quad \forall u \in (H^2(\Omega) + V_h), \; \forall v \in V_h.$$

The proof of such property follows from a combination of Cauchy-Schwarz inequalities,

$$a_h(u, v)$$

$$\leq |u|_{1,\Omega}|v|_{1,\Omega} + \gamma\|u\|_{\frac{1}{2}, h, \partial\Omega}\|v\|_{\frac{1}{2}, h, \partial\Omega} + \|v\|_{\frac{1}{2}, h, \partial\Omega}\|\partial_n u\|_{-\frac{1}{2}, h, \partial\Omega}$$

$$\quad + \|u\|_{\frac{1}{2}, h, \partial\Omega}\|\partial_n v\|_{-\frac{1}{2}, h, \partial\Omega}$$

$$\lesssim |u|_{1,\Omega}|v|_{1,\Omega} + \gamma\|u\|_{\frac{1}{2}, h, \partial\Omega}\|v\|_{\frac{1}{2}, h, \partial\Omega} + \|v\|_{\frac{1}{2}, h, \partial\Omega}\|\partial_n u\|_{-\frac{1}{2}, h, \partial\Omega} + \|u\|_{\frac{1}{2}, h, \partial\Omega}|v|_{1,\Omega}$$

$$\lesssim |||u|||_{1,h,\Omega}\|v\|_{1,h,\Omega}, \quad \forall u \in (H^2(\Omega) + V_h), \; \forall v \in V_h.$$

Combining consistency, stability and boundedness, we are able to perform the error analysis of Nitsche's method. We remind that the finite element space V_h satisfies a well known approximation property in the H^1 norm, which can be easily extended to the mesh dependent norm $||| \cdot |||_{1,h,\Omega}$ owning to inverse inequalities, in particular for any $v \in H^{k+1}(\Omega)$,

$$\inf_{v_h \in V_h} |||v - v_h|||_{1,h,\Omega} \lesssim h^k \|v\|_{k+1,\Omega}.$$

Then, Strang's lemma allows us to conclude that given $u \in H^{k+1}(\Omega)$ with $k \geq 1$ the weak solution of $-\Delta u = f$ in Ω with $u = g$ on $\partial\Omega$ and given u_h the solution of Nitsche's method with γ large enough, the following a-priori error estimate holds true,

$$\|u - u_h\|_{1,h,\Omega} \lesssim \inf_{v_h \in V_h} |||u - v_h|||_{1,h,\Omega} \lesssim h^k \|u\|_{k+1,\Omega}, \qquad (17)$$

and in case of self-adjoint problems and $s = 1$, exploiting Aubin-Nitsche's Lemma one obtains,

$$\|u - u_h\|_{0,\Omega} \lesssim h^{k+1} \|u\|_{k+1,\Omega}. \qquad (18)$$

The optimality of approximation properties in the L^2-norm show the advantage of Nitsche's method with respect to the penalty technique, because the latter scheme turns out to be slightly suboptimal in this norm. Indeed, the analysis of [7] shows that, provided $u \in H^4(\Omega)$, for piece-wise linear elements on polygonal domains with perfectly fitted boundaries the optimal penalty choice is $\varepsilon \sim h^{\frac{1}{3}}$ and it leads to

$$\|u - u_h\|_{1,\Omega} \lesssim h\|u\|_{4,\Omega}, \quad \|u - u_h\|_{0,\Omega} \lesssim h^{\frac{5}{3}} \|u\|_{4,\Omega}.$$

For quadratic Lagrangian elements with the choice $\varepsilon \sim h^2$, it is possible to prove that the penalty method satisfies the following error estimates,

$$\|u - u_h\|_{1,\Omega} \lesssim h^2 \|u\|_{5,\Omega}, \quad \|u - u_h\|_{0,\Omega} \lesssim h^2 \|u\|_{5,\Omega},$$

which, under the strengthened regularity assumption $u \in H^5(\Omega)$, are optimal for the H^1-norm case, but suboptimal when the convergence is measured in the L^2-norm.

Conversely, the Lagrange multipliers method provides optimal convergence rates with respect to h. More precisely, we assume that the spaces V_h, Λ_h satisfy the following approximation properties respectively,

$$\inf_{v_h \in V_h} \|v - v_h\|_{1,\Omega} \lesssim h^k \|v\|_{k+1,\Omega}, \quad \inf_{\mu_h \in \Lambda_h} \|\mu - \mu_h\|_{0,\partial\Omega} \lesssim h^{l+1} \|\mu\|_{l+1,\partial\Omega},$$

for regular functions $v \in H^{k+1}(\Omega)$, $\mu \in H^{l+1}(\partial\Omega)$. Then, provided that conditions (7)- (8) hold true for V_h, Λ_h, the following error estimates are satisfied, see [39–41],

$$\|u - u_h\|_{1,\Omega} + \|\lambda - \lambda_h\|_{-\frac{1}{2},h,\partial\Omega} \lesssim h^k \|u\|_{k+1,\Omega} + h^{l+\frac{3}{2}} \|\mu\|_{l+1,\partial\Omega}.$$

Thanks to the property $\lambda + \partial_n u = 0$, the Lagrange multipliers method has the advantage to simultaneously provide an approximation of the solution u and of its flux at the boundary. For Nitsche's method, the calculation of fluxes can be achieved after the solution of the problem determining u_h. It is interesting to observe that an accurate flux reconstruction involves both the normal gradient of the numerical solution and the penalty term. Indeed, multiplying (1) with homogeneous Dirichlet boundary data $g = 0$ by a test function $v_h \in V_h$, integrating over Ω and applying

Green's formula, we straightforwardly obtain

$$(\nabla u, \nabla v_h)_\Omega - (\partial_n u, v_h)_{\partial\Omega} = (f, v_h), \quad \forall v_h \in V_h.$$

Subtracting Nitsche's scheme from previous equation we obtain,

$$(\partial_n u, v_h)_{\partial\Omega} = (\partial_n u_h, v_h)_{\partial\Omega} + s\,(\partial_n v_h, u_h)_{\partial\Omega}$$
$$- \gamma \sum_{E \in \mathscr{B}_h} h_E^{-1}\,(u_h, v_h)_E + (\nabla(u - u_h), \nabla v_h)_\Omega$$

and by selecting $v_h = 1$ we obtain the following flux reconstruction formula,

$$\int_{\partial\Omega} \partial_n u = \int_{\partial\Omega} \partial_n u_h - \gamma \sum_{E \in \mathscr{B}_h} h_E^{-1} \int_E u_h.$$

1.3 Nitsche's Method for Interface Problems

The aim of this section is to briefly illustrate the application of Nitsche's method to a prototype of the interface problem. This subject has been and still is an active field of research, and the topics addressed here represent a summary of the seminal works by Hansbo et al, [8, 31].

Our simplified multi-domain problem consists of two non overlapping polygonal subdomains, Ω_i, $i = 1, 2$, with interface $\Gamma := \overline{\Omega}_1 \cap \overline{\Omega}_2$. We aim to find $u_i \in H^1(\Omega_i)$ that weakly satisfy,

$$\begin{cases} -\Delta u_i = f, & \text{in } \Omega_i, \\ u_i = 0, & \text{on } \partial\Omega \cap \partial\Omega_i, \\ u_1 - u_2 = 0, & \text{on } \Gamma, \\ \partial_n u_1 - \partial_n u_2 = 0, & \text{on } \Gamma, \end{cases} \tag{19}$$

where $\mathbf{n}$ denotes a unit normal vector associated to Γ and $\partial_n u := \nabla u \cdot \mathbf{n}$, where $\mathbf{n}$ on Γ can be either chosen as $\mathbf{n} := \mathbf{n}_1$ or equivalently $\mathbf{n} := \mathbf{n}_2$. Such ambiguity does not affect the application of Nitsche's method. To proceed, we define jumps and averages of quantities across the interface Γ. In particular, given a function $v : \overline{\Omega}_1 \cup \overline{\Omega}_2 \to \mathbb{R}$, its jump across the interface is defined as $[\![v]\!] := v_1 - v_2$, according to the sign of the vector $\mathbf{n}$, which is here selected as $\mathbf{n} = \mathbf{n}_1$, while the average is $\{v\} := \frac{1}{2}(v_1 + v_2)$. Problem (19) can be rewritten more conveniently as follows,

$$\begin{cases} -\Delta u_i = f, & \text{in } \Omega_i, \\ u_i = 0, & \text{on } \partial\Omega \cap \partial\Omega_i, \\ [\![u]\!] = 0, & \text{on } \Gamma, \\ [\![\partial_n u]\!] = 0, & \text{on } \Gamma. \end{cases} \qquad (20)$$

As an instance of the rich family of mortar methods for interface problems, the peculiarity of Nitsche's scheme is to provide an approximation $u_h := [u_{h,1}, u_{h,2}]$ of (20) that is non conforming with $H^1(\Omega)$, as alternative to most popular domain decomposition techniques, such as Dirichlet-Neumann splitting.

For the discretisation of (20) let $\mathcal{T}_{h,i}$ be a family of shape-regular, quasi-uniform triangulations of Ω_i. Note that $\mathcal{T}_{h,i}$ with $i = 1, 2$ may be non conforming at the interface. Let $\mathcal{B}_{h,i}$ and $\mathcal{G}_{h,i}$ the collections of the faces/edges at the boundary and at the interface respectively. We look for discrete functions $[u_{h,1}, u_{h,2}] \in V_h :=$ $V_{h,1} \times V_{h,2}$, where $V_{h,i}$ are Lagrangian finite element spaces on $\mathcal{T}_{h,i}$.

A weak formulation of the multi-domain problem that is prone to discretisation by Nitsche's method is obtained by multiplying (20)$_a$ with a test function $v_i \in$ $H^1\Omega_i$ and applying integration by parts, such that

$$\sum_{i=1,2} \left(\int_{\Omega_i} \nabla u \cdot \nabla v - \int_{\partial\Omega_i} \nabla u \cdot \mathbf{n}_i v \right)$$

$$= \sum_{i=1,2} \left(\int_{\Omega_i} \nabla u \cdot \nabla v - \int_{\partial\Omega_i \setminus \Gamma} \nabla u \cdot \mathbf{n}_i v \right) - \int_{\Gamma} [\![\nabla u \cdot \mathbf{n}v]\!].$$

Interface conditions prescribing continuity of fluxes, i.e. $[\![\partial_n u]\!] = 0$, can be enforced in the bilinear form with the help of the following algebraic identity $[\![ab]\!] = [\![a]\!]\{b\} + [\![b]\!]\{a\}$, such that

$$[\![\nabla u \cdot \mathbf{n}v]\!] = [\![\nabla u \cdot \mathbf{n}]\!]\{v\} + \{\nabla u \cdot \mathbf{n}\}[\![v]\!] = \{\nabla u \cdot \mathbf{n}\}[\![v]\!] + \{\nabla v \cdot \mathbf{n}\}[\![u]\!]$$

where we exploit $[\![u]\!] = 0$ owing to the strong consistency. For interface conditions prescribing continuity of the solution at the interface, we exploit penalty,

$$\sum_{i=1,2} \left(\sum_{E \in \mathcal{G}_{h,i}} \frac{\gamma}{h_E} \int_E [\![u]\!][\![v]\!] + \sum_{E \in \mathcal{B}_{h,i}} \frac{\gamma}{h_E} \int_E uv \right)$$

where γ is the penalty parameter already introduced for the approximation of Poisson's problem. Then, the extension of Nitsche's method to interface conditions consists in finding $u_h := [u_{h,1}, u_{h,2}] \in V_h := V_{h,1} \times V_{h,2}$ such that

$$a_h(u_h, v_h) = F_h(v_h), \quad \forall v_h \in V_h \qquad (21)$$

with $a_i(u, v) := (\nabla u_i, \nabla v_i)_{\Omega_i}$ for any $u_i, v_i \in H^1(\Omega_i)$ and

$$a_h(u_h, v_h) := \sum_{i=1,2} \left(a_i(u_{h,i}, v_{h,i}) + \sum_{E \in \mathscr{G}_{h,i}} \gamma h_E^{-1} \left(\llbracket u_h \rrbracket, \llbracket v_h \rrbracket\right)_E \right)$$

$$- \left(\{\nabla u_h \cdot \mathbf{n}\}, \llbracket v_h \rrbracket\right)_\Gamma - \left(\{\nabla v_h \cdot \mathbf{n}\}, \llbracket u_h \rrbracket\right)_\Gamma$$

$$+ \sum_{i=1,2} \left(\sum_{E \in \mathscr{B}_{h,i}} \gamma h_E^{-1} (u_h, v_h)_E - (\nabla u_h \cdot \mathbf{n}_i, v_h)_{\partial \Omega_i \setminus \Gamma} \right.$$

$$\left. - (\nabla v_h \cdot \mathbf{n}_i, u_h)_{\partial \Omega_i \setminus \Gamma} \right),$$

$$F_h(v_h) := F(v_h) = \int_\Omega f v_h, \quad \text{since } u = 0 \text{ on } \partial \Omega,$$

where for simplicity we restrict the setting to the case $s = 1$. This turns out to be a Galerkin method with an approximation space that is not H^1-conformal on Ω. Indeed, u_h belongs to the broken Sobolev space $H^1(\Omega_1 \cup \Omega_2) := H^1(\Omega_1) \times H^1(\Omega_2)$ and the natural norms for the analysis of the problem read as follows,

$$\|v\|_{\pm\frac{1}{2},h,\mathscr{G}_{h,i}}^2 := \sum_{E \in \mathscr{G}_{h,i}} h_E^{\mp 1} \|v\|_{0,E}^2, \quad \forall v \in L^2(\Gamma),$$

$$\|v\|_{1,h,\Omega_1 \cup \Omega_2}^2 := \sum_{i=1,2} \left(|v_i|_{1,\Omega_i}^2 + \|v_i\|_{\frac{1}{2},h,\mathscr{B}_{h,i}}^2 + \|\llbracket v \rrbracket\|_{\frac{1}{2},h,\mathscr{G}_{h,i}}^2 \right), \quad \forall v_i \in H^1(\Omega_i).$$

Then, proceeding analogously to the case of a single domain, it is possible to verify that, if (20) admits a regular solution $u \in H^2(\Omega_1 \cup \Omega_2) \cap H_0^1(\Omega)$, then $a_h(u, v_h) = F_h(v_h)$ for any $v_h \in V_h$ and $a_h(u - u_h, v_h) = 0$ for any $v_h \in V_h$. Furthermore, $a_h(\cdot, \cdot)$ is bounded in the norm $\| \cdot \|_{1,h,\Omega_1 \cup \Omega_2}$ and also stable with a constant uniformly independent on the mesh characteristic size h. As a result of that, following the lines of Cea's lemma, we obtain an a priori estimate equivalent to (17).

We finally notice that Nitsche's multi-domain scheme can be easily decomposed into local problems, relative to each subdomain, and coupling terms that transfer information from one subdomain to others. In particular we write,

$$a_h(u_h, v_h) = \sum_{i=1,2} \sum_{j \neq i} \left[a_{h,i}(u_{h,i}, v_{h,i}) - c_{h,ij}(u_{h,j}, v_{h,i}) \right],$$

where each single term is defined as follows,

$$a_{h,i}(u_{h,i}, v_{h,i}) := a_i(u_{h,i}, v_{h,i}) + c_{h,ii}(u_{h,i}, v_{h,i}) + b_{h,i}(u_{h,i}, v_{h,i}),$$

$$c_{h,ii}(u_{h,i}, v_{h,i}) := \sum_{E \in \mathscr{G}_{h,i}} \gamma h_E^{-1} (u_{h,i}, v_{h,i})_E - \left(\frac{1}{2}\nabla u_{h,i} \cdot \mathbf{n}_i, v_{h,i}\right)_\Gamma$$

$$- \left(\frac{1}{2}\nabla v_{h,i} \cdot \mathbf{n}_i, u_{h,i}\right)_\Gamma,$$

$$c_{h,ij}(u_{h,j}, v_{h,i}) := \sum_{E \in \mathscr{G}_{h,i}} \gamma h_E^{-1} (u_{h,j}, v_{h,i})_E + \left(\frac{1}{2}\nabla u_{h,j} \cdot \mathbf{n}_i, v_{h,i}\right)_\Gamma$$

$$- \left(\frac{1}{2}\nabla v_{h,i} \cdot \mathbf{n}_i, u_{h,j}\right)_\Gamma,$$

$$b_{h,i}(u_{h,i}, v_{h,i}) := \sum_{i=1,2}\left[\sum_{E \in \mathscr{B}_{h,i}} \gamma h_E^{-1} (u_h, v_h)_E - (\nabla u_h \cdot \mathbf{n}_i, v_h)_{\partial\Omega_i \setminus \Gamma}\right.$$

$$\left. - (\nabla v_h \cdot \mathbf{n}_i, u_h)_{\partial\Omega_i \setminus \Gamma}\right].$$

Such decomposition suggests that, starting from problem (21), it is possible to devise an iterative splitting strategy that aims to decompose the solution of a multi-domain problem on Ω into a sequence of local problems on Ω_i. Indeed, owing to the introduction of the following relaxation operators, where the relaxation effect from one iteration to another is again obtained through a penalty term similar to the one of (11),

$$s_{h,i}^\sigma(u_{h,i}, v_{h,i}; u_{h,i}^{(old)}) := \sum_{E \in \mathscr{G}_{h,i}} \sigma h_E^{-1} \left(u_{h,i} - u_{h,i}^{(old)}, v_{h,i}\right)_E,$$

$$s_h^\sigma(u_h, v_h; u_h^{(old)}) := \sum_{i=1,2} s_{h,i}^\sigma(u_{h,i}, v_{h,i}; u_{h,i}^{(old)}).$$

The iterative method obtained by giving $u_{h,i}^0 \in V_{h,i}$ for $i = 1,2$ and looking for a sequence of approximations $u_{h,i}^k$ for any $k > 0$ such that,

$$a_{h,i}(u_{h,i}^k, v_{h,i}) + s_{h,i}^\sigma(u_{h,i}^k, v_{h,i}; u_{h,i}^{k-1}) - c_{h,ij}(u_{h,j}^{k-1}, v_{h,i}) = F_{h,i}(v_{h,i}), \ \forall v_{h,i} \in V_{h,i},$$
(22)

turns out to be convergent to $[u_{h,1}, u_{h,2}]$ provided that the relaxation parameter σ is large enough. Such technique has already been profitably applied to the approximation of advection dominated elliptic problems in [17] as well as to mixed problems in [19]. The convergence analysis of the iterative scheme is more easily performed if we rewrite it as follows

$$a_h(u_h^k, v_h) + s_h^\sigma(u_h^k, v_h; u_h^{k-1}) = F_h(v_h) - r_h(u_h^k - u_h^{k-1}, v_h),$$
(23)

$$r_h(u_h^k - u_h^{k-1}, v_h) := \sum_{i=1,2} \sum_{j \neq i} c_{h,ij}(u_{h,j}^k - u_{h,j}^{k-1}, v_{h,i}),$$

which is obtained from (22) by summing up the equations for $i = 1,2$ and introducing the new terms $\pm c_{h,ij}(u_{h,j}^k, v_{h,i})$. Notice that $r_h(u_h^k - u_h^{k-1}, v_h)$ plays the role of iteration residual and the interplay of s_h^σ with r_h is the key point to prove convergence of iterations. To this purpose, we look at the iteration error, that is $w_h^k := u_h - u_h^k$. By subtracting (23) from (21), we obtain an equation for w_h^k, precisely

$$a_h(w_h^k, v_h) + s_h^\sigma(w_h^k, v_h; w_h^{k-1}) = -r_h(w_h^k - w_h^{k-1}, v_h).$$

Then, convergence of u_h^k relies on the following inequality

$$\alpha \|w_h^k\|_{1,h,\Omega_1\cup\Omega_2}^2 + s_h^\sigma(w_h^k, w_h^k; w_h^{k-1}) \leq |r_h(w_h^k - w_h^{k-1}, w_h^k)|,$$

combined with the following estimates for s_h^σ and r_h,

$$r_h(w_h^k - w_h^{k-1}, w_h^k) = \sum_{i=1,2; j\neq i} \left[\left(\nabla(w_{h,j}^k - w_{h,j}^{k-1}) \cdot \mathbf{n}_i, w_{h,i}^k \right)_\Gamma \right.$$

$$\left. - \left(\nabla w_{h,i}^k \cdot \mathbf{n}_i, w_{h,j}^k - w_{h,j}^{k-1} \right)_\Gamma + \sum_{E\in\mathscr{G}_{h,i}} \gamma h_E^{-1} \left(w_{h,j}^k - w_{h,j}^{k-1}, w_{h,i}^k \right)_\Gamma \right],$$

$$\sum_{i=1,2} \left[\left(\nabla(w_{h,j}^k - w_{h,j}^{k-1}) \cdot \mathbf{n}_i, w_{h,i}^k \right)_\Gamma - \left(\nabla w_{h,i}^k \cdot \mathbf{n}_i, w_{h,j}^k - w_{h,j}^{k-1} \right)_\Gamma \right]$$

$$\lesssim \sum_{i=1,2; j\neq i} \left[\delta(\|w_{h,i}^k\|_{1,h,\Omega_i}^2 + \|w_{h,i}^{k-1}\|_{1,h,\Omega_i}^2) + \delta^{-1}\|w_{h,i}^k - w_{h,i}^{k-1}\|_{\frac{1}{2},h,\Gamma}^2 \right],$$

$$\sum_{i=1,2; j\neq i} \sum_{E\in\mathscr{G}_{h,i}} \gamma h_E^{-1} \left(w_{h,j}^k - w_{h,j}^{k-1}, w_{h,i}^k \right)_\Gamma$$

$$\lesssim \sum_{i=1,2} \gamma \left[\|w_{h,i}^k - w_{h,i}^{k-1}\|_{\frac{1}{2},h,\Gamma}^2 + \|[\![w_h^k]\!]\|_{\frac{1}{2},h,\Gamma}^2 \right] + s_h^\gamma(w_h^k, w_h^k; w_h^{k-1}),$$

$$s_h^\sigma(w_h^k, w_h^k; w_h^{k-1}) = \frac{\sigma}{2} \sum_{i=1,2} \left[\|w_{h,i}^k\|_{\frac{1}{2},h,\Gamma}^2 - \|w_{h,i}^{k-1}\|_{\frac{1}{2},h,\Gamma}^2 + \|w_{h,i}^k - w_{h,i}^{k-1}\|_{\frac{1}{2},h,\Gamma}^2 \right].$$

Together with suitable choices of δ and γ, the previous estimates can be suitably applied to obtain the following inequality

$$\beta \|w_h^k\|_{1,h,\Omega_1\cup\Omega_2}^2 + \xi \sum_{i=1,2} \|w_{h,i}^k - w_{h,i}^{k-1}\|_{\frac{1}{2},h,\Gamma}^2$$

$$\lesssim \varsigma \sum_{i=1,2} \left[\|w_{h,i}^{k-1}\|_{\frac{1}{2},h,\Gamma}^2 - \|w_{h,i}^k\|_{\frac{1}{2},h,\Gamma}^2 + \|w_{h,i}^{k-1}\|_{1,h,\Omega_i}^2 - \|w_{h,i}^k\|_{1,h,\Omega_i}^2 \right].$$

Summing up over the index k we conclude that there exists a constant $C > 0$ independent on k, but possibly depending on the initial state, such that

$$\sum_{k=1}^{\infty} \|w_h^k\|_{1,h,\Omega_1\cup\Omega_2} \leq C,$$

which implies convergence of the sequence u_h^k to u_h in the natural norm.

1.4 The Unfitted Version of Nitsche's Method

The increasing complexity of geometrical configurations in applications addressed by means of computational analysis has motivated the research of finite element schemes capable to handle the case where the computational mesh is not fitted to boundaries or interfaces. Instead, a physical domain with a possibly complex shape is embedded into a computational domain with simple shape that is easily partitioned into elements. Thanks to their flexibility in the treatment of Dirichlet boundary conditions, the method of Lagrange multipliers and Nitsche's scheme have been profitably applied to such purpose. We report here simple examples for such schemes, together with a brief discussion of their intrinsic drawbacks. We refer to Sect. 3 for a detailed development of suitable stabilisation techniques to obtain efficient and robust schemes for the approximation of problems on boundaries or interfaces that do not fit with the computational mesh.

For the set up of a finite element method with unfitted boundary, we denote by Ω the physical domain, embedded into a computational domain $\Omega_{\mathscr{T}}$ corresponding to a computational mesh $\mathscr{T}_h$. The basic restrictive assumption for the correct definition of unfitted boundary methods is the requirement that each element $K \in \mathscr{T}_h$ must have a non vanishing intersection with Ω and that the boundary $\partial\Omega$ is regular and intersects each element boundary ∂K at most twice and each open edge E at most once. We refer to Fig. 1 for an example of physical and computational domains. The approximation space consists on linear Lagrangian finite elements on $\Omega_{\mathscr{T}}$,

$$V_h := \{v_h \in C^0(\Omega_{\mathscr{T}}) : v_h|_K \in \mathbb{P}^1(K)\ \forall K \in \mathscr{T}_h\}.$$

Because of its simplicity, the penalty method turns out to be very attractive to build up finite element approximations on meshes not fitting the boundary of the physical domain. Under the assumption $\mathrm{dist}(\Omega, \Omega_{\mathscr{T}}) \lesssim h^2$, it is shown in [7] that the application of the simple penalty term $h^{-2}(u_h - g, v_h)_{\partial\Omega_{\mathscr{T}}}$ to a linear

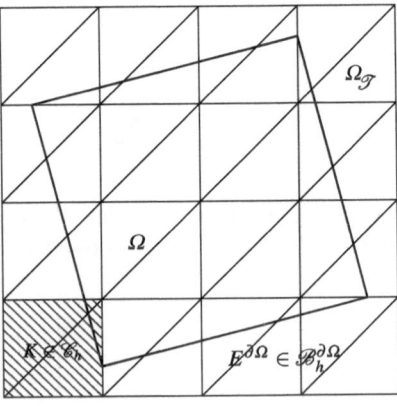

Fig. 1 A sketch of the physical domain, Ω, and the computational domain $\Omega_{\mathscr{T}}$, with the notation used to set up the fictitious domain method

finite element approximation without boundary constraints is sufficient to recover a discrete solution that satisfies the error estimates

$$\|u - u_h\|_{1,\Omega} \lesssim h\|u\|_{4,\Omega}, \quad \|u - u_h\|_{0,\Omega} \lesssim h^{\frac{3}{2}}\|u\|_{4,\Omega},$$

for Poisson's problem (or any other second-order self-adjoint variant) with regular solution $u \in H^4(\Omega)$.

To extend the method of Lagrange multipliers to unfitted meshes, a major difficulty is the construction of a suitable multiplier space for a boundary that does not coincide with the edges of the mesh. An effective and simple solution is studied in [26] where piecewise constant multipliers are applied on Ω. If the multiplier mesh is suitably coarser than the one relative to the primal unknown, the application of piecewise linear approximations with piecewise constant multipliers gives rise to a stable scheme. Unless the finite element spaces are chosen so that the discrete inf-sup condition is satisfied, some stabilisation must be introduced. One of the most popular stabilised methods was introduced by Hughes and Barbosa [5, 6]. In this case the difference between the discrete Lagrange multiplier and the discrete normal derivative is penalised. Such a method was proposed in the fictitious domain framework by Renard et al. [33]. Another recent stabilised method is based on the idea of interior penalty, where the stabilisation acts on the Lagrange multiplier alone and acts as a coarsening operator effectively penalising the distance of the discrete Lagrange multiplier to a stable subspace. We give an example of this later formulation taken from [15] below.

For the construction of such a space we assume that $\partial\Omega$ is a curved boundary without corners (for the extension to the polygonal case we refer to [15]) and we define the collection of all elements cut by the unfitted boundary as $\mathscr{C}_h := \{K \in \mathscr{T}_h : K \cap \partial\Omega \neq \emptyset\}$, then the space of multipliers is

$$\Lambda_h := \{v_h \in L^2(\mathscr{C}_h) : v_h|_K \in \mathbb{P}^0(K) \ \forall K \in \mathscr{C}_h\}.$$

For the approximation of problem (1) on an unfitted mesh we aim to find a couple $(u_h, \lambda_h) \in V_h \times \Lambda_h$ such that

$$\begin{cases} a(u_h, v_h) + b(\lambda_h, v_h) = F(v_h) & \forall v_h \in V_h, \\ b(\mu_h, u_h) - J(\lambda_h, \mu_h) = b(\mu_h, g) & \forall \mu_h \in \Lambda_h, \end{cases} \tag{24}$$

where the definitions of $a(\cdot, \cdot)$ and $b(\cdot, \cdot)$ do not change with respect to (6), while $J(\lambda_h, v_h)$ is a stabilisation term proposed in [15] and defined as follows,

$$J(\lambda_h, \mu_h) = \sum_{E \in \mathscr{E}_B} (\gamma h[\![\lambda_h]\!], [\![\mu_h]\!])_E,$$

where $\mathscr{E}_B$ is the set of edges or faces in $\mathscr{C}_h$ intersected by the boundary $\partial\Omega$, the jump of the piecewise constant function λ_h across such edges is denoted by $[\![\lambda_h]\!]$

and γ is a stabilisation parameter that should be selected large enough according to the analysis performed in [15]. Problem (24) features a remarkable advantage for unfitted boundaries, because the primal and the dual variables u_h and λ_h respectively, are defined on the same computational mesh $\mathscr{T}_h$, in contrast to a more classical choice for Lagrange multipliers that needs an independent partition of the boundary $\partial\Omega$.

The application of Nitsche's method to the case of unfitted boundary only requires a minor modification with respect to the standard case. We notice that the concept of edges or faces on the unfitted boundary is not properly defined yet. Then, instead of defining the penalty term on each edges, we simply consider $\gamma h^{-1}(u_h, v_h)_\Gamma$. As a result of that, Nitsche's method for an unfitted boundary requires to find $u_h \in V_h$ such that $a_h(u_h, v_h) = F_h(v_h)$ for any $v_h \in V_h$ with

$$a_h(u_h, v_h) := a(u_h, v_h) - (\partial_n u_h, v_h)_{\partial\Omega} - s(\partial_n v_h, u_h)_{\partial\Omega} + \gamma h^{-1}(u_h, v_h)_{\partial\Omega},$$

$$F_h(v_h) := F(v_h) + h^{-1}(g, v_h)_{\partial\Omega} - s(\partial_n v_h, g)_{\partial\Omega}.$$

The main drawback of such a scheme is its lack of robustness with respect to the position of the boundary. Indeed, an unfitted boundary Γ may cut the computational mesh such that some intersections of elements with the physical domain are very small and/or feature very large aspect ratios. In such cases, as illustrated in [12], the linear system arising from Nitsche's discrete problem may be ill posed. Let $\mathbf{x}_k$ be the vertexes of the computational mesh $\mathscr{T}_h$ and let $\mathscr{P}_k$ be the patch of elements relative to the vertex $\mathbf{x}_k$. Given a generic function $v_h \in V_h$, let $\mathbf{v}$ the vector of its degrees of freedom endowed with the Euclidean norm $\|\mathbf{v}\|$, namely these are the values of v_h in the vertexes $\mathbf{x}_k$ for linear Lagrangian finite elements. We denote with ν a non-dimensional parameter that quantifies the size of the minimal intersection of finite element patches with the physical domain. Precisely, we define

$$\nu = \min_k \frac{|\mathscr{P}_k \cap \Omega|}{|\mathscr{P}_k|},$$

where from now on $|\Omega|$ will denote the d-dimensional volume of $\Omega \subset \mathbb{R}^d$. According to the analysis developed in [12, 43], there exists a function $v_h^* \in V_h$ such that

$$\|v_h^*\|_{1,\varepsilon,\Omega}^2 \lesssim h^{d-2}\nu\|\mathbf{v}^*\|^2.$$

Denoting with A_h the stiffness matrix related to Nitsche's method, such an estimate directly implies that its spectral condition number admits the lower bound $K_2(A_h) \gtrsim \nu^{-1}h^{-2}$. For any boundary configuration such that $\nu \to 0$, matrix A_h becomes ill posed and almost singular. In conclusion, the development of stabilisation techniques to complement the unfitted Nitsche's scheme and make it fully robust with respect to any boundary configuration is a vivid field of research on which we will concentrate in Sect. 3.

2   A Modified Nitsche's Method for Large Contrast Problems

In the previous section we have studied how to apply Nitsche's method to enforce
interface conditions to couple second order problems of the same type on adjacent
domains. The purpose of this section is to extend this method to problems that
vary in character form one part of the domain to another. To be more precise,
we restrict to interface problems where the governing equations are similar on
adjacent subdomains, but they may be characterized by heterogeneous coefficients.
We refer to this large family of problems with the general name of large contrast
problems and we remark that they are encountered in relevant applications such as
computational mechanics, for the study of the deformation of heterogeneous bodies,
or geosciences, for the analysis of flow and mass transport in soils or aquifers.

Several authors have already successfully applied Nitsche's method to the
discretization of large contrast problems. For the case of computational mechanics
we refer for instance to [31], while for the analysis of a generic singularly perturbed
advection diffusion problem we refer to [17]. In this section we focus on the
latter case, in particular we study the coupling of a second order scalar problem
where one of the subproblems features a singularly perturbed behaviour. Typical
model problems are advection / diffusion equations with heterogeneous diffusion
coefficients between subregions, i.e.

$$-\nabla \cdot (\varepsilon \nabla u) + \beta \cdot \nabla u = f \text{ in } \Omega,$$

where ε denotes the diffusivity of a given medium and β is a given advective field,
which for simplicity we assume to be solenoidal. Provided that ε is a positive and
bounded function, the advection / diffusion problem turns out to be well posed
owing to a straightforward application of Lax-Milgram's lemma. In the case of
variable, possibly discontinuous diffusivity ε the interest in rewriting the problem as
a multi-domain problem, subdividing regions with uniform properties, arises from
the observation that internal layers of the solution may appear in the neighbourhood
of the interfaces where coefficients are discontinuous. In several applications, such
as heat or mass transfer problems, the configuration of such layers determine the
fluxes exchanged between different bodies, and thus a correct approximation of
them is necessary.

2.1   Approximation of Large Contrast Problems with Locally
Vanishing Diffusion

We consider for simplicity two non overlapping polygonal subdomains, Ω_i, $i =
1, 2$, with interface $\Gamma := \overline{\Omega}_1 \cup \overline{\Omega}_2$ as an instance of a more general multi-material
problem depicted in Figure 2. Furthermore, without significant loss of generality,
we restrict to the case of uniform coefficients on each subregion. In particular, given

Fig. 2 A general
multi-material problem (left)
restricted to a two-domain
case (right)

 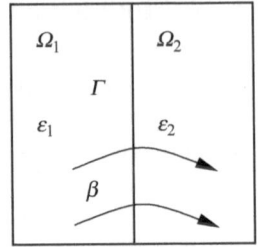

two constant parameters $\varepsilon_i > 0$, $i = 1, 2$ and $\beta \in [C^1(\Omega)]^d$ with $\nabla \cdot \beta = 0$, $|\beta| \simeq 1$, we aim to find u_i such that

$$\begin{cases} \nabla \cdot \left(- \varepsilon_i \nabla u_i + \beta u_i \right) = f_i, & \text{in } \Omega_i, \\ u_i = 0, & \text{on } \partial\Omega \cap \partial\Omega_i, \\ [\![u]\!] = 0, & \text{on } \Gamma, \\ [\![-\varepsilon \nabla u \cdot \mathbf{n} + \beta \cdot \mathbf{n} u]\!] = 0, & \text{on } \Gamma. \end{cases} \tag{25}$$

First of all, we notice that an internal layer may appear in the neighborhood of Γ when $\varepsilon_1 \neq \varepsilon_2$. This happens for instance if $\varepsilon_1 \ll \varepsilon_2$ and the interface Γ (or part of it) is an outflow region for the advective field β. In this case the internal layer is located upwind to the interface, in other words it is confined into the domain Ω_1. Moreover, in the singularly perturbed limit case, i.e. $\varepsilon_1 \to 0$, $\varepsilon_2 > 0$ the internal layer becomes thinner and stiffer, while the global solution u of (25) approaches a discontinuous function. In conclusion, under these particular conditions, the solution of the limit problem (25) with $\varepsilon_1 \to 0$ fails to be H^1-conformal. Thus, we focus on Nitsche's technique as a discretization method for interface problems pursuing the idea that only a H^1 non-conformal discretization technique can robustly approximate the problem under all possible conditions including the singularly perturbed limit.

For the discretization of problem (25) we could proceed in analogy with Poisson problem, already addressed in Sect. 1.3. Since problem (25) is written in divergence form, it is easy to extend the treatment of natural interface conditions of type $[\![\partial_n u]\!] = 0$ to the case of the conormal derivative $[\![-\varepsilon \nabla u \cdot \mathbf{n} + \beta \cdot \mathbf{n} u]\!]$. We will see later on that such an approach will only partially fulfill the objective to set up a robust discretization scheme for local singularly perturbed problems. To further improve the resulting scheme, we look at interface conditions with a bias to domain decomposition methods. Observing that in the limit case $\varepsilon_1 \to 0$ the sub-problem in Ω_1 tends to an hyperbolic problem coupled to an elliptic problem on Ω_2, we consider the set up of a new Nitsche method arising from a set of generalized interface conditions, introduced in [24] to couple both elliptic and hyperbolic problems, which give rise to the so called heterogeneous domain decomposition methods, see [42]. Our purpose is to obtain a weak coupling scheme that inherits the robustness of heterogeneous domain decomposition methods for the approximation of problems that vary in character form one part of the domain to another. As a result of that, such a method will turn out to be effective for problems whose solution features sharp internal layers.

The starting point of such a procedure is the definition and analysis of the coupling conditions between an advection / diffusion (elliptic) equation with a purely advective model (hyperbolic problem). Setting $\varepsilon_1 = 0$, $\varepsilon_2 > 0$ in (25), we identify Ω_1 as the hyperbolic and Ω_2 as the elliptic subregion. Let n_{hy} be the outward unit normal with respect to Ω_{hy} and let $\partial\Omega_{in} := \{x \in \partial\Omega_{hy} : \beta \cdot n_{hy} < 0\}$. According to this definition, the interface can be split in two parts $\Gamma_{in} := \Gamma \cap \partial\Omega_{in}$ and the complementary $\Gamma_{out} := \Gamma \setminus \Gamma_{in}$. We look for $u_{hy} = u_1$, $u_{el} = u_2$ such that

$$
\begin{cases}
\nabla\cdot(-\varepsilon\nabla u_{el} + \beta u_{el}) = f & \text{in } \Omega_{el}, \\
\nabla\cdot(\beta u_{hy}) = f & \text{in } \Omega_{hy}, \\
-\varepsilon\nabla u_{el} \cdot n + \beta \cdot n u_{el} = \beta \cdot n u_{hy} & \text{on } \Gamma, \\
u_{el} = u_{hy} & \text{on } \Gamma_{in}, \\
u_{el} = 0 & \text{on } \partial\Omega \cap \partial\Omega_{el}, \\
u_{hy} = 0 & \text{on } \partial\Omega \cap \partial\Omega_{in}.
\end{cases}
\tag{26}
$$

Comparing problem (25) with (26), we notice that interface conditions involving mass fluxes are naturally extended to the limit case $\varepsilon_1 = 0$, because the diffusive flux has disappeared from the right hand side of (26)$_c$. Conversely, interface conditions for the solution itself feature a singular behaviour in the vanishing viscosity case. Indeed, continuity of the solution is only enforced on the inflow part of the interface, referred to as the hyperbolic boundary, i.e. $\Gamma \cap \partial\Omega_{in}$, while on the complementary outflow interface the solutions u_{el} and u_{hy} do not conform; that is the global solution of the heterogeneous problem, u, can be discontinuous across this part of the interface. Nitsche's method turns out to be particularly effective to handle such conditions in a general setting. On the one hand, as for the aforementioned Poisson problem, the continuity of mass fluxes can be naturally handled by integrating by parts the local governing equations and exploiting the algebraic inequality $[\![ab]\!] = [\![a]\!]\{b\} + [\![b]\!]\{a\}$. On the other hand, the singular behaviour of the continuity of the solution u can be addressed by a suitable manipulation of the interface penalty term. Exploiting the flexibility of Nitsche's technique, we aim to set up a discrete interface problem that is strongly consistent with both problems (25), (26), resorting to a robust finite element scheme for the local singularly perturbed limit case. To perform this task we start from a unified formulation of continuity interface conditions for problems (25) and (26) in a sufficiently general setting that allows the extension of similar concepts to the case of fluid dynamics, like the case addressed in [19].

If ε_i were positive and quasi-uniform on Ω, the standard condition to enforce the continuity of the solution would be,

$$
\left[\frac{1}{2}|\beta \cdot n_\Gamma| + \{\varepsilon\} \right] [\![u]\!] = 0 \quad \text{on } \Gamma \setminus \partial\Omega.
$$

where we the factor $\left[\frac{1}{2}|\beta \cdot n_\Gamma| + \{\varepsilon\} \right]$ appears to modulate the intensity of the penalty term that weakly enforces the continuity requirement. In order to correct this condition in the case where ε_i significantly varies from region to region, we introduce an *heterogeneity factor*, which quantifies the variation of ε on the interface, $\lambda(x)|_\Gamma : \Gamma \to [-1, 1]$ such that

$$\lambda(x)|_\Gamma := \begin{cases} \frac{1}{2}\frac{[\![\varepsilon(x)]\!]}{\{\varepsilon(x)\}}, & \text{if } \{\varepsilon(x)\} > 0, \\ \\ 0, & \text{if } \{\varepsilon(x)\} = 0. \end{cases}$$

Then, starting from the case of uniform diffusivity considered above, we propose the following generalized interface conditions for the continuity of the solution,

$$\left[\frac{1}{2}|\beta \cdot n_\Gamma|\big(1 - \text{sign}(\beta \cdot n_\Gamma)\varphi_\Gamma(\lambda)\big) + \{\varepsilon\}\big(1 - \chi_\Gamma(\lambda)\big) \right][\![u]\!] = 0 \text{ on } \Gamma, \quad (27)$$

where $\varphi_\Gamma(\lambda)$ and $\chi_\Gamma(\lambda)$ are scaling functions that must satisfy the following requirements in order to make sure that in the limit case the continuity of the solution is enforced on $\Gamma \cap \partial\Omega_{in}$ solely. Precisely, we assume that they satisfy $|\chi_\Gamma(\lambda)| \le 1$, $|\varphi_\Gamma(\lambda)| \le 1$ and,

$$\begin{aligned} \chi_\Gamma(\lambda) &= 0 && \text{if } \lambda|_\Gamma = 0, \\ \chi_\Gamma(\lambda) &= 1 && \text{if } \lambda|_\Gamma = \pm 1, \\ \varphi_\Gamma(\lambda) &= 0 && \text{if } \lambda|_\Gamma = 0, \\ \varphi_\Gamma(\lambda) &= \mp 1 && \text{if } \lambda|_\Gamma = \pm 1. \end{aligned}$$

According to these properties, we further assume that $\chi_\Gamma(\lambda)$ is a symmetric function while $\varphi_\Gamma(\lambda)$ is skew-symmetric.

It is straightforward to verify that when $\varepsilon_1 = \varepsilon_2$ and thus $\lambda(\varepsilon) = 0$, condition (27) coincides with $\left[\frac{1}{2}|\beta \cdot n_\Gamma| + \{\varepsilon\}\right][\![u]\!] = 0$. In the vanishing viscosity case let us fix $\mathbf{n} = \mathbf{n}_{hy}$ as reference orientation of the interface. Then, we obtain

$$\lambda = \frac{\varepsilon_{hy} - \varepsilon_{el}}{\varepsilon_{hy} + \varepsilon_{el}} = -1$$

and by consequence $\varphi_\Gamma(\lambda = -1) = 1$. As a result of that, it turns out that, for the elliptic / hyperbolic coupling, condition (27) is equivalent to $\big(1 - \text{sign}(\beta \cdot \mathbf{n})\varphi_\Gamma(\lambda)\big) = 2$ on Γ_{in} and $\big(1 - \text{sign}(\beta \cdot \mathbf{n})\varphi_\Gamma(\lambda)\big) = 0$ on Γ_{out}, which coincides with the continuity condition of (26).

In conclusion, to set up Nitsche's method that suits problem (25) and (26) we start from the following general formulation that combines both,

$$\begin{cases} \nabla \cdot \big(-\varepsilon_i \nabla u_i + \beta u_i\big) = f_i & \text{in } \Omega_i, \\ \big(\frac{1}{2}|\beta \cdot \mathbf{n}_i| - \frac{1}{2}\beta \cdot \mathbf{n}_i + \varepsilon\big)u_i = 0 & \text{on } \partial\Omega \cap \partial\Omega_i, \\ [\![-\varepsilon\nabla u \cdot \mathbf{n} + \beta \cdot \mathbf{n}u]\!] = 0 & \text{on } \Gamma, \\ \left[\frac{1}{2}|\beta \cdot \mathbf{n}|\big(1 - \text{sign}(\beta \cdot \mathbf{n})\varphi_\Gamma(\lambda)\big) + \{\varepsilon\}\big(1 - \chi_\Gamma(\lambda)\big)\right][\![u]\!] = 0 & \text{on } \Gamma. \end{cases}$$

$$(28)$$

2.1.1 Variational Formulation and Analysis

To proceed with the variational formulation of problem (28), propaedeutic to the application of Nitsche's coupling technique, we integrate the governing equations on each sub-region and applying Green's formula (including the advective terms) we obtain,

$$
\sum_{i=1,2} \left[\int_{\Omega_i} \left(\varepsilon_i \nabla u_i \cdot \nabla v_i - \beta u_i \cdot \nabla v_i \right) + \int_{\partial\Omega_i} \left(-\varepsilon_i \nabla u_i \cdot \mathbf{n}_i v_i + \beta \cdot \mathbf{n}_i u_i v_i \right) \right]
$$

$$
= \sum_{i=1,2} \left[\int_{\Omega_i} \left(\varepsilon_i \nabla u_i \cdot \nabla v_i - \beta u_i \cdot \nabla v_i \right) + \int_{\partial\Omega_i \backslash \Gamma} \left(-\varepsilon_i \nabla u_i \cdot \mathbf{n}_i v_i + \beta \cdot \mathbf{n}_i u_i v_i \right) \right]
$$

$$
+ \int_{\Gamma} [\![-\varepsilon \nabla u \cdot \mathbf{n} v + \beta \cdot \mathbf{n} u v]\!] . \tag{29}
$$

The term $[\![-\varepsilon\nabla u \cdot \mathbf{n} v + \beta \cdot \mathbf{n} u v]\!]$ allows us to weakly enforce continuity of the conormal derivatives. However, to maintain strong consistency with problem (28), it is necessary to generalize the technique already described for Poisson's equation, to the case of weighted averages,

$$
\{v(x)\}_w := w_i(x) v_i(x) + w_j(x) v_j(x),
$$
$$
\{v(x)\}^w := w_j(x) v_i(x) + w_i(x) v_j(x),
$$

with $i = 1, 2$, $j \neq i$, where v is a regular function, $x \in \Gamma$ and the weights necessarily satisfy $w_1(x) + w_2(x) = 1$. We say that these averages are conjugate, because they fulfill the following identity,

$$
[\![ab]\!] = \{a\}_w [\![b]\!] + \{b\}^w [\![a]\!],
$$

that can be exploited to obtain,

$$
[\![(-\varepsilon\nabla u \cdot \mathbf{n} + \beta \cdot \mathbf{n} u) v]\!] = [\![-\varepsilon\nabla u \cdot \mathbf{n} + \beta \cdot \mathbf{n} u]\!] \{v\}^w + \{-\varepsilon\nabla u \cdot \mathbf{n} + \beta \cdot \mathbf{n} u\}_w [\![v]\!]
$$

$$
= [\![-\varepsilon\nabla u \cdot \mathbf{n} + \beta \cdot \mathbf{n} u]\!] \{v\}^w - \{\varepsilon\nabla u \cdot \mathbf{n}\}_w [\![v]\!] + \{\beta \cdot \mathbf{n} u\}_w [\![v]\!].
$$

First, the previous identity allows to weakly enforce the continuity of fluxes at the interface, by setting $[\![-\varepsilon\nabla u \cdot \mathbf{n} + \beta \cdot \mathbf{n} u]\!] = 0$. Second, it shows that the choice of the averaging weights w_i is not completely arbitrary. Indeed, to reproduce the interface condition $-\varepsilon\nabla u_{el} \cdot n + \beta \cdot n u_{el} = \beta \cdot n u_{hy}$ at the level of the variational formulation, that is to maintain strong consistency with problem (26), we have to make sure that the term $\{\varepsilon\nabla u \cdot \mathbf{n}\}_w [\![v]\!]$ vanishes when $\varepsilon_1 = \varepsilon_{hy} = 0$ while $\{\beta \cdot \mathbf{n} u\}_w [\![v]\!] = \beta \cdot n u_{hy} [\![v]\!]$. Such requirements correspond to the following constraint:

$$
w_1 = 1, \ w_2 = 0 \quad \text{when} \quad \varepsilon_1 \to 0
$$

We conclude that, for a strongly consistent treatment of the interface conditions with Nitsche's method, not only the intensity of the penalty terms, but also the averaging weights must suitably depend on the coefficients of the problem, and in particular on their heterogeneity. We will discuss later on suitable expressions for these problem dependent parameters.

Defining the following problem dependent penalty factors that modulate the enforcement of interface and boundary conditions, respectively,

$$\xi_\Gamma(\varepsilon, \beta) := \frac{1}{2}\left(|\beta \cdot \mathbf{n}| - \beta \cdot \mathbf{n}\varphi_\Gamma(\lambda|_\Gamma)\right) + \{\varepsilon\}\left(1 - \chi_\Gamma(\lambda|_\Gamma)\gamma h_E^{-1}\right),$$

$$\xi_{i,\partial\Omega}(\varepsilon, \beta) := \frac{1}{2}\left(|\beta \cdot \mathbf{n}_i| - \beta \cdot \mathbf{n}_i\right) + \varepsilon\gamma h_E^{-1},$$

where $\gamma > 0$ is a penalty parameter to be selected large enough in order to ensure stability of the resulting scheme, the bilinear form corresponding to Nitsche's method for the discretization of problem (28) is assembled adding the following penalty terms

$$\sum_{i=1,2}\left(\sum_{E\in\mathscr{G}_{h,i}}\xi_\Gamma(\varepsilon,\beta)\int_E [\![u_h]\!][\![v_h]\!] + \sum_{E\in\mathscr{B}_{h,i}}\xi_{i,\partial\Omega}(\varepsilon,\beta)\int_E u_h v_h\right),$$

to the equation arising from (29) after weak enforcement of flux continuity. Exploiting the same finite element approximation defined for Poisson's problem, see Sect. 1.3, we aim to find discrete functions $[u_{h,1}, u_{h,2}] \in V_h := V_{h,1} \times V_{h,2}$, where $V_{h,i}$ are Lagrangian finite element spaces on $\mathscr{T}_{h,i}$ relative to each subregion Ω_i, such that

$$a_h(u_h, v_h) = F_h(v_h), \quad \forall v_h \in V_h,$$

with

$$a_h(u_h, v_h) := \sum_{i=1,2}(\varepsilon_i \nabla u_{h,i} - \beta u_{h,i}, \nabla v_{h,i})_{\Omega_i}$$

$$+ \sum_{i=1,2}\left[\sum_{E\in\mathscr{G}_{h,i}}\xi_\Gamma(\varepsilon,\beta)\left([\![u_h]\!],[\![v_h]\!]\right)_E + \sum_{E\in\mathscr{B}_{h,i}}\xi_{i,\partial\Omega}(\varepsilon,\beta)\left(u_{h,i}, v_{h,i}\right)_E\right]$$

$$- \left(\{\varepsilon\nabla u_h \cdot \mathbf{n}\}_w, [\![v_h]\!]\right)_\Gamma - \left(\{\varepsilon\nabla v_h \cdot \mathbf{n}\}_w, [\![u_h]\!]\right)_\Gamma + \left(\{\beta \cdot \mathbf{n}u_h\}_w, [\![v_h]\!]\right)_\Gamma$$

$$- (\varepsilon_i \nabla u_{h,i} \cdot \mathbf{n}_i, v_{h,i})_{\partial\Omega_i\backslash\Gamma} - (\varepsilon_i \nabla v_{h,i} \cdot \mathbf{n}_i, u_{h,i})_{\partial\Omega_i\backslash\Gamma}$$

$$+ (\beta \cdot \mathbf{n}_i u_{h,i}, v_{h,i})_{\partial\Omega_i\backslash\Gamma},$$

$$F_h(v_h) := F(v_h) = \int_\Omega f v_h, \quad \text{if } u = 0 \text{ on } \partial\Omega.$$

Three remarks are in order. First, we restrict ourselves to homogeneous Dirichlet boundary conditions, but the corresponding schemes for non-homogeneous

Dirichlet or Neumann conditions can be obtained similarly. Secondly, we have applied the symmetrization technique already addressed for Poisson problem. For non symmetric problems such as advection / diffusion equations, also skew symmetrization turns out to be an interesting option. We will not dwell here on a detailed comparison of the two possibilities, but for a detailed discussion on the benefits of the non symmetric option we refer the interested reader to [38, 44]. Finally, we remind that the bilinear form $a_h(\cdot, \cdot)$ is not yet completely determined, because the scaling functions $\varphi_\Gamma(\lambda)$, $\chi_\Gamma(\lambda)$ and the averaging weights w_i still require a precise definition. Since there are infinitely many expressions that satisfy the aforementioned consistency requirements, we propose some criteria that allow to identify an admissible and effective choice for such parameters.

According to the usual practice for advection / diffusion equations, we split the bilinear form into its diffusive and advective components, denoted with $a_h^\varepsilon(\cdot, \cdot)$ and $a_h^\beta(\cdot, \cdot)$ respectively,

$$
\begin{aligned}
a_h^\varepsilon(u_h, v_h) := & \sum_{i=1,2} (\varepsilon_i \nabla u_{h,i}, \nabla v_{h,i})_{\Omega_i} \\
& + \sum_{i=1,2} \sum_{E \in \mathcal{G}_{h,i}} \left[\frac{1}{2} |\beta \cdot \mathbf{n}| + \{\varepsilon\}(1 - \chi_\Gamma(\lambda|_\Gamma)\gamma h_E^{-1}) \right] (\llbracket u_h \rrbracket, \llbracket v_h \rrbracket)_E \\
& + \sum_{i=1,2} \sum_{E \in \mathcal{B}_{h,i}} \left(\frac{1}{2} |\beta \cdot \mathbf{n}_i| + \varepsilon \gamma h_E^{-1} \right) (u_{h,i}, v_{h,i})_E \\
& - (\{\varepsilon \nabla u_h \cdot \mathbf{n}\}_w, \llbracket v_h \rrbracket)_\Gamma - (\{\varepsilon \nabla v_h \cdot \mathbf{n}\}_w, \llbracket u_h \rrbracket)_\Gamma \\
& - (\varepsilon_i \nabla u_{h,i} \cdot \mathbf{n}_i, v_{h,i})_{\partial \Omega_i \setminus \Gamma} - (\varepsilon_i \nabla v_{h,i} \cdot \mathbf{n}_i, u_{h,i})_{\partial \Omega_i \setminus \Gamma}, \\
a_h^\beta(u_h, v_h) := & \sum_{i=1,2} \left[-(\beta u_{h,i}, \nabla v_{h,i})_{\Omega_i} + \frac{1}{2}(\beta \cdot \mathbf{n}_i u_{h,i}, v_{h,i})_{\partial \Omega_i \setminus \Gamma} \right] \\
& + \left(\{\beta \cdot \mathbf{n} u_h\}_w - \frac{1}{2}\beta \cdot \mathbf{n}\varphi_\Gamma(\lambda|_\Gamma), \llbracket v_h \rrbracket \right)_\Gamma.
\end{aligned}
$$

The aforementioned assumption that $\chi_\Gamma(\lambda)$ is a symmetric function together with the choice of exploiting the symmetric Nitsche formulation, makes sure that the diffusion bilinear form $a_h^\varepsilon(\cdot, \cdot)$ respects the symmetry of the underlying operator. Correspondingly, we want to make sure that $a_h^\beta(\cdot, \cdot)$ is skew-symmetric, i.e. $a_h^\beta(u_h, v_h) = -a_h^\beta(v_h, u_h)$. Since the satisfaction of such property depends on $\varphi_\Gamma(\lambda)$, this is our criterion to determine the expression of this function. Exploiting integration by parts, we observe that $a_h^\beta(u_h, v_h)$ becomes skew-symmetric provided that the following equality holds true for any test function v_h,

$$
\{v_h\}^w + \frac{1}{2}\varphi_\Gamma(\lambda)\llbracket v_h \rrbracket = \{v_h\}_w - \frac{1}{2}\varphi_\Gamma(\lambda)\llbracket v_h \rrbracket,
$$

which is equivalent to define

$$\varphi_\Gamma(\lambda) := (w_i - w_j)$$

in the particular case when the reference normal vector on the interface Γ, namely **n**, points from Ω_i to Ω_j. Moreover, the following identity holds true,

$$\beta \cdot \mathbf{n}\{v_h\}_w - \frac{1}{2}\beta \cdot \mathbf{n}(w_i - w_j)[\![v_h]\!] = \beta \cdot \mathbf{n}\{v_h\},$$

and we notice that the advective bilinear form becomes,

$$a_h^\beta(u_h, v_h) = \sum_{i=1,2}\left[-(\beta u_{h,i}, \nabla v_{h,i})_{\Omega_i} + \frac{1}{2}(\beta \cdot \mathbf{n}_i u_{h,i}, v_{h,i})_{\partial\Omega_i \setminus \Gamma}\right] + (\{\beta \cdot \mathbf{n}u_h\}, [\![v_h]\!])_\Gamma,$$

which, together with the penalty term proportional to $\frac{1}{2}|\beta \cdot \mathbf{n}_i|$, corresponds to the treatment of advective fluxes through the interface by means of a standard upwind method.

For the identification of a suitable function $\chi_\Gamma(\lambda)$ and of the weights w_i in terms of ε and β, we formulate some technical requirements that will facilitate the proof of coercivity of $a_h^\varepsilon(\cdot, \cdot)$ in the forthcoming analysis of the scheme. First, we select $\chi_\Gamma(\lambda)$ such that,

$$\{\varepsilon\}(1 - \chi_\Gamma(\lambda)) = \{\varepsilon\}_w.$$

Noticing that for any regular function v it holds $\{v\}_w = \{v\} - (w_j - w_i)[\![v]\!]$ and reminding of the definition of the heterogeneity factor $\lambda = [\![\varepsilon]\!]/(2\{\varepsilon\})$, we conclude that the aforementioned requirement for $\chi_\Gamma(\lambda)$ corresponds to set,

$$\chi_\Gamma(\lambda|_\Gamma) = 1 - \frac{\{\varepsilon\}_w}{\{\varepsilon\}} = (w_j - w_i)\lambda.$$

Finally, the weights w_i are conveniently selected in order to satisfy the following equality for any test function v,

$$\{\varepsilon v\}_w = \{\varepsilon\}_w\{v\},$$

that implies that $2\{\varepsilon\}_w = \varepsilon_i w_i = \varepsilon_j w_j$ being equivalent to set

$$w_i = \frac{\varepsilon_j}{\varepsilon_i + \varepsilon_j}, \quad w_j = \frac{\varepsilon_i}{\varepsilon_i + \varepsilon_j} \quad \text{and} \quad \{\varepsilon\}_w = \frac{2\varepsilon_i\varepsilon_j}{\varepsilon_i + \varepsilon_j}.$$

We observe that the aforementioned requirement $w_1 = 1$, $w_2 = 0$ when $\varepsilon_1 \to 0$ is satisfied and that the term $\{\varepsilon\nabla u \cdot \mathbf{n}\}_w[\![v]\!] = \{\varepsilon\}_w\{\varepsilon\nabla u \cdot \mathbf{n}\}[\![v]\!]$ vanishes together with the diffusivity parameter.

With these particular choices of scaling functions and weights, the stability of the discrete scheme, i.e. the consistency of its bilinear form, is readily proved. Indeed, it is sufficient to consider the diffusive (symmetric) part $a_h^\varepsilon(\cdot, \cdot)$, because we have shown that $a_h^\beta(\cdot, \cdot)$ is skew symmetric and it does not contribute to the energy of the system. First of all, we straightforwardly verify that,

$$
a_h^\varepsilon(v_h, v_h) = \sum_{i=1,2} \|\varepsilon^{\frac{1}{2}} \nabla v_h\|_{0,\Omega_i}^2
$$

$$
+ \sum_{i=1,2} \sum_{E \in \mathscr{G}_{h,i}} \|(\frac{1}{2}|\beta \cdot \mathbf{n}| + \gamma \{\varepsilon\}_w h_E^{-1})^{\frac{1}{2}} [\![v_h]\!]\|_{0,E}^2
$$

$$
+ \sum_{i=1,2} \sum_{E \in \mathscr{B}_{h,i}} \|(\frac{1}{2}|\beta \cdot \mathbf{n}| + \varepsilon \gamma h_E^{-1})^{\frac{1}{2}} v_h\|_{0,E}^2
$$

$$
- 2 (\{\varepsilon \nabla v_h\}_w \cdot \mathbf{n}, [\![v_h]\!])_\Gamma - 2 (\varepsilon \nabla v_h \cdot \mathbf{n}, v_h)_{\partial \Omega},
$$

where the first three terms on the right hand side represent the energy norm that is applied for the stability and convergence analysis of the scheme. For the remaining terms of $a_h^\varepsilon(\cdot, \cdot)$, we exploit that $\{\varepsilon\}_w = 2 w_i \varepsilon_i \le 2 \varepsilon_i$ to obtain the following upper bound,

$$
2 (\{\varepsilon \nabla v_h\}_w \cdot \mathbf{n}, [\![v_h]\!])_\Gamma + 2 (\varepsilon \nabla v_h \cdot \mathbf{n}, v_h)_{\partial \Omega}
$$

$$
= \sum_{i=1,2} 2 (\varepsilon_i w_i \nabla v_{h,i} \cdot \mathbf{n}, [\![v_h]\!])_\Gamma + 2 (\varepsilon \nabla v_h \cdot \mathbf{n}, v_h)_{\partial \Omega}
$$

$$
\le \sum_{i=1,2} \sum_{E \in \mathscr{G}_{h,i}} \left[\delta h_E \|(\varepsilon_i)^{\frac{1}{2}} \nabla v_{h,i} \cdot \mathbf{n}\|_{0,E}^2 + \frac{1}{\delta h_E} \|\{\varepsilon\}_w^{\frac{1}{2}} [\![v_h]\!]\|_{0,E}^2 \right]
$$

$$
+ \sum_{i=1,2} \sum_{E \in \mathscr{B}_{h,i}} \left[\delta h_E \|\varepsilon_i^{\frac{1}{2}} \nabla v_{h,i} \cdot \mathbf{n}\|_{0,E}^2 + \frac{1}{\delta h_E} \|\varepsilon^{\frac{1}{2}} v_h\|_{0,E}^2 \right]
$$

$$
\lesssim \sum_{i=1,2} \delta \|\varepsilon^{\frac{1}{2}} \nabla v_h\|_{0,\Omega_i}^2 + \frac{1}{\delta} \|\{\varepsilon\}_w^{\frac{1}{2}} [\![v_h]\!]\|_{\frac{1}{2},h,\Gamma}^2 + \frac{1}{\delta} \|\varepsilon^{\frac{1}{2}} v_h\|_{\frac{1}{2},h,\partial\Omega}^2.
$$

Then, $a_h^\varepsilon(\cdot, \cdot)$ turns out to be coercive for a sufficiently small δ and large γ.

In [23] the scheme has been extended to Problem (25) with an anisotropic symmetric positive definite diffusion tensor $K : \Omega \to \mathbb{R}^{d \times d}$ replacing the scalar diffusivity ε, under the practical assumption that K is a constant on each sub-region denoted with K_i. With the aforementioned choice of the scaling function $\chi_\Gamma(\lambda|_\Gamma) = (w_j - w_i) \lambda$, the diffusive part of the bilinear form becomes

$$
a_h^K(u_h, v_h) := \sum_{i=1,2} (K_i \nabla u_{h,i}, \nabla v_{h,i})_{\Omega_i}
$$

$$
+ \sum_{i=1,2} \sum_{E \in \mathscr{G}_{h,i}} \left[\frac{1}{2}|\beta \cdot \mathbf{n}| + \gamma \{\kappa\}_w h_E^{-1}) \right] ([\![u_h]\!], [\![v_h]\!])_E
$$

$$+ \sum_{i=1,2} \sum_{E \in \mathscr{B}_{h,i}} \left(\frac{1}{2} |\beta \cdot \mathbf{n}_i| + \gamma \kappa_i h_E^{-1} \right) (u_{h,i}, v_{h,i})_E$$

$$- \left(\{ \mathbf{n}^T K \nabla u_h \cdot \mathbf{n} \}_w, [\![v_h]\!] \right)_\Gamma - \left(\{ \mathbf{n}^T K \nabla v_h \cdot \mathbf{n} \}_w, [\![u_h]\!] \right)_\Gamma$$

$$- \left(\mathbf{n}_i^T K_i \nabla u_{h,i} \cdot \mathbf{n}_i, v_{h,i} \right)_{\partial \Omega_i \setminus \Gamma} - \left(\mathbf{n}_i^T K_i \nabla v_{h,i} \cdot \mathbf{n}_i, u_{h,i} \right)_{\partial \Omega_i \setminus \Gamma},$$

where the averaging weights are selected as follows

$$\kappa_i := \mathbf{n}^T K_i \mathbf{n}, \quad w_i = \frac{\kappa_j}{\kappa_i + \kappa_j}, \quad w_j = \frac{\kappa_i}{\kappa_i + \kappa_j}, \quad \text{and} \quad \{\kappa\}_w = \frac{2\kappa_i \kappa_j}{\kappa_i + \kappa_j}.$$

2.1.2 Stabilized Galerkin Methods for Singularly Perturbed Equations

The aforementioned Nitsche technique allows to robustly enforce interface conditions among second order elliptic problems with discontinuous diffusion coefficients, but such technique does not cure the intrinsic instability of any standard Galerkin approximation applied to singularly perturbed equations. For this reason, the previously developed scheme should be complemented with a stabilisation technique acting on each subregion Ω_i where the local Péclét number is large.

It is not our aim to review here the wide area of numerical schemes devoted to stabilisation of Galerkin method for transport dominated problems. We will simply present two options that suitably fit the present discretisation framework and are also related to Nitsche's idea.

Following [17], the first stabilisation strategy that we consider is suited for locally H^1-conforming approximations. More precisely, we use standard Lagrangian finite elements on each subdomain and obtain stability for high Péclét numbers by adding a penalty term on the gradient jumps over element faces. Combined with the previously presented Nitsche interface conditions, it will result in a robust continuous / discontinuous approximation of large contrast problems, where the discontinuous approximation functions are localized only along the discontinuities of problem coefficients. Denoting by $\mathscr{E}_{h,i}$ the collection of interior edges belonging to elements of $\mathscr{T}_{h,i}$, the stabilisation effect is then obtained by complementing the bilinear form $a_h(\cdot, \cdot)$ with the following additional terms on each Ω_i,

$$J_i(u_h, v_h) := \sum_{E \in \mathscr{E}_{h,i}} \left(\gamma_{cip} h_E^2 \| \beta \cdot n \|_{L^\infty(E)} [\![\nabla u_h \cdot \mathbf{n}_E]\!], [\![\nabla v_h \cdot \mathbf{n}_E]\!] \right)_E,$$

proposed and thoroughly analysed in [11, 13, 14], which consist of interior penalty forms controlling the jumps in the gradient over *interior* faces of each sub-domain Ω_i. Since the finite element approximation to which this stabilisation is applied involves continuous functions, the resulting scheme has been called *continuous interior penalty* (CIP). The main idea behind the stabilisation based on the jump in the gradient between adjacent elements is to introduce a least squares control over the part of the convective derivative that is not in the finite element space. A key result is the following property of the Oswald quasi-interpolant

$$\pi_h^* : \{v \in L^2(\Omega) : v|_K \in \mathbb{P}^k(K), \forall K \in \mathscr{T}_h\} \to \{v \in C^0(\Omega) : v|_K \in \mathbb{P}^1(K), \forall K \in \mathscr{T}_h\}$$

$$\pi_h^* v(x_j) := \frac{1}{n_j} \sum_{\{K : x_j \in K\}} v|_K(x_j), \quad \forall v \in \{v \in L^2(\Omega) : v|_K \in \mathbb{P}^k(K)\},$$

where x_j are the nodes of the local finite element meshes $\mathscr{T}_{h,i}$, and n_j is the number of elements containing x_j as a node. Let β_h be the piecewise affine Lagrange interpolant of β and let $u_h \in V_{h,i}$. Then there exists a constant $\gamma_{cip} \geq c_0 > 0$, depending only on the local mesh geometry, such that

$$\|h^{\frac{1}{2}}(\beta_h \cdot \nabla u_h - \pi_h^*(\beta_h \cdot \nabla u_h))\|_{0,\Omega_i}^2 \leq J_i(u_h, u_h).$$

Assuming that $\beta \in [W^{1,\infty}(\Omega)]^d$ with $\nabla \cdot \beta = 0$, $\varepsilon \in L^\infty(\Omega)$ and that the exact solution of the multi-domain problem satisfies $u \in H^s(\Omega_1 \cup \Omega_2) \cap H_0^1(\Omega)$ with $s \geq k+1 \geq 2$ it has been shown in [17] that the following error estimate holds true,

$$|||u - u_h|||_{1,h,\Omega_1 \cup \Omega_2} \lesssim \left(\|\varepsilon\|_{L^\infty(\Omega)}^{\frac{1}{2}} \mathscr{H}(0,u) + \|\beta\|_{L^\infty(\Omega)}^{\frac{1}{2}} \mathscr{H}(1,u) \right)$$

where for any $v \in H^s(\Omega_1 \cup \Omega_2) \cap H_0^1(\Omega)$

$$|||v|||_{1,h,\Omega_1 \cup \Omega_2}^2 := \sum_{i=1,2} \left(|\varepsilon_i^{\frac{1}{2}} v_i|_{1,\Omega_i}^2 + \|\varepsilon_i^{\frac{1}{2}} v_i\|_{\frac{1}{2},h,\mathscr{B}_{h,i}}^2 + \|\{\varepsilon\}_w^{\frac{1}{2}}[\![v]\!]\|_{\frac{1}{2},h,\mathscr{G}_{h,i}}^2 + J_i(v,v) \right),$$

and

$$\mathscr{H}(\alpha,u) = \left(\sum_{i=1}^N \sum_{K \in \mathscr{T}_{h,i}} h_K^{2k+\alpha} \|u\|_{k+1,K}^2 \right)^{\frac{1}{2}}.$$

The CIP stabilisation is a suitable method when heterogeneities of the diffusion coefficient appear at a scale that is much larger than the element size. Conversely, if the bulk is so fractured that the diffusivity varies at the scale of single elements, the following approach, based on a fully discontinuous approximation space, seems to be more appropriate. The main idea consists in exploiting the robustness of the proposed Nitsche's method for the enforcement of transmission conditions, combined with the observation that fully discontinuous finite elements provide stable approximation of transport problems. This turns out to transform the previous continuous / discontinuous approximation of multi-domain (25) into a fully discontinuous approximation where each element plays the role of a domain, giving rise to an instance of the so called interior penalty discontinuous Galerkin methods, [1]. Different variants of such a method have been applied to the discretization of elliptic, possibly singularly perturbed problems, [2]. Because of the application of weighted averages, we will denote the scheme proposed here as *weighted interior penalty method* (WIPG) and we will compare it with similar formulations such as the symmetric interior penalty (SIPG) and the non symmetric interior penalty (NIPG). We refer the interested reader to [2] for a broad review of literature and to [22,23,47] for further details about the present approach. Consistently with the fact that this new approximation scheme is stable also for transport dominated problems, we notice that the continuous interior penalty term on the gradient jumps vanishes since there are no interior faces in the element-based subdomains. To set up such discontinuous Galerkin scheme we reformulate problem (28) at the level of single elements $K \in \mathcal{T}_h$,

$$
\begin{cases}
-\varepsilon \Delta u + \beta \cdot \nabla u = f & \text{in } K, \\
[\![-\varepsilon \nabla u + \beta u]\!]_{\partial K} \cdot n_{\partial K} = 0 & \text{on } \partial K \setminus \partial \Omega, \\
\gamma_{h,E}(\varepsilon, \beta)[\![u]\!]_{\partial K} = 0 & \text{on } \partial K \setminus \partial \Omega, \\
\gamma_{h,\partial \Omega}(\varepsilon, \beta) u = 0 & \text{on } \partial K \cap \partial \Omega,
\end{cases}
$$

and proceeding as for Nitsche's method we look for $u_h \in V_h := \{v_h \in L^2(\Omega) : v_h|_K \in \mathbb{P}^k, \ \forall K \in \mathcal{T}_h\}$ such that

$$
\begin{aligned}
a_h^{(DG)}(u_h, v_h) :=& \sum_{K \in \mathcal{T}_h} \left((\varepsilon \nabla u_h - \beta u_h), \nabla v_h \right)_K \\
&+ \sum_{E \in \mathscr{E}_h} \Big[(\{\beta u_h\}_w \cdot \mathbf{n}_E, [\![v_h]\!])_E \\
&- (\{\varepsilon \nabla u_h\}_w \cdot \mathbf{n}_E, [\![v_h]\!])_E - (\{\varepsilon \nabla v_h\}_w \cdot \mathbf{n}_E, [\![u_h]\!])_E \\
&+ \left(\frac{1}{2} |\beta \cdot \mathbf{n}_E| - \frac{1}{2} \beta \cdot \mathbf{n}_E (w_E^- - w_E^+) + \gamma \{\varepsilon\}_w h_E^{-1} \right) ([\![u_h]\!], [\![v_h]\!])_E \Big] \\
&+ \sum_{E \in \mathscr{B}_h} \Big[\left(\frac{1}{2} \beta \cdot \mathbf{n}_E u_h, v_h \right)_E - (\varepsilon \nabla u_h \cdot \mathbf{n}_E, v_h)_E - (\varepsilon \nabla v_h \cdot \mathbf{n}_E, u_h)_E \\
&+ \left(\frac{1}{2} |\beta \cdot n| + \varepsilon \gamma h_E^{-1} \right) (u_h, v_h)_E \Big] = F(v_h),
\end{aligned}
$$

where $\mathscr{E}_h$ is the collection of interior edges, $\mathbf{n}_E$ denotes the reference unit normal vector to each inter-element interface and w_E^-, w_E^+ represent the weights relative to the inner element $(-)$ and outer element $(+)$ neighbouring the edge E, with respect to the reference direction $\mathbf{n}_E$, as depicted in Fig. 3.

For the numerical validation of the robustness of weighted Nitsche's transmission conditions in presence of locally singularly perturbed problems, we will apply the element-wise version.

Fig. 3 A sketch of the
element setting for the
extension of Nitsche's
method to a discontinuous
Galerkin scheme

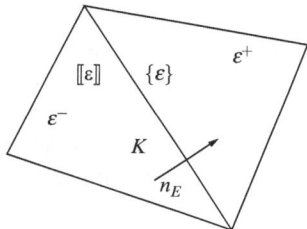

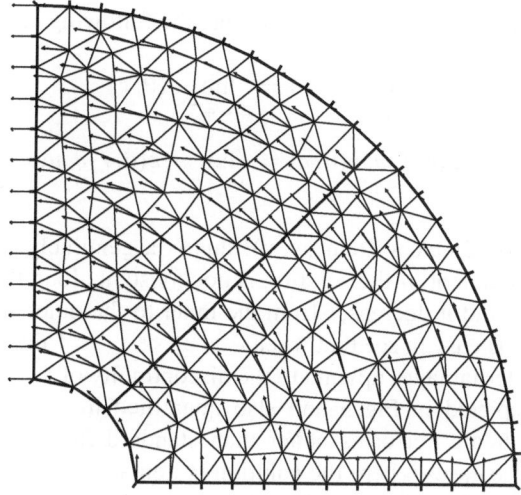

Fig. 4 The domain Ω and the subregions Ω_1, Ω_2 together with the computational mesh T_h and
the advective field β

2.1.3 Numerical Results and Discussion

To conclude this section, we will compare the efficiency of the proposed Nitsche
technique for singularly perturbed problems (WIPG) with the symmetric interior
penalty method (SIPG) and the non symmetric version (NIPG). Such methods are
obtained from WIPG by setting $w_E^{\pm} = \frac{1}{2}$, not depending on the diffusivity parameter.
The latter NIPG variant has the advantage that it only requires the condition $\gamma > 0$
to ensure stability. Consequently, we will set $\gamma = 2 \; 10^{-2}$ for NIPG while $\xi = 2$
for SIPG and WIPG, to study how this parameter influences the accuracy when ε is
vanishing.

To set up a test problem, featuring discontinuous coefficients, that allows us
to analytically compute the exact solution we consider a domain $\Omega \subset \mathbb{R}^2$
corresponding to the rectangle $\hat{\Omega} = (0, \pi/2) \times (1 - \pi/4, 1)$ in polar coordinates
(θ, r). We split $\hat{\Omega}$ into two subregions, $\hat{\Omega}_1 = (0, \pi/4) \times (1 - \pi/4, 1)$, $\Omega_2 = (\pi/4, \pi/2) \times (1 - \pi/4, 1)$. Then the domain Ω is split into Ω_1 and Ω_2, see Fig. 4,

owing to the mapping from polar to Cartesian coordinates. The viscosity $\varepsilon(x, y)$ is a discontinuous function across the interface between Ω_1 and Ω_2, namely the segment $x - y = 0$ with $x \in ((1 - \pi/4) \cos \pi/4, \cos \pi/4)$. Precisely, we will consider a constant $\varepsilon(x, y)$ in each subregion with several values of ε_1 in Ω_1 and a fixed $\varepsilon_2 = 1.0$ in Ω_2. Moreover, we set $\beta = [\beta_x = -y(x^2 + y^2)^{-1}, \beta_y = x(x^2 + y^2)^{-1}]$, $f = 0$ and the boundary conditions $u(x, y = 0) = 1$, $u(x = 0, y) = 0$ and $\nabla u \cdot n = 0$ otherwise. Then, the exact solution of the problem on each subregion $\hat{\Omega}_1, \hat{\Omega}_2$ can be expressed in polar coordinates as an exponential function with respect to θ independently from r. The global solution $u(\theta, r)$ is provided by choosing the value at the interface $\theta = \pi/4$ in order to ensure the following matching conditions,

$$\lim_{\theta \to \frac{\pi}{4}^-} u(\theta, r) = \lim_{\theta \to \frac{\pi}{4}^+} u(\theta, r),$$

$$\lim_{\theta \to \frac{\pi}{4}^-} -\varepsilon(\theta, r)\partial_\theta u(\theta, r) = \lim_{\theta \to \frac{\pi}{4}^+} -\varepsilon(\theta, r)\partial_\theta u(\theta, r).$$

In the Cartesian coordinate system (x, y), this is a genuinely 2-dimensional test case, because the gradient of the solution is not constant along the interface where ε is discontinuous, and it decreases from the inner to the outer side of the domain Ω. Furthermore, it is easy to see that when $0 \simeq \varepsilon_1 \ll \varepsilon_2 = 1$ the global solution, u, features a sharp internal layer upwind to the discontinuity of ε.

The results, depicted in Fig. 5 and also quantified in Table 1, give evidence that the WIPG scheme performs better than standard interior penalty methods, particularly in those cases where the solution is non smooth and at the same time the computational mesh with $h = 0.0654$ is not completely adequate to capture the singularities. From the analysis of Fig. 5, it is possible to identify three regimens where the numerical methods behave differently. The first one consists of the diffusive region, where all methods provide similar results. For the intermediate value of ε a transition takes place, because the computational mesh is not adequate anymore to capture the sharp internal layer that originates upwind to the discontinuity of ε. Initially, the error relative to each method increases when ε is reduced, but this trend is inverted for the WIPG method solely, after the threshold $\varepsilon = 10^{-6}$, while the error monotonically increases for SIPG and NIPG. Finally, the smallest value of ε_1 corresponds to the hyperbolic regimen. In the limit case $\varepsilon_1 \to 0$, the discontinuities of the global solution u are aligned with those of ε. However, we observe that the standard interior penalty schemes (SIPG or NIPG equivalently) provide solutions that are almost continuous, as reported in Fig. 5. This behaviour promotes the instability of the approximate solution in the neighborhood of the boundary layer, because the computational mesh is not adequate to smoothly approximate the high gradients across the interface. Conversely, the WIPG method is more effective, thanks to the consistency with the elliptic/hyperbolic limit case, because it replaces the part of the boundary layer with a jump that cannot be captured by the computational mesh.

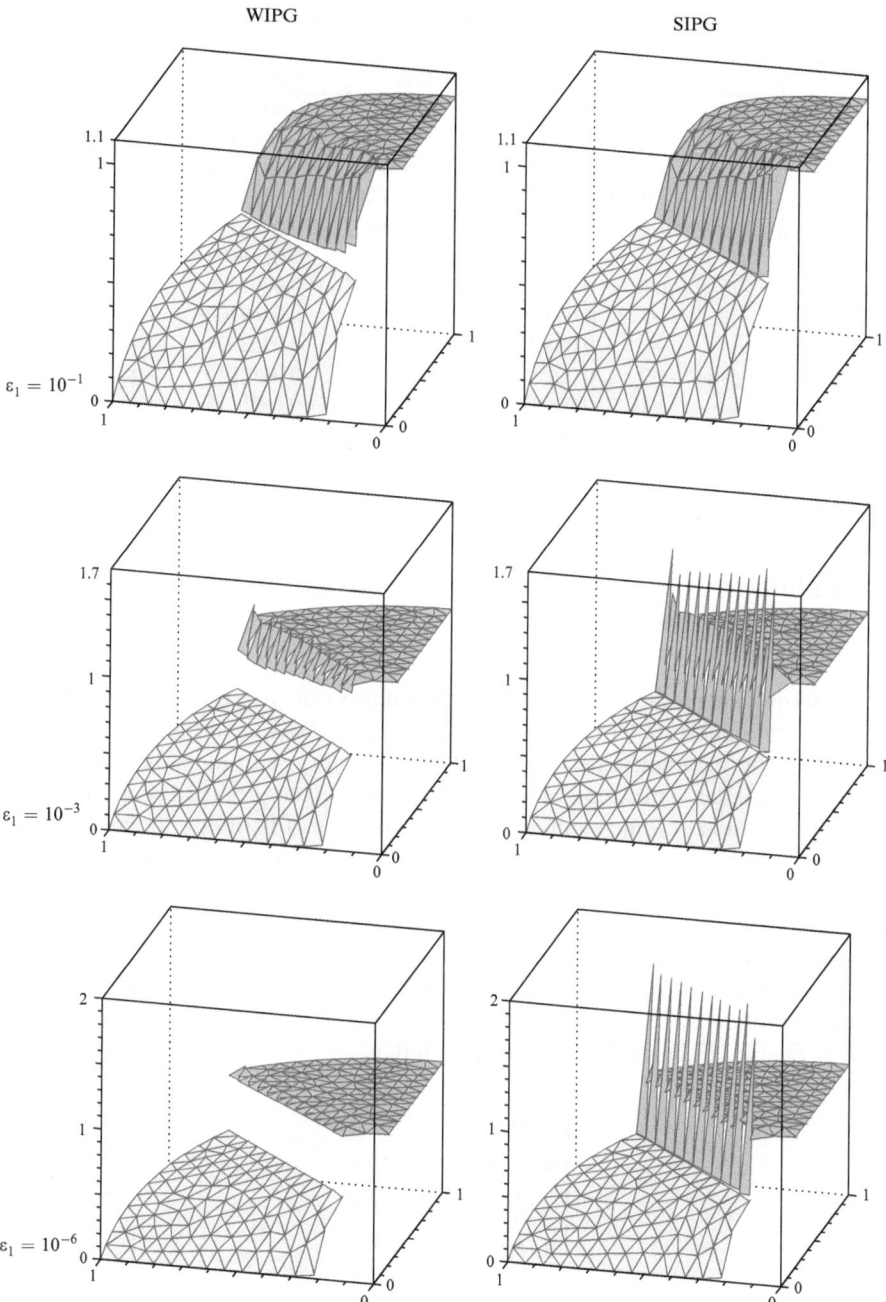

Fig. 5 A comparison of Nitsche's method with (WIPG) and without (SIPG) the application weighing technique

Table 1 The L^2 norm error for WIPG, SIPG, NIPG for different values of $\varepsilon_1 = 2^{-i}$ and a fixed value of $\varepsilon_2 = 1$ in the test problem depicted in Fig. 4

$\|u - u_h\|_{L^2}$	i	$\varepsilon_1 = 2^{-i}$	SIPG	NIPG	WIPG
Diffusive region	0	1	0.00101853	0.00123078	0.00101853
	−1	0.5	0.00123921	0.00134626	0.00121052
	−2	0.25	0.00200825	0.00167944	0.00182993
	−3	0.125	0.00393471	0.00333855	0.00315595
Transition region	−4	0.0625	0.0079422	0.00703319	0.00532886
	−5	0.03125	0.0144257	0.0130603	0.00780319
	−6	0.015625	0.0224454	0.0207315	0.00908097
	−7	0.0078125	0.0307374	0.0289709	0.00831401
	−8	0.00390625	0.0380299	0.0363924	0.00655286
Hyperbolic region	−9	0.00195312	0.0429129	0.0414616	0.0049148
	−10	0.000976562	0.0452834	0.0440218	0.00329726
	−11	0.000488281	0.0463316	0.0452286	0.00204598
	−12	0.000244141	0.0468732	0.0458791	0.00143603
	−13	0.00012207	0.0471628	0.0462332	0.00122399

3 Stabilized Nitsche's Method for Unfitted Boundaries and Interfaces

Fictitious domain methods turn out to be particularly effective for the approximation of boundary value problems on domains of complex shape and for free interface problems. The parametric description of the boundary with the subsequent mesh generation and the application of interface tracking techniques represent difficulties for the application of finite element methods. The idea of fictitious domain schemes consists in embedding the physical domain into a larger domain with reasonably simple shape. However, as discussed in [36], to preserve the accuracy of the selected finite element method, it is necessary to restrict the integration of the discrete variational formulation to the physical domain.

To illustrate the limitations of standard finite element approximations of unfitted interface problems, let us split the interval $\Omega := (0, 1)$ in two parts $\Omega_1 := (0, \Gamma)$, $\Omega_2 := (\Gamma, 1)$ and look for $u(x)$ such that,

$$\begin{cases} -\varepsilon_i u_i'' = 1 & \text{in } \Omega_i, \\ u_1 = u_2 & \text{on } \Gamma, \\ \varepsilon_1 u_1' = \varepsilon_2 u_2' & \text{on } \Gamma, \\ u_1 = u_2 = 0 & \text{on } \partial\Omega. \end{cases}$$

Let us approximate u with piecewise linear finite elements on a uniform partition of width h. For any positive $\varepsilon_1 \neq \varepsilon_2$ we have $u \notin H^2(\Omega)$. Then optimal convergence cannot be expected. In particular, as confirmed by numerical results reported in Table 2, sub-optimal convergence is verified if we select Γ such that it never coincides with a vertex of the partitions underlying the finite element space.

Table 2 Convergence rate of the error $\|u-u_h\|_{0,\Omega}$ of linear finite elements for an unfitted interface problem. The physical domain $\Omega = [0, 1]$ is divided in two subdomains $\Omega_1 = [0, \frac{1}{\sqrt{5}}]$ and $\Omega_2 = [\frac{1}{\sqrt{5}}, 1]$. The exact solution is of the form $u_i(x) = -\frac{x^2}{2\varepsilon_i} + b_i x + c_i$. The coefficient b_i and c_i are chosen such that the functions u_i satisfy the boundary conditions and the continuity conditions at interface

h	$\varepsilon_1 = \varepsilon_2 = 1$	$\varepsilon_1 = 1, \varepsilon_2 = 10^{-2}$
5.00×10^{-2}	2.28×10^{-4}	4.60×10^{-2}
2.50×10^{-2}	5.70×10^{-5}	3.49×10^{-2}
1.25×10^{-2}	1.42×10^{-5}	3.23×10^{-2}
6.25×10^{-3}	3.56×10^{-6}	3.15×10^{-2}
3.12×10^{-3}	8.91×10^{-7}	2.92×10^{-2}
p	1.99	0.15

Since the boundary and the interface do not necessarily conform with the mesh, an optimally convergent finite element method must be defined on sub-elements. In the case of interface problems, this additional difficulty can be taken into account by enriching the approximation space with additional basis functions that lie on a portion of the mesh elements. Such a technique is often called the extended finite element method (XFEM) and has been successfully applied to different applications such as crack propagation problems [21] and free interface problems in fluid dynamics [27, 43].

The approximation of elliptic problems with unfitted boundary or interface has already been investigated in recent works, we mention for instance [18, 20, 25, 32, 37]. The discretisation schemes that we consider are closely related to [28, 30], where an extended finite element method has been combined with a Nitsche technique to enforce the matching conditions between contiguous sub-regions. However, the application of Nitsche's method for the treatment of boundary or interface conditions may give rise to numerical instabilities in presence of small element cuts. More precisely, it has been observed in [12, 15, 16, 43] that the stability and the condition number of the finite element scheme depend on how the interface cuts the computational mesh. To cure them, the application of interior penalty stabilisation techniques has been successfully considered in a sequel of papers [12, 15, 16]. The idea of such stabilisation methods is to introduce in the discrete formulation a minimum of artificial diffusion to ensure the positivity of the discrete bilinear form for any configuration of the boundary or interface.

For interface problems, the need to introduce additional finite element basis functions lying on sub-elements to restore optimal convergence represents a second source of instability. Following the approach proposed in [43], we study the H^1 stability of the extended finite element space in the case of piecewise linear approximation. We analyse the condition number of the corresponding mass and stiffness matrices in presence of small sub-elements and we conclude that their spectrum is affected by how elements are cut.

Finally, we will apply Nitsche's method to enforce transmission conditions in the extended finite element space for interface problems governed by symmetric elliptic equations with large contrast between diffusion coefficients. We aim to develop a scheme that is robust with respect to the configuration of sub-elements as well as the heterogeneity of the diffusion coefficients.

3.1 The Unfitted Nitsche Method for Boundary Conditions

We recall and analyse the Nitsche's method for the approximation of boundary conditions on a computational mesh that does not fit the physical domain. Let $\mathscr{T}_h^0$ be a given admissible computational mesh whose elements entirely cover the physical domain Ω. We also assume that all elements of $\mathscr{T}_h^0$ have non-empty intersection with Ω. Let $\Omega_{\mathscr{T}}$ be the domain covered by $\mathscr{T}_h^0$. To improve this possibly coarse approximation, we will also consider a family of shape regular, quasi-uniform triangulations, $\mathscr{T}_h$, built by recursive refinement of $\mathscr{T}_h^0$, omitting any elements whose intersection with Ω is empty. As previously mentioned in Sect. 1.4, to keep the analysis of the schemes as simple as possible, we consider linear Lagrangian finite elements

$$V_h := \{v_h \in C^0(\Omega_{\mathscr{T}}) : v_h|_K \in \mathbb{P}^1(K) \ \forall K \in \mathscr{T}_h\}.$$

Referring to Poisson problem (1) with homogeneous boundary conditions i.e. $g = 0$, Nitsche's method requires to find $u_h \in V_h$ such that $a_h(u_h, v_h) = F_h(v_h)$ for any $v_h \in V_h$ with

$$a_h(u_h, v_h) := a(u_h, v_h) - (\partial_n u_h, v_h)_{\partial\Omega} - s\,(\partial_n v_h, u_h)_{\partial\Omega} + \gamma h^{-1}\,(u_h, v_h)_{\partial\Omega}\,, \quad (30)$$

$$F_h(v_h) := F(v_h).$$

where $s = \pm 1$ gives rise to the symmetric or non symmetric formulations.

Although formally equivalent to the case of fitted boundary, the treatment of the unfitted case hides some additional difficulties for the set up of the discrete problem.

Firstly, for the assembly of mass and stiffness matrices, integrals over cut elements must be computed, such as

$$\int_{K\cap\Omega} u_h \cdot v_h, \quad \int_{K\cap\Omega} \nabla u_h \cdot \nabla v_h,$$

where $K \cap \Omega$ is a portion of a triangle or a tetrahedron. In two or three space dimensions, $K \cap \Omega$ may not be a simplex. For these reasons, the computation of these integrals requires particular attention, and the fact that $|K \cap \Omega|$ may vanish affects the condition number of mass and stiffness matrices, as it will be discussed in the forthcoming sections.

Secondly, the assembly of boundary terms involves integrals over manifolds that do not coincide with edges or faces. To automatically perform such calculations, some approximation of the boundary configuration is necessary. For instance, the boundary can be represented by means of the level set of a discrete distance function. This means that there exists a discrete implicit surface (or hyper-surface when $d = 3$) $\varphi_h \in V_h$ that defines $\partial\Omega$ as its zero level set. Coherently with the notation adopted in the previous sections, we apply here a two-dimensional notation and we denote quantities related to element edges of faces with E.

We denote with $\mathscr{C}_h := \{K \in \mathscr{T}_h : |K \cap \partial\Omega|_{\mathbb{R}^{d-1}} > 0\}$ a *crust* of elements with non vanishing intersection with the boundary, measured in $\mathbb{R}^{d-1}$ topology. Assuming that each element K is an open set, with the piecewise linear description of the boundary we observe that for all $K \in \mathscr{C}_h$ the set $\partial\Omega \cap \partial K$ consists on two points (in the two-dimensional case) and the portion of $\partial\Omega$ that connects them is a straight line (or a planar surface in three dimensions). An example is illustrated in Fig. 1. If $\partial\Omega$ lies on an entire edge of an element K, then such element does not belong to $\mathscr{C}_h$. We denote with $E^{\partial\Omega} := K \cap \partial\Omega$ the cut edges and with $\mathscr{B}_h^{\partial\Omega}$ their collection, see Fig. 1. For a fixed regular mesh $\mathscr{T}_h$, the size of any $E^{\partial\Omega} \in \mathscr{B}_h^{\partial\Omega}$ is upper bounded by the mesh characteristic size, h, but it can become arbitrarily small. For this reason, the penalty term cannot be scaled with respect of the size of edges or faces lying on $\partial\Omega$, but it has been taken inversely proportional to the characteristic mesh size.

In order to analyse how the configuration of the boundary with respect to the mesh affects the stability of the scheme, we introduce the following indicator,

$$\nu' := \min_{K \in \mathscr{C}_h} \frac{|K \cap \Omega|}{|K|},$$

that corresponds to the minimum relative intersection of an element with the physical domain Ω. Since the unfitted Nitsche's method requires to evaluate integrals over cut elements or cut edges, we expect that the parameter ν' may affect the stability properties of the scheme.

Before addressing the analysis of the present unfitted Nitsche method, it is useful to recall some norms and related discrete inequalities as the basis for the forthcoming investigation. Concerning the norms, we notice that the definition of $\|\cdot\|_{\pm\frac{1}{2},h,\partial\Omega}$ should be adapted to the present scheme as follows,

$$\|v\|_{\pm\frac{1}{2},h,\partial\Omega} := h^{\mp\frac{1}{2}}\|v\|_{0,\partial\Omega},$$

while the definition of the energy and augmented norms is unchanged,

$$\|v\|_{1,h,\Omega}^2 := |v|_{1,\Omega}^2 + \|v\|_{\frac{1}{2},h,\partial\Omega}^2,$$

$$|||v|||_{1,h,\Omega}^2 := \|v\|_{1,h,\Omega}^2 + \|\partial_n v_h\|_{-\frac{1}{2},h,\partial\Omega}^2.$$

Exploiting inverse inequalities we easily prove that,

$$\|v_h\|_{-\frac{1}{2},h,\partial\Omega} = \sum_{E^{\partial\Omega} \in \mathscr{D}_h^{\partial\Omega}} h\|v_h\|_{E^{\partial\Omega}}^2 \lesssim \sum_{K \in \mathscr{C}_h} \|v_h\|_K^2 \lesssim \|v_h\|_{0,\Omega_{\mathscr{T}}}^2,$$

which is not satisfactory for our purpose, because the right hand side involves the entire computational domain and not the physical domain Ω solely. Proceeding similarly, the desired right hand side can be obtained,

$$\|v_h\|_{-\frac{1}{2},h,\partial\Omega} \lesssim \max_{K \in \mathscr{C}_h} \frac{|K|}{|K \cap \Omega|} \sum_{K \in \mathscr{C}_h} \|v_h\|_{0,K \cap \Omega}^2 \lesssim (v')^{-1}\|v_h\|_{0,\Omega}^2. \tag{31}$$

Since, given Ω, it is possible to construct a triangulation $\mathscr{T}_h$ with an arbitrarily small v, (31) shows that unfitted Nitsche's method is not robust with respect to the configuration of the boundary. Precisely, we say that a scheme is robust with respect to the parameter v if the spectrum of the discrete problem admits lower and upper bounds that are independent on the parameter itself.

Our main purpose is to study how small cut elements affect the fundamental properties of the numerical scheme. We perform such analysis simultaneously for symmetric ($s = 1$) and non symmetric ($s = -1$) schemes. To quantify the stability of the scheme, we look at the coercivity of the bilinear form and we exploit (31) to observe that

$$a_h(v_h, v_h) = |v_h|_{1,\Omega}^2 + \gamma\|v_h\|_{\frac{1}{2},h,\partial\Omega}^2 - (s+1)(\partial_n v_h, v_h)_{\partial\Omega}$$

$$\gtrsim \left(1 - (\delta_1 + \delta_2(s+1))(v')^{-1}\right)|v_h|_{1,\Omega}^2 + \delta_1\|\partial_n v_h\|_{-\frac{1}{2},h,\partial\Omega}^2$$

$$+ \left(\gamma - (s+1)\delta_2^{-1}\right)\|v_h\|_{\frac{1}{2},h,\partial\Omega}^2,$$

where δ_1, δ_2 are positive constants to be suitably chosen.

Three conclusions come out immediately. For the non symmetric case, i.e. $s = -1$, coercivity of $a_h(\cdot,\cdot)$ holds in the energy norm $\|\cdot\|_{1,h,\Omega}$ with $\delta_1 = 0$ and for any positive δ_2 and γ. As a result of that, the stability estimate of the non symmetric variant is robust with respect to the configuration of the interface. This is never true for the symmetric case because $s + 1 = 2$. If we analyse coercivity in the norm $\|\cdot\|_{1,h,\Omega}$, we can set $\delta_1 = 0$, but to make sure that the first term on the right hand side is positive it is necessary to satisfy $\delta_2 \lesssim v'$. Such a restriction entails that $\gamma \gtrsim (v')^{-1}$, which is unsatisfactory because the penalty term depends on the interface configuration and it becomes arbitrarily large for small element cuts. Finally, neither the non symmetric nor the symmetric cases feature robust stability properties in the augmented norm $\|\|\cdot\|\|_{1,h,\Omega}$. Indeed, coercivity of $a_h(\cdot,\cdot)$ in this norm can be only proved under the condition $\delta_1 \lesssim v'$, but in this case the control on the additional term $\delta_1\|\partial_n v_h\|_{-\frac{1}{2},h,\partial\Omega}^2$ is lost for small element cuts.

Concerning the boundedness of the bilinear form for discrete test functions, the consistency term $(\partial_n u_h, v_h)_{\partial\Omega}$ must be controlled by means of the energy or the augmented norms. To this purpose, the choice of the norm makes a significant difference. Since $|||v_h|||_{1,h,\Omega}$ directly controls $\|\partial_n v_h\|_{-\frac{1}{2},h,\partial\Omega}$, owing to a Cauchy-Schwarz inequality it is straightforward to conclude that

$$(\partial_n u_h, v_h)_{\partial\Omega} \leq \|\partial_n u_h\|_{-\frac{1}{2},h,\partial\Omega} \|v_h\|_{+\frac{1}{2},h,\partial\Omega} \leq |||u_h|||_{1,h,\Omega} |||v_h|||_{1,h,\Omega}.$$

Conversely, if we perform our analysis in the energy norm $\| \cdot \|_{1,h,\Omega}$, resorting to inverse inequality (31) is necessary to obtain an upper bound of the consistency term,

$$(\partial_n u_h, v_h)_{\partial\Omega} \leq \|\partial_n u_h\|_{-\frac{1}{2},h,\partial\Omega} \|v_h\|_{+\frac{1}{2},h,\partial\Omega}$$

$$\lesssim (\nu')^{-1} \|\nabla u_h\|_{0,\Omega} \|v_h\|_{+\frac{1}{2},h,\partial\Omega} \lesssim (\nu')^{-1} \|u_h\|_{1,h,\Omega} \|v_h\|_{1,h,\Omega}.$$

In the latter case, however, the fact that the continuity constant is proportional to $(\nu')^{-1}$ spoils the robustness of the scheme.

In conclusion, such analysis shows that both the symmetric and non symmetric variants of the unfitted Nitsche's method are unsatisfactory if we aim to set up a scheme that is fully robust with respect to the configuration of the computational mesh with respect to the boundary and the possibility to produce small element cuts. For this reason, in the forthcoming section we will propose a stabilisation technique to override this limitation of Nitsche's method.

3.2   The Ghost Penalty Stabilisation Method

In order to design a fully robust fictitious domain method, stability must be obtained in a norm at least as strong as the norm $|||u_h|||_{1,h,\Omega}$. This can be achieved by modifying the bilinear form in the interface zone. The idea is to add a penalty term that improves the stability in the elements cut by the interface and distributes the coercivity to the parts of the triangulation outside the physical domain. This added term must guarantee stability but at the same time be weakly consistent to the right order. Since the nodes outside the physical domain are often referred to as ghost nodes, this term is called the ghost penalty term.

Below we will follow the approach proposed in [16] with the higher order generalisation of [12]. For the proofs of the results we refer to these references. Recalling the definitions of (30) we propose the formulation: find $u_h \in V_h$ such that

$$a_h(u_h, v_h) + g_h(u_h, v_h) = F_h(v_h), \quad \forall v_h \in V_h. \tag{32}$$

where $g_h(\cdot, \cdot)$ is the ghost penalty stabilisation term. Define the set of element edges in the boundary zone by

$$\mathscr{E}_B := \{F = K \cap K', \text{ where either } K \in \mathscr{C}_h \text{ or } K' \in \mathscr{C}_h\}.$$

A possible ghost penalty term for piecewise affine approximations is then given by a penalty on the jumps of the gradients over the element edges of $\mathscr{E}_B$,

$$g_h(u_h, v_h) := \sum_{E \in \mathscr{E}_B} (\gamma_g h_E [\![\nabla u_h \cdot \mathbf{n}_E]\!], [\![\nabla v_h \cdot \mathbf{n}_E]\!])_E.$$

We also introduce the discrete $H^1(\Omega_{\mathscr{T}_h})$-norm

$$\|v_h\|_{1,h,\Omega_{\mathscr{T}}}^2 := \|\nabla v_h\|_{0,\Omega_{\mathscr{T}}}^2 + \|v_h\|_{\frac{1}{2},h,\partial\Omega}^2 \text{ with } \|v_h\|_{0,\Omega_{\mathscr{T}}}^2 := \sum_{K \in \mathscr{T}_h} \|v_h\|_{0,K}^2.$$

The enhanced stability obtained by adding $g_h(\cdot, \cdot)$, is reflected in the coercivity estimate

$$|||v_h|||_{1,h,\Omega}^2 \lesssim \|v_h\|_{1,h,\Omega_{\mathscr{T}_h}}^2 \lesssim a_h(v_h, v_h) + g_h(v_h, v_h), \quad \forall v_h \in V_h. \tag{33}$$

The first inequality is a consequence of the discrete trace inequality

$$\|\nabla v_h \cdot \mathbf{n}\|_{-\frac{1}{2},h,\partial\Omega} \lesssim \|\nabla v_h\|_{0,\mathscr{C}_h}$$

and the second holds thanks to the following fundamental property of the ghost penalty term

$$\|\nabla v_h\|_{0,\Omega_{\mathscr{T}}}^2 \lesssim \|\nabla v_h\|_{0,\Omega}^2 + g_h(v_h, v_h). \tag{34}$$

For piecewise affine Lagrangian finite element approximations, the idea in [16] to prove such an inequality is to observe that the gradient over any cut element $K \in \mathscr{C}_h$ is a piecewise constant function that is bounded from above by the gradient on another element $K' \notin \mathscr{C}_h$ plus the jumps of gradients across all elements that should be crossed to connect K with K'. Indeed, the ghost penalty stabilisation provides control on the additional terms involving jumps.

Under regularity assumptions on Ω, for all $v \in H^2(\Omega)$, we may introduce an extension operator $\mathbb{E} : H^2(\Omega) \mapsto H^2(\Omega_{\mathscr{T}})$ such that $\mathbb{E}v|_\Omega = v$ and $\|\mathbb{E}v\|_{H^2(\Omega_{\mathscr{T}})} \lesssim \|v\|_{H^2(\Omega)}$. It is then convenient to introduce an interpolation operator $i_h : H^2(\Omega) \mapsto V_h$ by $i_h v := I_h \mathbb{E}v$ where I_h is the standard nodal Lagrange interpolator. It is then straightforward to show that

$$|||v - i_h v|||_g := |||v - i_h v|||_{1,h,\Omega} + \sqrt{g_h(\mathbb{E}v - i_h v, \mathbb{E}v - i_h v)} \lesssim h\|v\|_{H^2(\Omega)}. \tag{35}$$

For the convergence analysis we need the following continuity result, that is a straightforward application of Cauchy-Schwarz inequalities and local trace inequalities. For all $v \in H^2(\Omega)$ and $w_h \in V_h$ there holds

$$|a_h(v - i_h v, w_h) + g_h(v - i_h v, w_h)|$$

$$\lesssim |||v - i_h v|||_g \left(\sum_{K \in \mathcal{T}_h} \|\nabla w_h\|_K^2 + \|w_h\|_{\frac{1}{2}, h, \partial \Omega}^2 \right)^{1/2}. \quad (36)$$

The optimal convergence estimate

$$|||u - u_h|||_{1, h, \Omega} \lesssim h \|u\|_{H^2(\Omega)}$$

is an immediate consequence of (33), (35) and (36). Using a duality argument one may also prove that

$$\|u - u_h\|_{0, \Omega} \lesssim h^2 \|u\|_{H^2(\Omega)}.$$

In a similar fashion exploiting the uniform upper and lower bounds of the bilinear form one may show that the condition number of the system matrix is robust with respect to the interface position. We will give some detail on this analysis in Sect. 3.3 in the case of multi-domain problems with large contrast.

In the case of high order approximations a penalty on the normal gradient is insufficient. Either one has to resort to a multi-penalty method or a stabilisation of local projection type, [12]. For example if we instead consider a Lagrangian finite element space where the polynomials are of degree k, the following multi-penalty operator will allow for a similar analysis in the high order case:

$$g_h(u_h, v_h) := \sum_{E \in \mathcal{E}_B} \sum_{i=1}^{k} (\gamma_g h_E^{2i-1} [\![\partial_n^i u_h]\!], [\![\partial_n^i v_h]\!])_E,$$

where $\partial_n^i u$ denotes the i-th order normal derivative of u across the edge E. Since such a quantity is combined with the jump across E, the orientation of the unit normal vector is irrelevant for the definition of $[\![\partial_n^i u_h]\!]$.

The role of this multi-penalty operator for the stabilisation of the unfitted method is better understood if we look at its connection with local projection operators. For any given element $K \in \mathcal{C}_h$, let $\mathcal{P}_K$ be the patch containing the shortest piecewise linear path connecting all the centres of mass to move from the centre of mass of K to the centre of $K' \notin \mathcal{C}_h$. Let $\mathcal{E}_K$ be the set of edges cut by such path. It is straightforward to verify that for any element $K \in \mathcal{C}_h$ and any corresponding patch $\mathcal{P}_K$ the ratio

$$\frac{|\mathcal{P}_K|}{|\mathcal{P}_K \cap \Omega|}$$

is uniformly bounded with respect to the position of the interface. Furthermore, for shape-regular and quasi-uniform meshes we expect that the number of individual elements contained in $\mathcal{P}_K$ is uniformly bounded from above. As a result of that, we have

$$\sum_{K \in \mathscr{C}_h} \|v_h\|^2_{0,\mathscr{P}_K} \simeq \sum_{K \in \mathscr{C}_h} \|v_h\|^2_{0,K},$$

where $a \simeq b$ means that there exist two constants c, C, uniformly independent of the mesh characteristic size h, such that $ca \leq b \leq Ca$.

Let us define by $\pi_h : L^2(\mathscr{P}_K) \to \mathbb{P}^k(\mathscr{P}_K)$ the local L^2 projection onto $\mathbb{P}^k(\mathscr{P}_K)$. The following equivalence can be proven, see [12] and references therein,

$$(v_h - \pi_h v_h, v_h)_{\mathscr{P}_K} \simeq \sum_{i=1}^{k} \sum_{E \in \mathscr{E}_K} \int_E [\![\partial_n^i v_h]\!]^2.$$

We will then use the local projection operator to prove that the following local counterpart of (34) holds true for any polynomial order $k \geq 1$,

$$\|v_h\|^2_{0,\mathscr{P}_K} \lesssim \|v_h\|^2_{0,\mathscr{P}_K \cap \Omega} + h^{-2} \int_{\mathscr{P}_K} \left(v_h - \pi_h v_h\right)^2. \tag{37}$$

To prove (37) we look at the restriction on $\mathscr{P}_K$ of any $v_h \in V_h$ and we split it as $v_h = \pi_h v_h + r_k$ where $r_k = v_h - \pi_h v_h$. We notice that either $\pi_h v_h = 0$ on the entire patch, or $\pi_h v_h \neq 0$ on any subset of $\mathscr{P}_k$ with non zero measure. As a result of that we obtain,

$$\|\nabla \pi_h v_h\|^2_{0,\mathscr{P}_K} \lesssim \frac{|\mathscr{P}_K|}{|\mathscr{P}_K \cap \Omega|} \|\nabla \pi_h v_h\|^2_{0,\mathscr{P}_K \cap \Omega}. \tag{38}$$

When the residual $r_k \neq 0$, exploiting (38), we notice that

$$\|\nabla v_h\|^2_{0,\mathscr{P}_K} \lesssim \|\nabla \pi_h v_h\|^2_{0,\mathscr{P}_K} + \|\nabla r_k\|^2_{0,\mathscr{P}_K} \lesssim \|\nabla \pi_h v_h\|^2_{0,\mathscr{P}_K \cap \Omega} + \|\nabla r_k\|^2_{0,\mathscr{P}_K}.$$

Owing to the inverse inequality we observe that,

$$\|\nabla r_k\|^2_{0,\mathscr{P}_K \cap \Omega} \lesssim \|\nabla r_k\|^2_{0,\mathscr{P}_K} \lesssim h^{-2} \|r_k\|^2_{0,\mathscr{P}_K}$$

and combining the previous estimates we conclude that,

$$|\nabla v_h\|^2_{0,\mathscr{P}_K} \lesssim \|\nabla \pi_h v_h\|^2_{0,\mathscr{P}_K \cap \Omega} - \|\nabla r_k\|^2_{0,\mathscr{P}_K \cap \Omega} + 2h^{-2} \|r_k\|^2_{0,\mathscr{P}_K}$$

$$\lesssim \|\nabla \pi_h v_h + \nabla r_k\|^2_{0,\mathscr{P}_K \cap \Omega} + 2h^{-2} \|r_k\|^2_{0,\mathscr{P}_K}$$

$$\lesssim \|\nabla v_h\|^2_{0,\mathscr{P}_K \cap \Omega} + 2h^{-2} \int_{\mathscr{P}_K} \left(v_h - \pi_h v_h\right)^2.$$

In conclusion, summing up over all elements $K \in \Omega_{\mathscr{T}}$, applying the previous estimate for any element cut by the interface $K \in \mathscr{C}_h$ and exploiting the equivalence between multi-penalty and the local projection we conclude that,

$$\|\nabla v_h\|_{\Omega_{\mathscr{T}}}^2 \lesssim \|\nabla v_h\|_{\Omega}^2 + \sum_{i=1}^{k} \sum_{E \in \mathscr{E}_B} \int_E [\![\partial_n^i v_h]\!]^2,$$

which generalizes (34) to high order Lagrangian finite elements.

3.3 The Unfitted Nitsche Method for Large Contrast Problems

We now place ourselves in the setting of Sect. 2, considering two non-overlapping subdomains, Ω_i, $i = 1, 2$, with interface $\Gamma := \overline{\Omega}_1 \cap \overline{\Omega}_2$. This time, however, the mesh will not be fitted to the interface. For simplicity we assume that Γ is a plane separating the two domains. The problem that we will study is (25), but this time we let $\beta = 0$. We recall the equations here for convenience

$$\begin{cases} \nabla \cdot \left(-\varepsilon_i \nabla u_i \right) = f_i, & \text{in } \Omega_i, \\ u_i = 0, & \text{on } \partial\Omega \cap \partial\Omega_i, \\ [\![u]\!] = 0, & \text{on } \Gamma, \\ [\![-\varepsilon\nabla u \cdot \mathbf{n}]\!] = 0, & \text{on } \Gamma. \end{cases} \tag{39}$$

We let $\mathscr{T}_{hi}$ denote a triangulation fitted to $\partial\Omega_i \setminus \Gamma$, but not to Γ. Let $\mathscr{T}_{h1}$ and $\mathscr{T}_{h2}$ match across the interface so that $\mathscr{T}_{h1} \cup \mathscr{T}_{h2}$ is a conforming triangulation of Ω. Let

$$V_{h,i} := \{v_h \in C^0(\overline{\Omega}_{\mathscr{T}_{hi}}) : v_h|_K \in \mathbb{P}_1(K), \text{ for all } K \in \mathscr{T}_{hi}; v_h|_{\partial\Omega} = 0\},$$

with $\mathbb{P}_1(K)$ denoting the set of polynomials of degree less than or equal to 1 on K. We denote by $\mathscr{T}_h := \mathscr{T}_{h1} \cup \mathscr{T}_{h2}$ the triangulation of the physical domain and by V_h^{Ω} the corresponding piecewise affine finite element space. Here we have for simplicity included the boundary conditions in the approximation space.

We may then write the following formulation, similar to that of Sect. 2. Find $[u_{h,1}, u_{h,2}] \in V_h := V_{h,1} \times V_{h,2}$, such that

$$a_h(u_h, v_h) = F_h(v_h), \quad \forall v_h \in V_h \tag{40}$$

where now

$$a_h(u_h, v_h) := \sum_{i=1,2} (\varepsilon_i \nabla u_{h,i}, \nabla v_{h,i})_{\Omega_i} + \gamma\xi(\varepsilon)h^{-1} ([\![u_h]\!], [\![v_h]\!])_{\Gamma}$$

$$- (\{\varepsilon\nabla u_h \cdot \mathbf{n}\}_w, [\![v_h]\!])_{\Gamma} - (\{\varepsilon\nabla v_h \cdot \mathbf{n}\}_w, [\![u_h]\!])_{\Gamma}, \tag{41}$$

$$F_h(v_h) := F(v_h) = (f, v_h)_{\Omega}.$$

Table 3 Convergence rate of (40) with linear finite elements for the test case already considered for Table 2

h	$\|u - u_h\|_{0,\Omega}$		$\|u - u_h\|_{1,h,\Omega}$	
	$\varepsilon_1 = \varepsilon_2 = 1$	$\varepsilon_1 = 1, \varepsilon_2 = 10^{-2}$	$\varepsilon_1 = \varepsilon_2 = 1$	$\varepsilon_1 = 1, \varepsilon_2 = 10^{-2}$
5.00×10^{-2}	2.24×10^{-4}	1.69×10^{-2}	1.48×10^{-2}	1.08×10^{-1}
2.50×10^{-2}	5.65×10^{-5}	4.20×10^{-3}	7.30×10^{-3}	5.41×10^{-2}
1.25×10^{-2}	1.41×10^{-5}	1.00×10^{-3}	3.60×10^{-3}	2.70×10^{-2}
6.25×10^{-3}	3.55×10^{-6}	2.64×10^{-4}	1.80×10^{-3}	1.35×10^{-2}
3.12×10^{-3}	8.90×10^{-7}	6.62×10^{-5}	9.02×10^{-4}	6.70×10^{-3}
p	1.99	1.99	1.00	1.00

The main advantage of method (40) consists in the fact that it restores the optimal convergence rate that is lost for the approximation of problem (39) with standard Lagrangian finite elements when the mesh does not fit with the interface. Indeed, for the test case already addressed for Table 2, we observe the convergence rates reported in Table 3, where $\|v\|_{1,h,\Omega}^2 := \sum_{i=1,2} \|\varepsilon_i^{1/2}\nabla v\|_{0,\Omega}^2 + \|\{\varepsilon\}_w^{1/2}[\![v]\!]\|_{\frac{1}{2},h,\Gamma}^2$ being $\|v\|_{\pm\frac{1}{2},h,\Gamma} := h^{\mp 1/2}\|v\|_{0,\Gamma}$ as for Nitsche's fictitious domain method.

However, this method has two major drawbacks that will be discussed thoroughly. Firstly, the corresponding matrix may become ill conditioned in case $|K \cap \Omega_i|$ is small for all the triangles in the support of a basis function. Secondly, the scheme cannot be simultaneously robust with respect to small cut elements and large contrast problems. A partial remedy exploiting the arbitrary choice of the averaging weights w_i will be proposed below, but a robust stability estimate can be only achieved with the help of a stabilisation term.

3.3.1 Stability Analysis of the Discrete Space with Cut Elements

The objective of this section is to reformulate the definition of $V_h = V_{h,1} \times V_{h,2}$ as an approximation space of functions with support on the physical domain Ω. As discussed in [43, 48], this allows us to exhibit and analyse the instabilities arising from the presence of small cut elements.

We consider the alternative representation of V_h proposed in [43], which exploits a hierarchical representation in terms of a standard Lagrangian finite element space, enriched with additional basis functions over cut elements. We start by defining the following restriction operator :

$$R_i : L^2(\Omega) \to L^2(\Omega), \quad R_i v := \begin{cases} v|_{\Omega_i} & \text{in } \Omega_i, \\ 0 & \text{in } \Omega \setminus \Omega_i. \end{cases}$$

Mimicking the Nitsche's fictitious domain method, we denote by $\mathscr{C}_h := \{K \in \mathscr{T}_h : |K \cap \Gamma|_{\mathbb{R}^{d-1}} > 0\}$ the crust of elements with non vanishing intersection with the interface. Let $\mathscr{I}$ be the set of indexes numbering the nodes associated to V_h^Ω and let

$\{x_k\}_{k \in \mathscr{I}}$ be the corresponding set of points on Ω. We define collections of nodes neighbouring the interface and we apply them to construct enrichment spaces,

$$\mathscr{I}_i^\Gamma := \{k \in \mathscr{I} : x_k \in \Omega_j,\ \mathrm{supp}(\phi_k) \cap \mathscr{C}_h \neq \emptyset\},\ \forall i, j = 1, 2,\ j \neq i$$

$$V_{h,i}^\Gamma := \mathrm{span}\{R_i \phi_k : k \in \mathscr{I}_i^\Gamma\},$$

where ϕ_k denotes the hat basis function associated to the node x_k. Owing to Theorem 2 in [43], the following direct decomposition holds:

$$V_h = V_h^\Omega \oplus V_{h,1}^\Gamma \oplus V_{h,2}^\Gamma,$$

i.e. any function $v \in V_h$ can be uniquely decomposed as $v = v^\Omega + v_1^\Gamma + v_2^\Gamma$ with $v^\Omega \in V_h^\Omega$, $v_i^\Gamma \in V_{h,i}^\Gamma$. We notice that the spaces $V_{h,1}^\Gamma, V_{h,2}^\Gamma$ are L^2-orthogonal on Ω, because their basis functions have disjoint supports.

Owing to this decomposition, finite element matrices in V_h feature the following block structure that can be exploited in their analysis. Let us denote with $M \in \mathbb{R}^{N_h \times N_h}$ and $L \in \mathbb{R}^{N_h \times N_h}$ the standard mass and stiffness matrices in the finite element space V_h,

$$\mathbf{v}' M \mathbf{w} = (v, w)_{0,\Omega},\quad \mathbf{v}' L \mathbf{w} = (\nabla v, \nabla w)_{0, \cup \Omega_i},\quad \forall v, w \in V_h$$

which can be rearranged as follows

$$M = \begin{bmatrix} M^\Omega & M_1^{\Omega\Gamma} & M_2^{\Omega\Gamma} \\ (M_1^{\Omega\Gamma})' & M_1^\Gamma & 0 \\ (M_2^{\Omega\Gamma})' & 0 & M_2^\Gamma \end{bmatrix} \quad L = \begin{bmatrix} L^\Omega & L_1^{\Omega\Gamma} & L_2^{\Omega\Gamma} \\ (L_1^{\Omega\Gamma})' & L_1^\Gamma & 0 \\ (L_2^{\Omega\Gamma})' & 0 & L_2^\Gamma \end{bmatrix}.$$

To quantify how the presence of small cut elements affects the spectrum of finite element mass and stiffness matrices, we introduce the following mesh dependent indicators. Let $x_k \in \mathscr{I}_i^\Gamma$ be any vertex associated to the enrichment spaces $V_{h,i}^\Gamma$, let ϕ_k be the corresponding basis function and $\mathscr{P}_k$ be its patch. The indicators that affect the conditioning of a finite element method with respect to small sub-elements can be defined as

$$\underline{v}_i := \min_{k \in \mathscr{I}_i^\Gamma} \frac{|\mathscr{P}_k \cap \Omega_i|}{|\mathscr{P}_k|},\quad \overline{v}_i := \max_{k \in \mathscr{I}_i^\Gamma} \frac{|\mathscr{P}_k \cap \Omega_i|}{|\mathscr{P}_k|},$$

$$v := \min_i \min_{k \in \mathscr{I}_i^\Gamma} \frac{|\mathscr{P}_k \cap \Omega_i|}{|\mathscr{P}_k|}.$$

Furthermore, we assume that for any index k, the corresponding patch satisfies $\mathscr{P}_k \cap (\Omega \setminus \Omega_\Gamma) \neq \emptyset$, i.e. there exists at least one element in the patch that is not cut by the interface.

Under the previous assumption on the mesh and owing to Lemma 2 of [43] the following strengthened Cauchy-Schwarz inequality hold true for any $v^\Omega \in V_h^\Omega$, $v^\Gamma \in V_{h,1}^\Gamma \oplus V_{h,2}^\Gamma$. There exist constants $0 < c_{cs}^0, c_{cs}^1 < 1$ such that

$$(v^\Omega, v^\Gamma)_\Omega \le c_{cs}^0 \|v^\Omega\|_{0,\Omega} \|v^\Gamma\|_{0,\Omega},$$

$$(\nabla v^\Omega, \nabla v^\Gamma)_{\cup \Omega_i} \le c_{cs}^1 \|\nabla v^\Omega\|_{0,\cup\Omega_i} \|\nabla v^\Gamma\|_{0,\cup\Omega_i}.$$

Then, exploiting the decomposition $v = v^\Omega + v_1^\Gamma + v_2^\Gamma$ together with Pythagoras' theorem, straightforward computations show that

$$(1 - c_{cs}^0)(\|v^\Omega\|_{0,\Omega}^2 + \|v_1^\Gamma\|_{0,\Omega_1}^2 + \|v_2^\Gamma\|_{0,\Omega_2}^2)$$
$$\le \|v\|_{0,\Omega}^2 \le 2(\|v^\Omega\|_{0,\Omega}^2 + \|v_1^\Gamma\|_{0,\Omega_1}^2 + \|v_2^\Gamma\|_{0,\Omega_2}^2),$$
$$(1 - c_{cs}^1)(\|\nabla v^\Omega\|_{0,\Omega}^2 + \|\nabla v_1^\Gamma\|_{0,\Omega_1}^2 + \|\nabla v_2^\Gamma\|_{0,\Omega_2}^2)$$
$$\le \|\nabla v\|_{0,\cup\Omega_i}^2 \le 2(\|\nabla v^\Omega\|_{0,\Omega}^2 + \|\nabla v_1^\Gamma\|_{0,\Omega_1}^2 + \|\nabla v_2^\Gamma\|_{0,\Omega_2}^2).$$

The previous inequalities directly imply that the mass and stiffness matrices are spectrally equivalent to their block diagonals,

$$\mathbf{v}'M\mathbf{v} \simeq (\mathbf{v}^\Omega)'M^\Omega \mathbf{v}^\Omega + (\mathbf{v}_1^\Gamma)'M_1^\Gamma \mathbf{v}_1^\Gamma + (\mathbf{v}_2^\Gamma)'M_2^\Gamma \mathbf{v}_2^\Gamma, \tag{42}$$

$$\mathbf{v}'L\mathbf{v} \simeq (\mathbf{v}^\Omega)'L^\Omega \mathbf{v}^\Omega + (\mathbf{v}_1^\Gamma)'L_1^\Gamma \mathbf{v}_1^\Gamma + (\mathbf{v}_2^\Gamma)'L_2^\Gamma \mathbf{v}_2^\Gamma. \tag{43}$$

Since the spectral properties of M^Ω and L^Ω are well known, we focus on the analysis of M_i^Γ, L_i^Γ. As shown in [43], Lemma 3, for any $v_i^\Gamma \in V_{h,i}^\Gamma$ there exist positive constants $\underline{c}_0^\Omega, \overline{c}_0^\Omega$, independent on how the interface Γ cuts the mesh $\mathscr{T}_h$, such that

$$\underline{c}_0^\Omega \sum_{k \in \mathscr{I}_i^\Gamma} \left(\beta_k^i\right)^2 \|R_i \phi_k\|_{0,\Omega_i}^2 \le \|v_i^\Gamma\|_{0,\Omega_i}^2 \le \overline{c}_0^\Omega \sum_{k \in \mathscr{I}_i^\Gamma} \left(\beta_k^i\right)^2 \|R_i \phi_k\|_{0,\Omega_i}^2. \tag{44}$$

The extension of this analysis to the H^1-norm holds true due to the fact that gradients of the local basis functions on $V_{h,i}^\Gamma$ are linearly independent functions. Indeed, for any $v_i^\Gamma \in V_{h,i}^\Gamma$ there exist positive constants $\underline{c}_1^\Omega, \overline{c}_1^\Omega$, independent on how the interface Γ cuts the mesh $\mathscr{T}_h$, such that

$$\underline{c}_1^\Omega \sum_{k \in \mathscr{I}_i^\Gamma} \left(\beta_k^i\right)^2 \|R_i \nabla \phi_k\|_{0,\Omega_i}^2 \le \|\nabla v_i^\Gamma\|_{0,\Omega_i}^2 \le \overline{c}_1^\Omega \sum_{k \in \mathscr{I}_i^\Gamma} \left(\beta_k^i\right)^2 \|R_i \nabla \phi_k\|_{0,\Omega_i}^2. \tag{45}$$

Let $\mathbf{v}$ denote the vector of degrees of freedom that identify a generic function $v \in V_h$ and let $\|\mathbf{v}\|$ be its Euclidean norm. Let $\mathbf{v}_i^\Gamma$ and $\mathbf{v}^\Omega$ be the vectors relative to $v_i^\Gamma \in V_{h,i}^\Gamma$ and $v^\Omega \in V_h^\Omega$, respectively. For any $v_i^\Gamma \in V_{h,i}^\Gamma$ there exist positive constants $\underline{c}_0^\Gamma, \overline{c}_0^\Gamma$, independent on ν, h, such that

$$\underline{c}_0^\Gamma h^d \underline{\nu}_i^{2/d+1} \|\mathbf{v}_i^\Gamma\|^2 \le \|v_i^\Gamma\|_{0,\Omega_i}^2 \le \overline{c}_0^\Gamma h^d \overline{\nu}_i^{2/d+1} \|\mathbf{v}_i^\Gamma\|^2, \qquad (46)$$

and there exists $\underline{v}_i^\Gamma \in V_{h,i}^\Gamma$ such that

$$\|\underline{v}_i^\Gamma\|_{0,\Omega_i}^2 \le \overline{c}_0^\Gamma h^d \underline{\nu}_i^{2/d+1} \|\underline{\mathbf{v}}_i^\Gamma\|^2. \qquad (47)$$

To prove (46) we have to estimate the smallest $\|R_i \phi_k\|_{0,\Omega_i}^2$. We split the integrals over the elements that belong to the patch of $R_i \phi_k$ and we apply a suitable quadrature formula. We notice that the measure of the support where the integrals are evaluated is proportional to $h^d \underline{\nu}_i$ while the pointwise evaluations of the function to be integrated can be at most equivalent to $(\nu_i^{1/d})^2$. The upper bound is obtained replacing the smallest $\|R_i \phi_k\|_{0,\Omega_i}^2$ with the largest. By the same argument, (47) holds true if we select $\underline{v}_i^\Gamma := R_i \phi_k$ corresponding to $\min_{k \in \mathscr{I}_i^\Gamma} \|R_i \phi_k\|_{0,\Omega}^2$.

By means of the same reasoning applied to (45), a similar result can be shown for gradients of discrete functions, with a different scaling with respect to ν, because $R_i \nabla \phi_k$ are constant functions proportional to h^{-1} and thus $\|R_i \nabla \phi_k\|_{0,K_i}^2 \simeq h^{d-2} \nu_i^K$. As a result of that, there exist $\underline{c}_1^\Gamma, \overline{c}_1^\Gamma > 0$, independent on ν, h such that

$$\underline{c}_1^\Gamma h^{d-2} \underline{\nu}_i \|\mathbf{v}_i^\Gamma\|^2 \le \|\nabla v_i^\Gamma\|_{0,\Omega}^2 \le \overline{c}_1^\Gamma h^{d-2} \overline{\nu}_i \|\mathbf{v}_i^\Gamma\|^2. \qquad (48)$$

Furthermore, for the same $\underline{v}_i^\Gamma \in V_{h,i}^\Gamma$ of (47) we have

$$\|\nabla \underline{v}_i^\Gamma\|_{0,\Omega}^2 \le \overline{c}_1^\Gamma h^{d-2} \underline{\nu}_i \|\underline{\mathbf{v}}^\Gamma\|^2. \qquad (49)$$

Inequalities (47) and (49) show that minimal eigenvalues of $M_i^\Gamma L_i^\Gamma$ become arbitrarily small in presence of small element cuts. This clearly influences the conditioning of the finite element scheme, which will be affected by a factor ν^{-1}. However, the present analysis immediately points out a cure for this drawback. Indeed, combining (42) and (43) with (47) and (49) we conclude that the mass and stiffness matrices of the enriched finite element space V_h are spectrally equivalent to

$$M \simeq \begin{bmatrix} M^\Omega & 0 & 0 \\ 0 & \mathrm{diag}(M_1^\Gamma) & 0 \\ 0 & 0 & \mathrm{diag}(M_2^\Gamma) \end{bmatrix} \quad L \simeq \begin{bmatrix} L^\Omega & 0 & 0 \\ 0 & \mathrm{diag}(L_1^\Gamma) & 0 \\ 0 & 0 & \mathrm{diag}(L_2^\Gamma) \end{bmatrix}.$$

where, given a real square matrix B, we denote with $\mathrm{diag}(B)$ its diagonal. This shows that solving a finite element scheme in the enriched space V_h is computationally equivalent to solving it in the standard space V_h^Ω, because the only genuinely stiff block is L^Ω. Another way to formulate this conclusion is based on the *optimal condition number* of the problem, see [46]. More precisely, given $A \in \mathbb{R}^{N \times N}$ the optimal condition number is

$$K_{opt}(A) := \min_{D \in \mathbb{R}^{N \times N}} K_2(DAD)$$

and the previous analysis shows that $K_{opt}(\alpha_M M + \alpha_L L) = K_{opt}(\alpha_M M^\Omega + \alpha_L L^\Omega)$ for any positive constants α_M, α_L.

3.3.2 Stability Issues for the Unfitted Nitsche Method

An important question concerning the stability of the scheme is how to choose the averages in the interface terms. In [29], Hansbo and Hansbo proposed a method for which they could prove stability and optimal convergence. In their analysis they chose mesh dependent weights,

$$w_i|_K = \frac{|K \cap \Omega_i|}{|K|}, \quad \text{and} \quad \xi(\varepsilon) = \max\{\varepsilon_1, \varepsilon_2\}.$$

As a result of that, integrals of the normal derivative on the interface on elements with a very small fraction intersecting one of the physical domains will get a small weight, which will balance the factor of order $|K|$ appearing after taking the trace inequality. In Sect. 2 we showed that

$$w_1 = \frac{\varepsilon_2}{\varepsilon_1 + \varepsilon_2}, \quad w_2 = \frac{\varepsilon_1}{\varepsilon_1 + \varepsilon_2} \quad \text{and} \quad \xi(\varepsilon) = \{\varepsilon\}_w = \frac{2\varepsilon_1 \varepsilon_2}{\varepsilon_1 + \varepsilon_2}. \quad (50)$$

leads to robustness with respect to the jump in the diffusivities.

This poses a situation in which the unfitted character of the method requires a certain set of weights, and the large contrast character requires another set. For instance, these contradictory requirements clearly appear in the following estimate,

$$\int_{\partial K \cap \Gamma} \{\varepsilon\}_w (\nabla v_{h,i})^2 \lesssim h^{-1} w_i \left(1 + \frac{(\varepsilon_j w_j)|_K}{(\varepsilon_i w_i)|_K}\right) \frac{|K|}{|K \cap \Omega_i|} \|\varepsilon_i^{\frac{1}{2}} \nabla v_{h,i}\|_{0, K \cap \Omega_i}^2,$$

which is needed to quantify an upper bound for the spectrum of the discrete problem. Indeed, the constant

$$\left(1 + \frac{(\varepsilon_j w_j)|_K}{(\varepsilon_i w_i)|_K}\right) \frac{|K|}{|K \cap \Omega_i|},$$

may become arbitrarily large for some configuration of the interface or highly heterogeneous weights. A partial remedy consists selecting the weights w_i to minimise the dominating effect. If the worse case comes from the way the interface is cut, then we define

$$w_i = \frac{|K \cap \Omega_i|}{|K|} \quad \text{satisfying} \quad w_1 + w_2 = 1,$$

otherwise, when the heterogeneity of coefficients is dominating, we choose

$$w_i = \frac{\varepsilon_j}{\varepsilon_i + \varepsilon_j} \quad \text{such that} \quad \left(1 + \frac{(\varepsilon_j w_j)|_K}{(\varepsilon_i w_i)|_K}\right) = 2.$$

Nevertheless, this technique does not work in situations where both difficulties arise simultaneously. To handle both effects at the same time, we may draw from the fictitious domain formulation proposed in the previous section. The introduction of a ghost penalty term on the interface elements both in $\mathcal{T}_{h1}$ and $\mathcal{T}_{h2}$ gives the same extended coercivity as in the fictitious domain case and we are then allowed to choose the weights so as to control the large contrast in diffusivity.

3.3.3 The Stabilized Unfitted Nitsche Method

The stabilised method that we propose takes the form: find $[u_{h,1}, u_{h,2}] \in V_h :=$ $V_{h,1} \times V_{h,2}$, such that

$$a_h(u_h, v_h) + g_h(u_h, v_h) = F_h(v_h), \quad \forall v_h \in V_h, \tag{51}$$

where the weights have been chosen as in (50) and

$$g_h(u_h, v_h) := \sum_{i=1}^{2} \sum_{E \in \mathcal{E}_{B_i}} (\gamma_g \varepsilon_i h_E [\![\nabla u_{h,i} \cdot \mathbf{n}_E]\!], [\![\nabla v_{h,i} \cdot \mathbf{n}_E]\!])_E,$$

with

$$\mathcal{E}_{B_i} := \{E = K \cap K' : K \in \mathcal{T}_{hi}, K' \in \mathcal{T}_{hi} \text{ where either } K \cap \Gamma \neq \emptyset \text{ or } K' \cap \Gamma \neq \emptyset\}.$$

For the analysis of Nitsche's method for unfitted interfaces, we introduce the norms

$$\|v_h\|_{1,h,\Omega_{\mathcal{T}}}^2 := \sum_{i=1}^{2} \sum_{K \in \mathcal{T}_{hi}} \|\varepsilon_i^{\frac{1}{2}} \nabla v_{h,i}\|_{0,K}^2 + \|\{\varepsilon\}_w^{\frac{1}{2}} [\![v_{h,i}]\!]\|_{\frac{1}{2},h,\Gamma}^2$$

and

$$|||v_h|||_{1,h,\Omega}^2 := \sum_{i=1}^{2} \|\varepsilon_i^{\frac{1}{2}} \nabla v_{h,i}\|_{0,\Omega}^2 + \|\{\varepsilon\}_w^{\frac{1}{2}} \{\nabla v_h \cdot \mathbf{n}\}\|_{-\frac{1}{2},h,\Gamma}^2 + \|\{\varepsilon\}_w^{\frac{1}{2}} [\![v_h]\!]\|_{+\frac{1}{2},h,\Gamma}^2.$$

To obtain a robust stability estimate, we use the extended coercivity obtained thanks to the ghost penalty term combined with the inverse inequality to conclude that,

$$|||v_h|||_{1,h,\Omega}^2 \lesssim \|v_h\|_{1,h,\Omega_{\mathcal{T}}}^2 \lesssim a_h(v_h, v_h) + g_h(v_h, v_h), \quad \forall v_h \in V_h. \tag{52}$$

This is obtained in the same fashion as the analogous result for the fictitious domain method. The boundedness of the stabilised bilinear form is also guaranteed by means of standard arguments, see [12],

$$a_h(u_h, v_h) + g_h(u_h, v_h) \lesssim |||u_h|||_{1,h,\Omega} \, |||v_h|||_{1,h,\Omega}$$

$$\lesssim \|u_h\|_{1,h,\Omega_{\mathscr{I}}} \|v_h\|_{1,h,\Omega_{\mathscr{I}}} \quad \forall u_h, v_h \in V_h.$$

To proceed with the convergence analysis, we introduce extension operators $\mathbb{E}_i :$ $H^2(\Omega_i) \mapsto H^2(\Omega_{\mathscr{I}i})$ such that $\mathbb{E}_i v|_{\Omega_i} = v|_{\Omega_i}$ and $\|\mathbb{E}_i v\|_{H^2(\Omega_{\mathscr{I}i})} \lesssim \|v\|_{H^2(\Omega_i)}$. In a similar fashion as above, we define an interpolation operator $i_h : H^2(\Omega_1) \times H^2(\Omega_2) \mapsto V_h$ by $i_h v := [I_h \mathbb{E}_1 v, I_h \mathbb{E}_2 v]$ where I_h is the standard nodal Lagrange interpolator. It is straightforward to show that

$$|||v - i_h v|||_g := |||v - i_h v|||_{1,h,\Omega} + \sqrt{g_h (\mathbb{E}v - i_h v, \mathbb{E}v - i_h v)} \lesssim h \|v\|_{H^2(\Omega_1 \cup \Omega_2)}.$$

For the convergence analysis we need the following continuity result, that is a straightforward application of Cauchy-Schwarz inequalities and local trace inequalities. For all $v \in H^2(\Omega)$ and $w_h \in V_h$ there holds

$$|a_h(v - i_h v, w_h) + g_h(v - i_h v, w_h)|$$

$$\lesssim |||v - i_h v|||_g \Big(\sum_{i=1,2} \sum_{K \in \mathscr{T}_{h,i}} \|\nabla w_{h,i}\|_K^2 + \|w_h\|_{\frac{1}{2},h,\partial\Omega}^2 \Big)^{1/2}.$$

Then, the optimal convergence estimate $|||u - u_h|||_{1,h,\Omega} \lesssim h \|u\|_{H^2(\Omega)}$, is an immediate consequence of (52).

3.3.4 Bounded Condition Number

In this section we will show that the choice of weights (50) together with the use of ghost penalty term leads to a method with a system matrix whose condition number, after diagonal scaling with the diffusivity, has the same asymptotic scaling as the standard Galerkin method for the Poisson problem with fitted mesh and constant coefficients. This means that the conditioning is independent both of the interface configuration and the jump of the diffusivities. To fix the ideas let $\varepsilon_1 = 1$ and $0 < \varepsilon_2 < \varepsilon_1$. Other configurations can be obtained by scaling.

Let $\{\phi_{k,i}\}$ denote the nodal basis of $V_{h,i}$ with $i = 1, 2$. Consequently we may write $u_{h,i} \in V_{h,i}$ in the form $u_{h,i} := \sum_{k=1}^{N_i} U_{k,i} \phi_{k,i}$. The formulation (41) may then be written as the linear system

$$\begin{bmatrix} \varepsilon_1 \mathbf{A}_{11} + \{\varepsilon\}_w \mathbf{A}_{11}^{\Gamma} & \{\varepsilon\}_w \mathbf{A}_{12}^{\Gamma} \\ \{\varepsilon\}_w \mathbf{A}_{21}^{\Gamma} & \varepsilon_2 \mathbf{A}_{22} + \{\varepsilon\}_w \mathbf{A}_{22}^{\Gamma} \end{bmatrix} \begin{bmatrix} \mathbf{U}_1 \\ \mathbf{U}_2 \end{bmatrix} = \begin{bmatrix} \mathbf{F}_1 \\ \mathbf{F}_2 \end{bmatrix}, \tag{53}$$

where for symmetry it holds that $\mathbf{A}_{12}^\Gamma = \left(\mathbf{A}_{21}^\Gamma\right)^T$ and the vectors are defined by

$$\mathbf{U}_i := \{U_{k,i}\}_{k=1}^{N_i}, \quad \text{and} \quad \mathbf{F}_i := \{F_h(\phi_{k,i})\}_{k=1}^{N_i}$$

and the weight function $\{\varepsilon\}_w$ is given in (50). Examining formulation (40) we see that the matrices are given by

$$\mathbf{A}_{ii} := \{(\varepsilon_i \nabla \phi_{k,i}, \nabla \phi_{l,i})_{\Omega_i} + \varepsilon_i g_i (\phi_{k,i}, \phi_{l,i})\}_{k,l=1}^{N_i}, \quad i = 1, 2$$

where $g_i(\phi_{k,i}, \phi_{l,i})$ denotes a ghost penalty term on the subdomain Ω_i,

$$\mathbf{A}_{ii}^\Gamma := \{-\left(\nabla \phi_{k,i} \cdot \mathbf{n} + h^{-1} \gamma \phi_{k,i}, \phi_{l,i}\right)_\Gamma\}_{k,l=1}^{N_i} \quad i = 1, 2$$

which is independent of ε_i, and

$$\mathbf{A}_{ij}^\Gamma := \left\{\frac{1}{2}\left(\nabla \phi_{k,i} \cdot \mathbf{n} + h^{-1} \gamma \phi_{k,i}, \phi_{l,j}\right)_\Gamma + \frac{1}{2}\left(\nabla \phi_{l,j} \cdot \mathbf{n} + h^{-1} \gamma \phi_{l,j}, \phi_{k,i}\right)_\Gamma\right\},$$

with $k = 1, \ldots, N_i$, $l = 1, \ldots, N_j$ and $i, j = 1, 2$, $i \neq j$. After diagonal symmetric scaling, the system matrix takes the form

$$\mathbf{A}_{scal} := \begin{bmatrix} \mathbf{A}_{11} + \dfrac{\{\varepsilon\}_w}{\varepsilon_1} \mathbf{A}_{11}^\Gamma & \dfrac{\{\varepsilon\}_w}{\sqrt{\varepsilon_1 \varepsilon_2}} \mathbf{A}_{12}^\Gamma \\[2ex] \dfrac{\{\varepsilon\}_w}{\sqrt{\varepsilon_1 \varepsilon_2}} \mathbf{A}_{21}^\Gamma & \mathbf{A}_{22} + \dfrac{\{\varepsilon\}_w}{\varepsilon_2} \mathbf{A}_{22}^\Gamma. \end{bmatrix}.$$

To study the behaviour of the unfitted Nitsche method in the case of highly heterogeneous coefficients, we notice that the matrix $\mathbf{A}_{scal}$ converges to

$$\lim_{\varepsilon_2 \to 0} \mathbf{A}_{scal} = \begin{bmatrix} \mathbf{A}_{11} & \mathbf{0} \\[2ex] \mathbf{0} & \mathbf{A}_{22} + \mathbf{A}_{22}^\Gamma. \end{bmatrix}$$

in the limit $\varepsilon_2 \to 0$.

Under the assumption that Γ is a planar interface we know that for each subdomain Ω_i it holds $\partial \Omega_i \cap \partial \Omega \neq \emptyset$. Furthermore, since the homogeneous Dirichlet boundary conditions are strongly enforced in the finite element space, we conclude that the stiffness matrices $\mathbf{A}_{ii}$ are symmetric positive definite. Owing to an inverse inequality and with the stabilisation parameter γ large enough, this conclusion holds true even though we extend it to $\mathbf{A}_{ii} + \mathbf{A}_{ii}^\Gamma$. This illustrates that the stabilised Nitsche's method with symmetric diagonal scaling becomes robust with respect to the heterogeneity of coefficients provided that the averaging weights are selected as in (50).

To quantify the robustness of the scheme with respect to the configuration of the interface, we exploit the strengthened coercivity ensured by the ghost penalty term. Under the aforementioned assumption on the stiffness matrices, inverse and Poincaré inequalities imply that

$$\sum_{i=1}^{2} \sum_{K \in \mathcal{T}_{hi}} \|\varepsilon_i v_{h,i}\|_{0,K}^2 \lesssim \|v_h\|_{1,h,\Omega_{\mathcal{T}_h}}^2 \lesssim a_h(v_h, v_h) + g_h(v_h, v_h)$$

$$\lesssim \|v_h\|_{1,h,\Omega_{\mathcal{T}_h}}^2 \lesssim h^{-2} \sum_{i=1}^{2} \sum_{K \in \mathcal{T}_{hi}} \|\varepsilon_i v_{h,i}\|_{0,K}^2, \quad \forall v_h \in V_h.$$

The matrix of system (53) is thus spectrally equivalent to a block diagonal matrix that is uniformly independent of the configuration of the interface. Furthermore, when diagonal scaling is applied to such matrix, the equivalent system becomes independent of the heterogeneity for diffusion coefficients.

3.3.5 Asymptotic Convergence to the Fictitious Domain Method

The aim of this section is to show that unfitted Nitsche method for interface problems coincides with the corresponding unfitted boundary method in the case that the diffusion coefficient on one of the subdomains becomes arbitrarily large.

This property has two interesting consequences. On the one hand, it shows that the choice of the balancing weights proposed for large contrast problems is consistent with the unfitted boundary case. On the other hand, it allows to exploit the unfitted interface formulation as a fictitious domain method, where the choice of the computational domain $\Omega_{\mathcal{T}}$ is completely arbitrary with respect to the physical domain Ω, provided that in the complementary domain $\Omega_{\mathcal{T}} \setminus \Omega$ a sufficiently large diffusivity is applied.

To fix the ideas, we assume that $\varepsilon_1 = 1$ and study the case $\varepsilon_2 \to \infty$. Accordingly, we denote the fictitious domain bilinear forms (30) defined on Ω_1 by $a_{fd}(\cdot, \cdot)$, $g_{fd}(\cdot, \cdot)$ and denote the domain decomposition bilinear forms (41) by $a_{dd}(\cdot, \cdot)$, $g_{dd}(\cdot, \cdot)$. Let $u_{h,fd}$ denote the solution of (32) and $u_{h,dd}$ the solution of (40) with *ghost penalty* stabilisation. We assume that the penalty parameters for both formulations are set to the same values. We are interested in the behaviour of the discrete error between the two formulations in the limit as $\varepsilon_2 \to \infty$. We therefore define $e_h := (u_{h,fd} - u_{h,dd})|_{\Omega_1}$. Using the coercivity of the formulation (32) we have

$$\|e_h\|_{1,h,\Omega_{\mathcal{T}_{h1}}}^2 \leq a_{fd}(e_h, e_h) + g_{dd}(e_h, e_h).$$

Note that since $e_h|_{\mathcal{T}_{h2}} = 0$, $g_{fd}(e_h, e_h) = g_{dd}(e_h, e_h)$. By the definition of the discrete problems we have

$$\|e_h\|^2_{1,h,\Omega_{\mathcal{T}_{h1}}} \leq F_h(e_h) - a_{fd}(u_{h,dd},e_h) - g_{dd}(u_{h,dd},e_h)$$
$$= a_{dd}(u_{h,dd},e_h) - a_{fd}(u_{h,dd},e_h).$$

It is straightforward to show that

$$a_{dd}(u_{h,dd},e_h) - a_{fd}(u_{h,dd},e_h) = \left((1 - \frac{\varepsilon_2}{\varepsilon_2 + 1})\partial_n u_{h,dd}|_{\partial\Omega_1}, e_h|_{\partial\Omega_1}\right)_{\partial\Omega_1}$$
$$+ \left((1 - \frac{\varepsilon_2}{\varepsilon_2 + 1})\partial_n e_h|_{\partial\Omega_1}, u_{h,dd}|_{\partial\Omega_1}\right)_{\partial\Omega_1}$$
$$- \left((\frac{\varepsilon_2}{\varepsilon_2 + 1})\partial_n u_{h,dd}|_{\partial\Omega_2}, e_h|_{\partial\Omega_1}\right)_{\partial\Omega_1}$$
$$- \left((\frac{\varepsilon_2}{\varepsilon_2 + 1})\partial_n e_h|_{\partial\Omega_1}, u_{h,dd}|_{\partial\Omega_2}\right)_{\partial\Omega_1}$$
$$+ \left((1 - \frac{\varepsilon_2}{\varepsilon_2 + 1})\gamma_{bc}h^{-1}u_{h,dd}|_{\partial\Omega_1}, e_h|_{\partial\Omega_1}\right)_{\partial\Omega_1}$$
$$- \left(\frac{\varepsilon_2}{\varepsilon_2 + 1}\gamma h^{-1}u_{h,dd}|_{\partial\Omega_2}, e_h|_{\partial\Omega_1}\right)_{\partial\Omega_1}.$$

By repeated application of the mesh weighted Cauchy-Schwarz inequality and trace inequalities in the right hand side we arrive at the bound

$$|a_{dd}(u_{h,dd},e_h) - a_{fd}(u_{h,dd},e_h)|$$
$$\lesssim \left(1 - \frac{\varepsilon_2}{\varepsilon_2 + 1}\right)\left(\sum_{K\in\mathcal{T}_{h1}}(1 + h_K^{-1})\|\nabla u_{h,dd}\|_K^2\right)^{\frac{1}{2}}\|e_h\|_{1,h,\Omega_{\mathcal{T}_{h1}}}$$
$$+ \frac{\varepsilon_2}{(\varepsilon_2 + 1)}\left(\sum_{K\in\mathcal{T}_{h2}}(1 + h_K^{-1})\|\nabla u_{h,dd}\|_K^2\right)^{\frac{1}{2}}\|e_h\|_{1,h,\Omega_{\mathcal{T}_{h1}}}.$$

Using the formulation (40), (41) with *ghost penalty* stabilisation, observing that there exists a positive constant C_F such that $F_h(v_h) \leq C_F\|v_h\|_{1,h,\Omega_{\mathcal{T}}}$ and exploiting the stability (52), we obtain the following estimates for $u_{h,dd}|_{\Omega_1}$ and $u_{h,dd}|_{\Omega_2}$,

$$\sqrt{\sum_{K\in\mathcal{T}_{h1}}\|\nabla u_{h,dd}\|_K^2} \leq C_F, \qquad \sqrt{\sum_{K\in\mathcal{T}_{h2}}\|\varepsilon^{\frac{1}{2}}\nabla u_{h,dd}\|_K^2} \leq C_F.$$

We conclude that we have the bound

$$\|e_h\|_{1,h,\Omega_{\mathcal{T}_{h1}}} \leq \frac{1}{(\varepsilon_2 + 1)}(1 + h^{-1})C_F$$

and that $\lim_{\varepsilon_2 \to \infty} \|e_h\|_{1,h,\Omega_{\mathcal{T}_{h1}}} = 0$. Hence, the unfitted Nitsche interface method reduces to the fictitious domain method in the limit of infinite diffusivity.

Acknowledgements The authors acknowledge the support of the project *5 per Mille Junior* *"Computational models for heterogeneous media. Application to microscale analysis of tissue engineered constructs"*, CUP D41J10000490001, Politecnico di Milano.

References

1. Arnold, D.N.: An interior penalty finite element method with discontinuous elements. SIAM J. Numer. Anal. **19**(4), 742–760 (1982). DOI 10.1137/0719052. URL http://dx.doi.org/10.1137/0719052
2. Arnold, D.N., Brezzi, F., Cockburn, B., Marini, L.D.: Unified analysis of discontinuous Galerkin methods for elliptic problems. SIAM J. Numer. Anal. **39**(5), 1749–1779 (2001/02). DOI 10.1137/S0036142901384162. URL http://dx.doi.org/10.1137/S0036142901384162
3. Babuška, I.: The finite element method with Lagrangian multipliers. Numer. Math. **20**, 179–192 (1972/73)
4. Babuška, I.: The finite element method with penalty. Math. Comp. **27**, 221–228 (1973)
5. Barbosa, H.J.C., Hughes, T.J.R.: The finite element method with Lagrange multipliers on the boundary: circumventing the Babuška-Brezzi condition. Comput. Methods Appl. Mech. Engrg. **85**(1), 109–128 (1991). DOI 10.1016/0045-7825(91)90125-P. URL http://dx.doi.org/10.1016/0045-7825(91)90125-P
6. Barbosa, H.J.C., Hughes, T.J.R.: Boundary Lagrange multipliers in finite element methods: error analysis in natural norms. Numer. Math. **62**(1), 1–15 (1992). DOI 10.1007/BF01396217. URL http://dx.doi.org/10.1007/BF01396217
7. Barrett, J.W., Elliott, C.M.: Finite element approximation of the Dirichlet problem using the boundary penalty method. Numer. Math. **49**(4), 343–366 (1986). DOI 10.1007/BF01389536. URL http://dx.doi.org/10.1007/BF01389536
8. Becker, R., Hansbo, P., Stenberg, R.: A finite element method for domain decomposition with non-matching grids. M2AN Math. Model. Numer. Anal. **37**(2), 209–225 (2003). DOI 10.1051/m2an:2003023. URL http://dx.doi.org/10.1051/m2an:2003023
9. Bramble, J.H.: The Lagrange multiplier method for Dirichlet's problem. Math. Comp. **37**(155), 1–11 (1981). DOI 10.2307/2007496. URL http://dx.doi.org/10.2307/2007496
10. Brenner, S.C., Scott, L.R.: The mathematical theory of finite element methods, *Texts in Applied Mathematics*, vol. 15, third edn. Springer, New York (2008). DOI 10.1007/978-0-387-75934-0. URL http://dx.doi.org/10.1007/978-0-387-75934-0
11. Burman, E.: A unified analysis for conforming and nonconforming stabilized finite element methods using interior penalty. SIAM J. Numer. Anal. **43**(5), 2012–2033 (electronic) (2005). DOI 10.1137/S0036142903437374. URL http://dx.doi.org/10.1137/S0036142903437374
12. Burman, E.: Ghost penalty. Comptes Rendus Mathematique **348**(21–22), 1217–1220 (2010). DOI DOI:10.1016/j.crma.2010.10.006
13. Burman, E., Ern, A.: Continuous interior penalty hp-finite element methods for advection and advection-diffusion equations. Math. Comp. **76**(259), 1119–1140 (electronic) (2007). DOI 10.1090/S0025-5718-07-01951-5. URL http://dx.doi.org/10.1090/S0025-5718-07-01951-5
14. Burman, E., Fernández, M.A., Hansbo, P.: Continuous interior penalty finite element method for Oseen's equations. SIAM J. Numer. Anal. **44**(3), 1248–1274 (electronic) (2006). DOI 10.1137/040617686. URL http://dx.doi.org/10.1137/040617686
15. Burman, E., Hansbo, P.: Fictitious domain finite element methods using cut elements: I. a stabilized lagrange multiplier method. Computer Methods in Applied Mechanics and Engineering **199**(41-44), 2680 – 2686 (2010). DOI DOI:10.1016/j.cma.2010.05.011

16. Burman, E., Hansbo, P.: Fictitious domain finite element methods using cut elements: Ii. a stabilized nitsche method. Applied Numerical Mathematics **In Press, Corrected Proof**, – (2011). DOI DOI:10.1016/j.apnum.2011.01.008
17. Burman, E., Zunino, P.: A domain decomposition method based on weighted interior penalties for advection-diffusion-reaction problems. SIAM J. Numer. Anal. **44**(4), 1612–1638 (electronic) (2006). DOI 10.1137/050634736. URL http://dx.doi.org/10.1137/050634736
18. Codina, R., Baiges, J.: Approximate imposition of boundary conditions in immersed boundary methods. Internat. J. Numer. Methods Engrg. **80**(11), 1379–1405 (2009). DOI 10.1002/nme.2662. URL http://dx.doi.org/10.1002/nme.2662
19. D'Angelo, C., Zunino, P.: A finite element method based on weighted interior penalties for heterogeneous incompressible flows. SIAM J. Numer. Anal. **47**(5), 3990–4020 (2009). DOI 10.1137/080726318. URL http://dx.doi.org/10.1137/080726318
20. Dolbow, J., Harari, I.: An efficient finite element method for embedded interface problems. Internat. J. Numer. Methods Engrg. **78**(2), 229–252 (2009). DOI 10.1002/nme.2486. URL http://dx.doi.org/10.1002/nme.2486
21. Dolbow, J., Moës, N., Belytschko, T.: An extended finite element method for modeling crack growth with frictional contact. Comput. Methods Appl. Mech. Engrg. **190**(51–52), 6825–6846 (2001). DOI 10.1016/S0045-7825(01)00260-2. URL http://dx.doi.org/10.1016/S0045-7825 (01)00260-2
22. Dryja, M.: On discontinuous Galerkin methods for elliptic problems with discontinuous coefficients. Comput. Methods Appl. Math. **3**(1), 76–85 (electronic) (2003). Dedicated to Raytcho Lazarov
23. Ern, A., Stephansen, A.F., Zunino, P.: A discontinuous Galerkin method with weighted averages for advection-diffusion equations with locally small and anisotropic diffusivity. IMA J. Numer. Anal. **29**(2), 235–256 (2009). DOI 10.1093/imanum/drm050. URL http://dx.doi.org/10.1093/imanum/drm050
24. Gastaldi, F., Quarteroni, A.: On the coupling of hyperbolic and parabolic systems: analytical and numerical approach. Appl. Numer. Math. **6**(1–2), 3–31 (1989/90). Spectral multi-domain methods (Paris, 1988)
25. Gerstenberger, A., Wall, W.A.: An embedded Dirichlet formulation for 3D continua. Internat. J. Numer. Methods Engrg. **82**(5), 537–563 (2010)
26. Girault, V., Glowinski, R.: Error analysis of a fictitious domain method applied to a Dirichlet problem. Japan J. Indust. Appl. Math. **12**(3), 487–514 (1995)
27. Groß, S., Reusken, A.: An extended pressure finite element space for two-phase incompressible flows with surface tension. J. Comput. Phys. **224**(1), 40–58 (2007). DOI 10.1016/j.jcp.2006.12.021. URL http://dx.doi.org/10.1016/j.jcp.2006.12.021
28. Hansbo, A., Hansbo, P.: An unfitted finite element method, based on Nitsche's method, for elliptic interface problems. Comput. Methods Appl. Mech. Engrg. **191**(47–48), 5537–5552 (2002). DOI 10.1016/S0045-7825(02)00524-8. URL http://dx.doi.org/10.1016/S0045-7825(02) 00524-8
29. Hansbo, A., Hansbo, P.: An unfitted finite element method, based on Nitsche's method, for elliptic interface problems. Comput. Methods Appl. Mech. Engrg. **191**(47–48), 5537–5552 (2002). DOI 10.1016/S0045-7825(02)00524-8. URL http://dx.doi.org/10.1016/S0045-7825(02)00524-8
30. Hansbo, A., Hansbo, P.: A finite element method for the simulation of strong and weak discontinuities in solid mechanics. Comput. Methods Appl. Mech. Engrg. **193**(33–35), 3523–3540 (2004). DOI 10.1016/j.cma.2003.12.041. URL http://dx.doi.org/10.1016/j.cma.2003.12.041
31. Hansbo, P.: Nitsche's method for interface problems in computational mechanics. GAMM-Mitt. **28**(2), 183–206 (2005)
32. Harari, I., Dolbow, J.: Analysis of an efficient finite element method for embedded interface problems. Comput. Mech. **46**(1), 205–211 (2010). DOI 10.1007/s00466-009-0457-5. URL http://dx.doi.org/10.1007/s00466-009-0457-5

33. Haslinger, J., Renard, Y.: A new fictitious domain approach inspired by the extended finite element method. SIAM J. Numer. Anal. **47**(2), 1474–1499 (2009). DOI 10.1137/070704435. URL http://dx.doi.org/10.1137/070704435

34. Juntunen, M., Stenberg, R.: Nitsche's method for general boundary conditions. Math. Comp. **78**(267), 1353–1374 (2009). DOI 10.1090/S0025-5718-08-02183-2. URL http://dx.doi.org/10.1090/S0025-5718-08-02183-2

35. Ladyzhenskaya, O.A.: The boundary value problems of mathematical physics, *Applied Mathematical Sciences*, vol. 49. Springer-Verlag, New York (1985). Translated from the Russian by Jack Lohwater [Arthur J. Lohwater]

36. Maury, B.: Numerical analysis of a finite element/volume penalty method. SIAM J. Numer. Anal. **47**(2), 1126–1148 (2009). DOI 10.1137/080712799. URL http://dx.doi.org/10.1137/080712799

37. Moës, N., Béchet, E., Tourbier, M.: Imposing Dirichlet boundary conditions in the extended finite element method. Internat. J. Numer. Methods Engrg. **67**(12), 1641–1669 (2006). DOI 10.1002/nme.1675. URL http://dx.doi.org/10.1002/nme.1675

38. Oden, J.T., Babuška, I., Baumann, C.E.: A discontinuous hp finite element method for diffusion problems. J. Comput. Phys. **146**(2), 491–519 (1998). DOI 10.1006/jcph.1998.6032. URL http://dx.doi.org/10.1006/jcph.1998.6032

39. Pitkäranta, J.: Boundary subspaces for the finite element method with Lagrange multipliers. Numer. Math. **33**(3), 273–289 (1979). DOI 10.1007/BF01398644. URL http://dx.doi.org/10.1007/BF01398644

40. Pitkäranta, J.: Local stability conditions for the Babuška method of Lagrange multipliers. Math. Comp. **35**(152), 1113–1129 (1980). DOI 10.2307/2006378. URL http://dx.doi.org/10.2307/2006378

41. Pitkäranta, J.: The finite element method with Lagrange multipliers for domains with corners. Math. Comp. **37**(155), 13–30 (1981). DOI 10.2307/2007497. URL http://dx.doi.org/10.2307/2007497

42. Quarteroni, A., Pasquarelli, F., Valli, A.: Heterogeneous domain decomposition: principles, algorithms, applications. In: Fifth International Symposium on Domain Decomposition Methods for Partial Differential Equations (Norfolk, VA, 1991), pp. 129–150. SIAM, Philadelphia, PA (1992)

43. Reusken, A.: Analysis of an extended pressure finite element space for two-phase incompressible flows. Comput. Vis. Sci. **11**(4-6), 293–305 (2008). DOI 10.1007/s00791-008-0099-8. URL http://dx.doi.org/10.1007/s00791-008-0099-8

44. Rivière, B., Wheeler, M.F., Girault, V.: A priori error estimates for finite element methods based on discontinuous approximation spaces for elliptic problems. SIAM J. Numer. Anal. **39**(3), 902–931 (2001). DOI 10.1137/S003614290037174X. URL http://dx.doi.org/10.1137/S003614290037174X

45. Stenberg, R.: On some techniques for approximating boundary conditions in the finite element method. J. Comput. Appl. Math. **63**(1-3), 139–148 (1995). DOI 10.1016/0377-0427(95)00057-7. URL http://dx.doi.org/10.1016/0377-0427(95)00057-7. International Symposium on Mathematical Modelling and Computational Methods Modelling 94 (Prague, 1994)

46. Strang, G., Fix, G.J.: An analysis of the finite element method. Prentice-Hall Inc., Englewood Cliffs, N. J. (1973). Prentice-Hall Series in Automatic Computation

47. Zunino, P.: Discontinuous Galerkin methods based on weighted interior penalties for second order PDEs with non-smooth coefficients. J. Sci. Comput. **38**(1), 99–126 (2009). DOI 10.1007/s10915-008-9219-3. URL http://dx.doi.org/10.1007/s10915-008-9219-3

48. Zunino, P., Cattaneo, L., Colciago, C.M.: An unfitted interface penalty method for the numerical approximation of contrast problems. Tech. rep., MOX - Department of Mathematics, Politecnico di Milano (2011)

Editorial Policy

1. Volumes in the following three categories will be published in LNCSE:

i)   Research monographs
ii) Tutorials
iii) Conference proceedings

Those considering a book which might be suitable for the series are strongly advised to contact the publisher or the series editors at an early stage.

2. Categories i) and ii). Tutorials are lecture notes typically arising via summer schools or similar events, which are used to teach graduate students. These categories will be emphasized by Lecture Notes in Computational Science and Engineering. **Submissions by interdisciplinary teams of authors are encouraged.** The goal is to report new developments – quickly, informally, and in a way that will make them accessible to non-specialists. In the evaluation of submissions timeliness of the work is an important criterion. Texts should be well-rounded, well-written and reasonably self-contained. In most cases the work will contain results of others as well as those of the author(s). In each case the author(s) should provide sufficient motivation, examples, and applications. In this respect, Ph.D. theses will usually be deemed unsuitable for the Lecture Notes series. Proposals for volumes in these categories should be submitted either to one of the series editors or to Springer-Verlag, Heidelberg, and will be refereed. A provisional judgement on the acceptability of a project can be based on partial information about the work: a detailed outline describing the contents of each chapter, the estimated length, a bibliography, and one or two sample chapters – or a first draft. A final decision whether to accept will rest on an evaluation of the completed work which should include

– at least 100 pages of text;
– a table of contents;
– an informative introduction perhaps with some historical remarks which should be accessible to readers unfamiliar with the topic treated;
– a subject index.

3. Category iii). Conference proceedings will be considered for publication provided that they are both of exceptional interest and devoted to a single topic. One (or more) expert participants will act as the scientific editor(s) of the volume. They select the papers which are suitable for inclusion and have them individually refereed as for a journal. Papers not closely related to the central topic are to be excluded. Organizers should contact the Editor for CSE at Springer at the planning stage, see *Addresses* below.

In exceptional cases some other multi-author-volumes may be considered in this category.

4. Only works in English will be considered. For evaluation purposes, manuscripts may be submitted in print or electronic form, in the latter case, preferably as pdf- or zipped ps-files. Authors are requested to use the LaTeX style files available from Springer at http:// www. springer.com/authors/book+authors?SGWID=0-154102-12-417900-0.

For categories ii) and iii) we strongly recommend that all contributions in a volume be written in the same LaTeX version, preferably LaTeX2e. Electronic material can be included if appropriate. Please contact the publisher.

Careful preparation of the manuscripts will help keep production time short besides ensuring satisfactory appearance of the finished book in print and online.

5. The following terms and conditions hold. Categories i), ii) and iii):

Authors receive 50 free copies of their book. No royalty is paid.
Volume editors receive a total of 50 free copies of their volume to be shared with authors, but no royalties.

Authors and volume editors are entitled to a discount of 33.3 % on the price of Springer books purchased for their personal use, if ordering directly from Springer.

6. Commitment to publish is made by letter of intent rather than by signing a formal contract. Springer-Verlag secures the copyright for each volume.

Addresses:

Timothy J. Barth
NASA Ames Research Center
NAS Division
Moffett Field, CA 94035, USA
barth@nas.nasa.gov

Michael Griebel
Institut für Numerische Simulation
der Universität Bonn
Wegelerstr. 6
53115 Bonn, Germany
griebel@ins.uni-bonn.de

David E. Keyes
Mathematical and Computer Sciences
and Engineering
King Abdullah University of Science
and Technology
P.O. Box 55455
Jeddah 21534, Saudi Arabia
david.keyes@kaust.edu.sa

and

Department of Applied Physics
and Applied Mathematics
Columbia University
500 W. 120 th Street
New York, NY 10027, USA
kd2112@columbia.edu

Risto M. Nieminen
Department of Applied Physics
Aalto University School of Science
and Technology
00076 Aalto, Finland
risto.nieminen@tkk.fi

Dirk Roose
Department of Computer Science
Katholieke Universiteit Leuven
Celestijnenlaan 200A
3001 Leuven-Heverlee, Belgium
dirk.roose@cs.kuleuven.be

Tamar Schlick
Department of Chemistry
and Courant Institute
of Mathematical Sciences
New York University
251 Mercer Street
New York, NY 10012, USA
schlick@nyu.edu

Editor for Computational Science
and Engineering at Springer:
Martin Peters
Springer-Verlag
Mathematics Editorial IV
Tiergartenstrasse 17
69121 Heidelberg, Germany
martin.peters@springer.com

Lecture Notes
in Computational Science
and Engineering

23. L.F. Pavarino, A. Toselli (eds.), *Recent Developments in Domain Decomposition Methods.*

24. T. Schlick, H.H. Gan (eds.), *Computational Methods for Macromolecules: Challenges and Applications.*

25. T.J. Barth, H. Deconinck (eds.), *Error Estimation and Adaptive Discretization Methods in Computational Fluid Dynamics.*

26. M. Griebel, M.A. Schweitzer (eds.), *Meshfree Methods for Partial Differential Equations.*

27. S. Müller, *Adaptive Multiscale Schemes for Conservation Laws.*

28. C. Carstensen, S. Funken, W. Hackbusch, R.H.W. Hoppe, P. Monk (eds.), *Computational Electromagnetics.*

29. M.A. Schweitzer, *A Parallel Multilevel Partition of Unity Method for Elliptic Partial Differential Equations.*

30. T. Biegler, O. Ghattas, M. Heinkenschloss, B. van Bloemen Waanders (eds.), *Large-Scale PDE-Constrained Optimization.*

31. M. Ainsworth, P. Davies, D. Duncan, P. Martin, B. Rynne (eds.), *Topics in Computational Wave Propagation.* Direct and Inverse Problems.

32. H. Emmerich, B. Nestler, M. Schreckenberg (eds.), *Interface and Transport Dynamics.* Computational Modelling.

33. H.P. Langtangen, A. Tveito (eds.), *Advanced Topics in Computational Partial Differential Equations.* Numerical Methods and Diffpack Programming.

34. V. John, *Large Eddy Simulation of Turbulent Incompressible Flows.* Analytical and Numerical Results for a Class of LES Models.

35. E. Bänsch (ed.), *Challenges in Scientific Computing - CISC 2002.*

36. B.N. Khoromskij, G. Wittum, *Numerical Solution of Elliptic Differential Equations by Reduction to the Interface.*

37. A. Iske, *Multiresolution Methods in Scattered Data Modelling.*

38. S.-I. Niculescu, K. Gu (eds.), *Advances in Time-Delay Systems.*

39. S. Attinger, P. Koumoutsakos (eds.), *Multiscale Modelling and Simulation.*

40. R. Kornhuber, R. Hoppe, J. Périaux, O. Pironneau, O. Wildlund, J. Xu (eds.), *Domain Decomposition Methods in Science and Engineering.*

41. T. Plewa, T. Linde, V.G. Weirs (eds.), *Adaptive Mesh Refinement – Theory and Applications.*

42. A. Schmidt, K.G. Siebert, *Design of Adaptive Finite Element Software.* The Finite Element Toolbox ALBERTA.

43. M. Griebel, M.A. Schweitzer (eds.), *Meshfree Methods for Partial Differential Equations II.*

44. B. Engquist, P. Lötstedt, O. Runborg (eds.), *Multiscale Methods in Science and Engineering.*

45. P. Benner, V. Mehrmann, D.C. Sorensen (eds.), *Dimension Reduction of Large-Scale Systems.*

46. D. Kressner, *Numerical Methods for General and Structured Eigenvalue Problems.*

47. A. Boriçi, A. Frommer, B. Joó, A. Kennedy, B. Pendleton (eds.), *QCD and Numerical Analysis III.*

48. F. Graziani (ed.), *Computational Methods in Transport.*

49. B. Leimkuhler, C. Chipot, R. Elber, A. Laaksonen, A. Mark, T. Schlick, C. Schütte, R. Skeel (eds.), *New Algorithms for Macromolecular Simulation.*

50. M. Bücker, G. Corliss, P. Hovland, U. Naumann, B. Norris (eds.), *Automatic Differentiation: Applications, Theory, and Implementations.*

51. A.M. Bruaset, A. Tveito (eds.), *Numerical Solution of Partial Differential Equations on Parallel Computers.*

52. K.H. Hoffmann, A. Meyer (eds.), *Parallel Algorithms and Cluster Computing.*

53. H.-J. Bungartz, M. Schäfer (eds.), *Fluid-Structure Interaction.*

54. J. Behrens, *Adaptive Atmospheric Modeling.*

55. O. Widlund, D. Keyes (eds.), *Domain Decomposition Methods in Science and Engineering XVI.*

56. S. Kassinos, C. Langer, G. Iaccarino, P. Moin (eds.), *Complex Effects in Large Eddy Simulations.*

57. M. Griebel, M.A Schweitzer (eds.), *Meshfree Methods for Partial Differential Equations III.*

58. A.N. Gorban, B. Kégl, D.C. Wunsch, A. Zinovyev (eds.), *Principal Manifolds for Data Visualization and Dimension Reduction.*

59. H. Ammari (ed.), *Modeling and Computations in Electromagnetics: A Volume Dedicated to Jean-Claude Nédélec.*

60. U. Langer, M. Discacciati, D. Keyes, O. Widlund, W. Zulehner (eds.), *Domain Decomposition Methods in Science and Engineering XVII.*

61. T. Mathew, *Domain Decomposition Methods for the Numerical Solution of Partial Differential Equations.*

62. F. Graziani (ed.), *Computational Methods in Transport: Verification and Validation.*

63. M. Bebendorf, *Hierarchical Matrices. A Means to Efficiently Solve Elliptic Boundary Value Problems.*

64. C.H. Bischof, H.M. Bücker, P. Hovland, U. Naumann, J. Utke (eds.), *Advances in Automatic Differentiation.*

65. M. Griebel, M.A. Schweitzer (eds.), *Meshfree Methods for Partial Differential Equations IV.*

66. B. Engquist, P. Lötstedt, O. Runborg (eds.), *Multiscale Modeling and Simulation in Science.*

67. I.H. Tuncer, Ü. Gülcat, D.R. Emerson, K. Matsuno (eds.), *Parallel Computational Fluid Dynamics 2007.*

68. S. Yip, T. Diaz de la Rubia (eds.), *Scientific Modeling and Simulations.*

69. A. Hegarty, N. Kopteva, E. O'Riordan, M. Stynes (eds.), *BAIL 2008 – Boundary and Interior Layers.*

70. M. Bercovier, M.J. Gander, R. Kornhuber, O. Widlund (eds.), *Domain Decomposition Methods in Science and Engineering XVIII.*

71. B. Koren, C. Vuik (eds.), *Advanced Computational Methods in Science and Engineering.*

72. M. Peters (ed.), *Computational Fluid Dynamics for Sport Simulation.*

73. H.-J. Bungartz, M. Mehl, M. Schäfer (eds.), *Fluid Structure Interaction II - Modelling, Simulation, Optimization.*

74. D. Tromeur-Dervout, G. Brenner, D.R. Emerson, J. Erhel (eds.), *Parallel Computational Fluid Dynamics 2008.*

75. A.N. Gorban, D. Roose (eds.), *Coping with Complexity: Model Reduction and Data Analysis.*

76. J.S. Hesthaven, E.M. Rønquist (eds.), *Spectral and High Order Methods for Partial Differential Equations.*

77. M. Holtz, *Sparse Grid Quadrature in High Dimensions with Applications in Finance and Insurance.*

78. Y. Huang, R. Kornhuber, O.Widlund, J. Xu (eds.), *Domain Decomposition Methods in Science and Engineering XIX.*

79. M. Griebel, M.A. Schweitzer (eds.), *Meshfree Methods for Partial Differential Equations V.*

80. P.H. Lauritzen, C. Jablonowski, M.A. Taylor, R.D. Nair (eds.), *Numerical Techniques for Global Atmospheric Models.*

81. C. Clavero, J.L. Gracia, F.J. Lisbona (eds.), *BAIL 2010 – Boundary and Interior Layers, Computational and Asymptotic Methods.*

82. B. Engquist, O. Runborg, Y.R. Tsai (eds.), *Numerical Analysis and Multiscale Computations.*

83. I.G. Graham, T.Y. Hou, O. Lakkis, R. Scheichl (eds.), *Numerical Analysis of Multiscale Problems.*

84. A. Logg, K.-A. Mardal, G.N. Wells (eds.), *Automated Solution of Differential Equations by the Finite Element Method.*

85. J. Blowey, M. Jensen (eds.), *Frontiers in Numerical Analysis - Durham 2010.*

For further information on these books please have a look at our mathematics catalogue at the following URL: www.springer.com/series/3527

Monographs in Computational Science and Engineering

For further information on this book, please have a look at our mathematics catalogue at the following URL: www.springer.com/series/7417

Texts in Computational Science and Engineering

For further information on these books please have a look at our mathematics catalogue at the following URL: www.springer.com/series/5151